Advances in Stereotactic and Functional Neurosurgery 10

*Proceedings of the 10ᵗʰ Meeting
of the European Society for Stereotactic
and Functional Neurosurgery,
Stockholm 1992*

Edited by
*B. A. Meyerson, G. Broggi, J. Martin-Rodriguez,
C. Ostertag, M. Sindou*

*Acta Neurochirurgica
Supplementum 58*

Springer-Verlag Wien GmbH

Assoc. Professor Dr. Björn A. Meyerson
Department of Neurosurgery, Karolinska Hospital, Stockholm, Sweden

Professor Dr. Giovanni Broggi
Istituto Neurologico "C. Besta", Milano, Italy

Professor Dr. Jose Martin-Rodriguez
Fundacion "Sixto Obrador", Hospital Ramon y Cajal, Madrid, Spain

Professor Dr. Christoph Ostertag
Neurochirurgische Universitätsklinik, Freiburg, Federal Republic of Germany

Professor Dr. Marc Sindou
Hôpital Neurologique Pierre Wertheimer, Lyon, France

With 85 Figures

© 1993 Springer-Verlag Wien
Originally published by Springer-Verlag Wien New York in 1993
Softcover reprint of the hardcover 1st edition 1993

Product Liability: The publisher can give no guarantee for information about drug dosage and application thereof contained in this book. In every individual case the respective user must check its accuracy by consulting other pharmaceutical literature. The use of registered names, trademarks, etc. in this publication does not inply, even in the absence of a specific statement, that such names are exempt from the relevant protective laws and regulations and therefore free for general use.

Typesetting: Thomson Press, New Delhi, India
Printed on acid-free and chlorine free bleached paper

ISSN 0065-1419 (Acta Neurochirurgica/Suppl.)
ISSN 0720-7972 (Advances in Stereotactic and Functional Neurosurgery)
ISBN 978-3-7091-9299-3   ISBN 978-3-7091-9297-9 (eBook)
DOI 10.1007/978-3-7091-9297-9

Preface

These proceedings from the Xth Congress of the European Society for Stereotactic and Functional Neurosurgery in Stockholm reflect the growing interest in these fields of neurosurgery. It is the most extensive volume in this series of publications and it contains a large number of original articles pertaining to the most recent advances in stereotactic and functional neurosurgery. Not long ago stereotactic neurosurgery was considered an esoteric subspeciality practised only by those involved in treating movement disorders and pain. In the last decade we have witnessed the incorporation of stereotactic methodology in the management of common neurosurgical diseases, and the stereotactic technique is now practised in all major neurosurgical centers. As with other surgical techniques and tools, however, the utilization of stereotactic methods requires special knowledge and training. This book comprises articles which give insight into new stereotactic applications and technology. For example, the usage of stereotaxis without a frame and the so-called navigator systems in open tumor surgery is dealt with in two papers.

The introduction and development of radiosurgery is closely linked to the advancement of stereotactic technique. Radiosurgical treatment of tumors and cerebrovascular diseases has been one of major achievements in modern neurosurgery. This publication contains several original reports illustrating the efficacy of radiosurgery in problematic neurosurgical diseases.

In the field of functional neurosurgery there has been an impressive development in the treatment of pain and many of the new treatment modalities in pain have been introduced and developed by neurosurgeons. Various forms of electric neurostimulation have become a routine in the management of certain pain conditions resistant to conventional therapy. Neuropathic forms of chronic pain are particularly problematic, and the physiological background for such pain has until recently been poorly understood. The pathophysiology of peripheral neuropathic pain is the topic of a thorough review by Clifford Woolf from London. There are also a number of other papers dealing with experimental and clinical studies of neuropathic pain.

Neural transplantation is a novel and fascinating approach to the treatment of parkinsonism. This form of therapy is unique in the sense that it aims at the restoration of diseased nervous tissue with specific neurotransmittor-releasing properties. Transplantation also gives hope as a therapy for other hitherto incurable neurological diseases. The role of trophic factors in neural transplantation is reviewed by Lars Olson, Stockholm, and a clinical update is given by Roy Bakay, Atlanta.

The classical neurosurgical treatment of tremor is thalamotomy which is still widely practised. However, in recent years it has been convincingly shown that the same tremor blocking effect can be achieved by electric stimulation in the thalamus. This new and non-destructive approach for dealing with incapacitating tremor in parkinsonism and other extrapyramidal diseases is reviewed by Alim Benabid, Grenoble.

These proceedings demonstrate that stereotactic and functional neurosurgery represent active fields of both clinical and basic neuroscience. The many original and new reports presented at the meeting in Stockholm in 1992 is a manifestation of the vitality of the European Society for Stereotactic and Functional Neurosurgery.

Björn Meyerson
President ESSFN

September 1993

Contents

Radiosurgery

Pain

Epilepsy

Neurotransplantation

Acta Neurochir (1993) [Suppl] 58: 3–7

Reparative Strategies in the Brain: Treatment Strategies Based on Trophic Factors and Cell Transfer Techniques

L. Olson

Department of Histology and Neurobiology, Karolinska Institutet, Sweden

Summary

Three reparative strategies based on transfer of genes, molecules, or cells to the central nervous system are reviewed. When neurons are already lost, they can sometimes be replaced by transfer to the target area of neurons or other cells compensating for the lost functions. This technique is undergoing clinical trials in Parkinson's disease. Before neurons have died, it may be possible to prevent "stressed" neurons from dying, and stimulate nerve terminal ramifications from remaining neurons using treatment with neurotrophic factors. Such approaches, with an emphasis on the NGF family of neurotrophins and their receptors, are reviewed. Finally, advances of molecular biology techniques suggest that it should be possible to transfer genes directly into non-dividing cells of the central nervous system. The three different approaches all aim at long-lasting counteractive and reparative measures in the central nervous system. It is predicted that they have general applicability, and may become important not only in neuro-degenerative diseases, but also in other common afflictions of the nervous system such as ischaemia, stroke and injury.

Keywords: Parkinson's disease; Alzheimer's disease; stroke; brain injury; neurotrophins; NGF; adrenal medullary cell transfer; foetal dopamine neuron transfer.

Introduction

It is a well-established fact that long nerve fiber pathways do not regenerate in the adult mammalian central nervous system, while axonal regeneration is an effective and clinically important mechanism in the peripheral nervous system. The reason for this difference long remained obscure, but the elegant experiments of Aguayo and collaborators[3] demonstrated many years ago that adult CNS axons are also able to elongate efficiently if given the appropriate environment such as a section of a peripheral nerve. The discrepancy between regenerative capabilities in CNS and PNS has at least three possible explanations: namely, the production of neurotrophic factors by Schwann cells but not oligodendroglial cells, or secondly, the production of nerve growth inhibitory factors by oligodendroglial but not Schwann cells, or thirdly, the formation of scar tissue to a greater extent in the CNS than in the PNS. It now appears as if all three possibilities operate. We are equipped with relatively effective regenerative mechanisms for peripheral nerve injury, but the cost for the extreme complexity of the central nervous system and perhaps for maintenance of "learned" functions is an absence of such regenerative possibilities in the central nervous system. Faced with the many reasons for CNS malfunction caused by diseases of modern living, and longevity, the need for reparative intervention in the brain and spinal cord is nevertheless obvious.

While intensive research during the last two decades has not yet given us methods to effectively stimulate regeneration of long fiber tracts in the central nervous system, it has led to the development of two other principal repair strategies. The first is a cell replacement strategy: Neurons that have been lost can sometimes be replaced by other cells such as embryonic neurons which are then implanted, not at the original site of such nerve cell bodies, but directly into the axon terminal area of the lost neurons. With this strategy one avoids the necessity of reforming a long axon pathway, obviously loses the original circuitry, but gains new nerve terminals. The second approach is applicable before neurons have died, and aims at preventing "stressed" neurons from dying, and stimulating neo-formation of nerve terminal ramifications from remaining neurons using treatment with neurotrophic factors. Obviously, these two principles can be combined. In the following, I shall discuss some of the recent animal background for treatment with grafts and growth factors as well as comment upon the ongoing clinical trials.

From a theoretical point of view one may look at reparative strategies as various ways of transfering molecules or cells to the brain to obtain long-lasting effects. Thus *molecular* and *cellular transfer techniques* differ from current treatment strategies of neurodegenerative diseases such as Parkinson's disease and Alzheimer's disease. Current treatments attempt to increase dopamine and acetylcholine neurotransmission, respectively, by pharmacological means such as administering a dopamine precursor or an acetylcholine esterase inhibitor respectively. These treatments may be effective as long as the drugs are taken regularly, but the positive effects disappear immediately upon drug withdrawal.

Neurotrophic Molecules

From a situation just a few years ago when nerve growth factor, NGF[8], was the only neurotrophic factor known, we now are faced with a very complex but also very promising situation in which we begin to realize that brain development and maintenance as well as anti-degenerative effects are all mediated via a large number of specific neurotrophic factors acting upon an array of specific receptors. It appears as if every "growth factor" that has been searched for in the central nervous system (e.g. NGF family, FGF family, PDGF family, TGF family, IGF family, CNTF) has been found to be expressed either by neurons or glial cells, and in many cases, physiological and/or pharmacological roles for these factors have been discovered.

The NGF Family of Neurotrophins

In order to attempt to understand the role of the neurotrophins and their receptors (see[4]) in the brain, one must know the precise site of synthesis and localization of the neurotrophins as well as of their receptors. The four neurotrophins together with their common so-called low-affinity receptor, p75, and the more specific so-called high-affinity receptors *trk*, *trk*B and *trk*C, respectively, encompass eight gene products that need to be localized. Sites of synthesis can be revealed by in situ hybridization or possibly by immunohistochemistry using antibodies against prosequences within the protein. Presence of mature protein can only be determined by immunohistochemistry which is particularly demanding in the case

of neurotrophins for two reasons: these proteins appear not to be stored, and are present in extremely low amounts. Secondly, the high degree of sequence homology between the proteins makes it necessary to generate antibodies against restricted sequences of the molecule in order to find neurotrophin-specific epitopes. The bulk of the neuroanatomical data has been obtained from in situ hybridization studies.

Once it was realized that brain-derived neurotrophic factor (BDNF) had a very high amino acid sequence homology with NGF[8], further members of the NGF family could be searched for using PCR techniques. Today we know four closely related members, NGF, BDNF, neurotrophin-3 (NT-3) and NT-4 (see[1] and [18]). All four neurotrophins are expressed in the brain. NGF, BDNF, and NT-3 are expressed by specific, partially overlapping, subsets of neurons in the hippocampal formation and other areas of the brain. BDNF appears to be the factor in this family with the most widespread distribution, including many neurons throughout the cerebral cortex, most pyramidal cells and granule cells of the hippocampal formation, as well as several subcortical areas. NGF has a more restricted expression involving scattered cells along the pyramidal cell layer of the hippocampal formation and large neurons in the hilar region of the dentate gyrus bordering the granular cells. NT-3 is the most restricted of the three being expressed in a specific subset of pyramidal cells in the medial CA1 area, in CA2 and in the granular cells of the hippocampal formation.

Localization of neurotrophin proteins has not been very successful so far for NGF and NT-3. However, we have been able to develop BDNF-specific antibodies[22] which have proved useful localizing BDNF-containing structures in the brain[20,21]. The neurotrophin receptors[2] are more abundant in the cells that make them and thus have proved somewhat easier to demonstrate using in situ hybridization as well as immunohistochemistry. There is a widespread expression of the low-affinity NGF receptor. The presumed high-affinity receptor for NGF, *trk*, is specifically located on cholinergic neurons which also contain low-affinity receptor. This fits well with the observed effects of NGF on central cholinergic neurons. *trk*B is much more widespread in the brain, *trk*B mRNA appears to be present also in neurons that contain BDNF mRNA, and this pattern suggests widespread local actions of BDNF of paracrine and perhaps also autocrine nature. It should also be noted that the neurotrophin receptors occur in different forms. For instance, there is a truncated form of the *trk*B receptor lacking the intracellular protein kinase domain which is even more widespread and probably also present on glial cells[19].

Neurotrophin Reactions to Injury and Disturbances

Summarizing several recent reports of effects of perturbations in the brain, one notes that many different kinds of disturbances such as mechanical lesions, ischaemia, hypoglycaemia, excitotoxic lesions and epileptogenic treatments cause rapid upregulation of NGF and BDNF, while NT-3 appears to change less or not at all. Interestingly, such upregulation can also be seen in the corresponding neurotrophin receptor mRNA levels. This is in contrast to neurotransmitter systems, the classical example being the dopamine system, in which there is an inverse relationship between receptor sensitivity and the amount of dopamine, tending to maintain homeostasis of dopamine neurotransmission. For the NGF and BDNF neurotrophins, both neurotrophin synthesis and receptor synthesis increase in parallel and rapidly, thus rapidly increasing neurotrophin efficacy in response to neuronal "stress". It has recently been shown that not only neurotrophins in the NGF family, but probably several other neurotrophic compounds such as members of the FGF family are also upregulated rapidly in response to injury. A tentative conclusion from these studies is that many neurotrophic factors act in a local para- or autocrine mode to protect neurons in jeopardy. Thus pharmacological treatment with such factors might be useful also clinically to rescue "stressed" neurons.

Transfer of Cells to the CNS

While terms such as "grafts" and "transplants" are used frequently in various experimental situations and in clinical trials, it is important both from a scientific standpoint and to avoid public misconception that the techniques involving transfer of living cells to the central nervous system are properly described as different from various transplant strategies used in other parts of the body. Thus, grafting to the brain and the spinal cord only involves cell suspensions or small tissue fragments. It is not, cannot be, and should not be transplantation of whole parts of brain tissue. Cell transfer can be used in several different ways: the most straightforward is the replacement therapy attempted in Parkinson's disease, in which the lost

adult dopamine neurons are replaced by implantation of embryonic dopamine neuroblasts into the target area (Fig. 1). This technique has proven efficient in animals and clinical results to date are relatively promising[10,16,23]. As one alternative in Parkinson's disease, one may transfer cells to the brain that secrete the lost signal substance, but are not themselves substantia nigra neurons. One possibility that has sustained extensive animal experimentation (see[12]) and is also currently on clinical trial is to use chromaffin adrenal medullary cells[11,13]. Being of neural crest origin, chromaffin cells are closely related to sympathetic neurons. When removed from the surrounding adrenal cortex and, particularly when supplemented with nerve growth factor, adult chromaffin adrenal medullary cells are able to transform into neuronal phenotypes, extend long neurites, and synthesize noradrenaline as well as dopamine. The main advantages with the adrenal medullary cells as compared to foetal substantia nigra cells include the fact that the patient is his or her own donor, which makes the procedure more easily organized and also makes immunosuppressive treatment unnecessary.

Animal experimentation has suggested other alternatives to foetal substantia nigra neuroblasts in Parkinson's disease: one approach aims at transfering

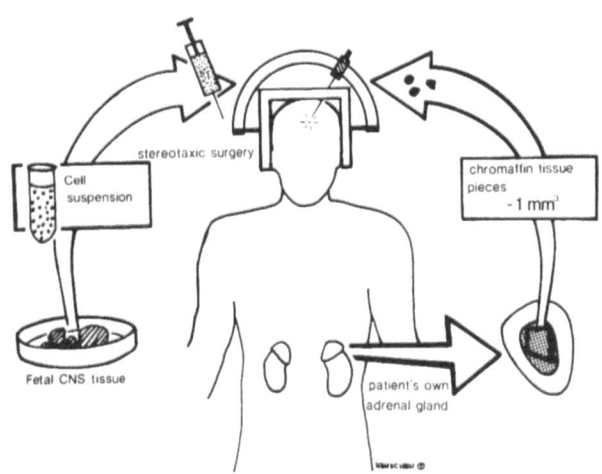

Fig. 1. Grafting procedures in Parkinson's disease. Schematic illustration of two cell transfer approaches currently undergoing clinical trials in Parkinson's disease. To the left is depicted the use of human fetal CNS tissue, obtained after early routine abortions. Mesencephalic tissue containing dopamine neurons is made into a cell suspension which is injected using stereotaxic surgery. To the right is depicted removal of the major part of one adrenal gland. Adrenal medullary tissue is then collected and similarly implanted using stereotactic surgery

the necessary enzymes for making L-dopa or dopamine into cell lines which could then be transplanted to serve as local sources of the lost neurotransmitter. If such cell lines are obtained from other sources, it is assumed that immunosuppression will again become necessary; however, if one could transfect primary cell lines, for instance fibroblasts from the patient, then immunosuppression could again be avoided. One problem with established cell lines is their tendency to continue to divide after transfer to the brain. There are now various ways of controlling unwanted mitogenic activity including the insertion of genes coding for products that make cell division temperature-sensitive. Thus, cell lines can be constructed that will only divide at temperatures lower than body temperature, and can thus be maintained in vitro, but will cease to divide after transfer to the brain. Another use of cell transfer techniques would be to transfect cells with neurotrophic factors. When introduced into the brain, such cells would presumably produce and secrete neurotrophic factors which would be released inside the blood-brain barrier to act at appropriate target systems. Again, animal experimentation suggests that this principle is valid[5,15,17]. However, there have been repeated problems with maintaining synthesis of neurotrophic molecules after transfer to the brain. Further experimentation using various promotors, perhaps inducible promotors, might overcome these problems.

As an alternative to transfected cells, it may also be useful to attempt grafting of cells that normally secrete neurotrophic factors. Examples of such cells are the cells of the striated tubules of the male mouse submandibular gland which secrete extremely high levels of NGF and Schwann cells, which after grafting, upregulate secretion of NGF and presumably several other neurotrophic factors. Schwann cells have been experimentally effective in supporting cografts of adrenal medullary tissue, an approach that has also been taken to clinical trials by combining adrenal medullary tissue pieces with peripheral nerve minces.

Transfer of Genes to the CNS

With the rapid advances of molecular biology techniques, it now begins to appear feasible to transfer genes also to non-dividing cells and obtain long-lasting effects. Thus, using retrovirus approaches or other techniques, it may be possible to transfer not cells, but rather genes into already existing neuronal or glial cells in the brain.

Treatment with Neurotrophic Factors

Based on animal experimentation, we have carried out initial clinical trials with intracerebral infusion of NGF in Parkinson's disease and in Alzheimer's disease. In the case of Parkinson's disease, NGF infusion into putamen has been carried out to support intraputaminal adrenal medullary autografts[13]. Our ongoing clinical results (three patients to date) suggests that one can indeed obtain better and more long-lasting effects of adrenal medullary implants by temporary NGF support.

In Alzheimer's disease our clinical trial with NGF infusion rests on the cholinergic hypothesis of cognitive dysfunction seen in Alzheimer's disease. We know that cholinergic neurons are NGF-sensitive and probably NGF-dependent. We also know that lesioning the cholinergic projections to the hippocampus and cortex cerebri will cause memory dysfunction and other cognitive disturbances. The cholinergic systems are severely degenerated in Alzheimer's disease. We have thus initiated clinical trials infusing NGF chronically into the lateral ventricle. The first patient received NGF via this route for three months, and we were able to note several changes towards normal including an improved cerebral blood flow and increased cerebral nicotine binding (both data obtained by PET), improvement of the EEG as well as certain cognitive improvements. These changes lasted as long as the NGF infusion or outlasted the NGF treatment for up to one year (EEG changes)[14].

Concluding Remarks

In this mini-review I have discussed various ways of transfering genes, neurotrophic molecules, or cells to the central nervous system in order to obtain long-lasting improvements in patients with neurodegenerative diseases. Clearly, the clinical trials are in a very early phase for both neurotrophic delivery and cell delivery, and have not been initiated in terms of gene delivery. There are many possible future improvements of the technologies. One interesting approach involves binding neurotrophins to molecules that will enable their transport across the blood-brain barrier, thus allowing for, e.g., i.v. injections of neurotrophins[7].

Taken together, the three approaches outlined above all aim at long-lasting counteractive and reparative measures in the central nervous system. They have general applicability and might become useful treatment strategies not only in neurodegenerative diseases such as Parkinson's disease, Alzheimer's disease, or ALS, but also in other common afflictions of the nervous system such as ischaemia, stroke and injury.

Acknowledgements

Supported by the Swedish Medical Research Council, the Swedish Natural Science Research Council, Karolinska Institutets Fonder, M. and M. Wallenbergs stiftelse and USPHS grants NS 09199 and AG04418.

References

1. Barde YA (1990) The nerve growth factor family. Prog Growth Factor Res 2: 237–248
2. Bothwell M (1991) Keeping track of neurotrophin receptors. Cell 65: 915–918
3. David S, Aguayo AJ (1981) Axonal elongation into peripheral nervous system "bridges" after central nervous system injury in adult rats. Science 214: 931–93
4. Ebendal T (1992) Function and evolution in the NGF family and its receptors. J Neurosci Res 32: 461–470
5. Ernfors P, Ebendal T, Olson L, Mouton P, Strömberg I, Persson H (1989) A cell line producing recombinant nerve growth factor evokes growth responses in intrinsic and grafted central cholinergic neurons. Proc Natl Acad Sci USA 86: 4756–4760
6. Freed CR, Breeze RE, Roxenberg NL, Schneck SA, Kriek E, Qi JX, Lone T, Zhang YB, Snyder JA, Wells TH, Ramig LO, Thompson L, Mazziotta JC, Huang SC, Grafton ST, Brooks D, Sawle G, Schroter G, Ansari AA (1992) Survival of implanted fetal dopamine cells and neurologic improvement 12 to 46 months after transplantation for Parkinson's disease. N Engl J Med 327: 1549–1555
7. Friden PM, Walus LR, Musso GF, Taylor MA, Malfroy B, Starzyk R (1991) Antitransferrin receptor antibody and antibody-drug conjugates cross the blood-brain barrier. Proc Natl Acad Sci USA 88: 4771–4775
8. Leibrock J, Lottspeich F, Hohn A, Hofer M, Hengerer B, Masiakowski P, Thoenen H, Barde YA (1989) Molecular cloning and expression of brain-derived neurotrophic factor. Nature 341: 149–152
9. Levi-Montalcini R (1987) The nerve growth factor 35 years later. Science 237: 1154–1162
10. Lindvall O, Rehncrona S, Brundin P, Gustavii B, Åstedt B, Widner H, Lindholm T, Björklund A, Leenders KL, Rothwell JC, Frackowiak R, Marsden CD, Johnels B, Steg G, Freedman R, Hoffer BJ, Seiger Å, Bygdeman M, Strömberg I, Olson L (1989) Human fetal dopamine neurons grafted into the striatum in two patients with severe Parkinson's disease: A detailed account of methodology and a 6-month follow-up. Arch Neurol 46: 615–631
11. Backlund E-O, Granberg PO, Hamberger B, Knutsson E, Mårtensson A, Sedvall G, Seiger Å, Olson L (1985) Transplantation of adrenal medullary tissue to striatum in Parkinsonism. First clinical trials. J Neurosurg 62: 169–173
12. Olson L (1988) Grafting in the mammalian central nervous system: Basic science with clinical promise. In: Magistretti P (ed) Discussions in neurosciences. FESN, Geneva, pp 1–73
13. Olson L, Backlund E-O, Ebendal T, Freedman R, Hamberger B, Hansson P, Hoffer B, Lindblom U, Meyerson B, Strömberg I, Sydow O, Seiger Å (1991) Intraputaminal infusion of nerve growth factor to support adrenal medullary autografts in Parkinson's disease: One-year follow-up of first clinical trial. Arch Neurol 48: 373–381
14. Olson L, Nordberg A, von Holst H, Bäckman L, Ebendal T, Alafuzoff I, Amberla K, Hartvig P, Herlitz A, Lilja A, Lundqvist H, Långström B, Meyerson B, Persson A, Viitanen M, Winblad B, Seiger Å (1992) Nerve growth factor affects ^{11}C-nicotine binding, blood flow, EEG, and verbal episodic memory in an Alzheimer patient (case report). J Neural Transm (Park Dis Dement Sect) 4: 79–95
15. Rosenberg MB, Friedmann T, Robertson RC, Tuszynski M, Wolff JA, Breakefield XO, Gage FH (1988) Grafting genetically modified cells to the damaged brain: Restorative effects of NGF expression. Science 242: 1575–1578
16. Spencer, DD, Robbins RJ, Naftolin F, Marek KL, Vollmer T, Leranth C, Roth RH, Price LH, Gjedde A, Bunney BS, Sass KJ, Elsworth JD, Kier EL, Makuch R, Hoffer PB, Redmond Jr DE (1992) Unilateral transplantation of human fetal mesencephalic tissue into the caudate nucleus of patients with Parkinson's disease. N Engl J Med 327: 1541–1548
17. Strömberg I, Wetmore CJ, Ebendal T, Ernfors P, Persson H, Olson L (1990) Rescue of basal forebrain cholinergic neurons after implantation of genetically modified cells producing recombinant NGF. J Neurosci Res 25: 405–411
18. Thoenen H (1991) The changing scene of neurotrophic factors. Trends Neurosci 14: 165–170
19. Wetmore C (1992) Brain-derived neurotrophic factor: Studies on the cellular localization and regulation of BDNF, related neurotrophins and their receptors at the mRNA and protein level. Ph.D. Thesis. Karolinska Institutet, Stockholm
20. Wetmore C, Cao Y, Pettersson RF, Olson L (1991) Brain-derived neurotrophic factor: Subcellular compartmentalization and interneuronal transfer as visualized with antipeptide antibodies. Proc Natl Acad Sci USA 88: 9843–9847
21. Wetmore C, Bean AJ, Olson L (1992) Regulation of cortical BDNF expression by non-NMDA glutamate receptors (manuscript)
22. Wetmore C, Cao Y, Pettersson RF, Olson L (1993) Brain-derived neurotrophic factor (BDNF) peptide antibodies: Characterization using a vaccinia virus expression system. J Histochem Cytochem 41: 521–534
23. Widner H, Tetrud J, Rehncrona St, Snow B, Brundin P, Gustavii B, Björklund A, Lindvall O, Langston JW. Bilateral foetal mesencephalic grafting in two patients with Parkinsonism induced by 1-methyl-4-phenyl-1,2,3,6-tetrahydropyridine (MPTP). N Engl J Med 327: 1556–1563

Correspondence: L. Olson, M.D., Department of Histology and Neurobiology, Karolinska Institutet, S-10401 Stockholm, Sweden.

Acta Neurochir (1993) [Suppl] 58: 8–16

Neurotransplantation: a Clinical Update

R. A. E. Bakay

Emory University School of Medicine and Yerkes Regional Primate Research Center, VAMC, Decatur, GA, U.S.A.

Summary

The use of grafts to correct neurological disorders has great promise. The progress toward this goal, as it relates to Parkinson's disease, is briefly reviewed. Although there are a number of questions that remain unanswered, recent reports of improvement in parkinsonian symptoms are very encouraging. In order to successfully evaluate the clinical trials, multicenter and/or standardized reporting techniques will be required. Future studies will need to concentrate on improving the graft survival and ability of the graft to reinnervate the host. Eventually alternative tissues may alleviate the need for fetal tissue. The use of neurotrophic factors should prove an important adjuvant to the repair and restoration of lost neurological function. As this technology is applied to other neurological diseases, it will be important to evaluate the appropriateness of grafting through extensive animal model studies and to balance the potential benefits of such therapy against the degree of risk from surgery and the severity of the disease.

Keywords: Parkinson's disease; transplantation and regeneration; foetal tissue; adrenal medulla.

Introduction

A decade has passed since the initial attempt at central nervous system (CNS) transplantation in humans. Progress has been slow and irregular[5,57,77]. Similar to other innovative surgical techniques, there have been pendular swings of great activity and interest to little activity and disappointment[55,92]. This has occurred despite the fact that activity in the neuroscientific community has been logarithmically expanding. The basic principles of CNS transplantation have become more sophisticated. We now can understand the reasons for the failures of the past and can plan how to make the future more successful.

History of Adrenal Medullary Grafting

The application of CNS transplantation for the treatment of Parkinson's disease was realized in Sweden by Erik-Olof Backlund on March 30, 1982[4]. The procedure was a bold application of the established scientific principles of the time. Fetal mesencephalic grafts placed in the striatum of unilateral nigrostriatal lesioned rats could correct parkinsonism-like behavioral abnormalities through anatomical reinnervation, and there was a suggestion that the chromaffin cells of the adrenal medulla could provide the same amelioration of symptoms[31]. In this first attempt to internalize the therapy for parkinsonism, two severely affected patients who responded to L-dopa but had severe dyskinesias were selected to have small fragments of adrenal medulla stereotactically placed into the caudate. Six months later, the patients' symptoms were unaffected by the procedure.

Based on the fact that the dopamine loss is greater in the putamen than the caudate, and on increasing evidence suggesting that the putamen is more important to the ultimate production of motor activity, Backlund and his colleagues stereotactically placed grafts in the putamen of two additional less severely affected patients[58]. Minor symptomatic improvements occurred but did not persist. Clinical evaluation of these patients was markedly enhanced by the contributions of a neurologist who specialized in movement disorders. Subsequently, it was demonstrated that, although neonatal adrenal tissue seemed to be effective in the rat model, adult tissue was not effective[29]. Clinical investigations at this point ceased, and basic science investigations of nerve growth factor (NGF) in the promotion of adult adrenal medullary survival were initiated[75,91].

Against this background, basic scientists viewed with great skepticism the report from Mexico in April 1987 by Madrazo *et al.*[66]. Using a transcortical intraventricular microsurgical approach, adrenal medullary grafting in the caudate produced a dramatic improvement in the parkinsonian symptoms of two patients. Madrazo attributed the success to the intraventricular placement, which in animal studies was demonstrated to be superior to intraparenchymal placement[30,86,99]. Although reported as preliminary and that "further work was necessary to see whether this procedure would be applicable over the long term in other types of patients with Parkinson's disease," the stimulus for clinical studies was enormous. Publication in a major medical journal heightened awareness in the neurosurgical community[66]; initiation of a clinical study in the U.S.[2] and the apparent confirmation independently in China[47] undoubtedly also contributed to the initiation of clinical studies worldwide[48,63,65,72,80,81]. By the end of 1987, in the U.S. alone 12 centers had operated on over 60 patients in an attempt to reproduce these findings.

The early reports that all patients improved regardless of age, and that morbidity was very low[66,67], could not be confirmed by investigators in the U.S. whether by reproducing exactly the Madrazo technique[17,35,37,56,73,78], modifying the technique to employ stereotactic guidance to a craniotomy[2,50,95], or employing stereotactic techniques alone[3,34]. There was indeed improvement in many of these patients. However, the degree of improvement was relatively small compared to that which had been anticipated, and the morbidity rate was markedly higher than anticipated. Two registries summarize this experience.

The United Parkinson Foundation (UPF) Registry developed a database from 13 centers on 61 patients evaluated by neurologists with experience in movement disorders using the United Parkinson's Disease Rating Scale (UPDRS). The Registry's overall assessment at one year (n = 52) reported that 57% of patients improved when examining only survivor data, or 44% of all patients improved[36]. Improvement was defined as a score better than baseline for the combined ON and OFF rating on the activity of daily living and motor components of the UPDRS. Morbidity at one year was 15% for neurological, 41% for psychiatric, 4% for cardiac, 2% for abdominal, and 2% for pulmonary. Six patients died, but none during the perioperative period. Even more discouraging were the results at 2 years (n = 56), which indicated a 32% improvement among the survivors and a 19% improvement in the total group of patients. The effectiveness was felt to be similar to the improvement in the amount of ON time achieved with new dopamine agonists.

The American Association of Neurological Surgeons GRAFT Project was designed to be far more inclusive[7]. Careful tracking of case reports in the U.S. recorded a total of 125 patients from the U.S. operated upon between 1987 and 1990. The GRAFT Project includes 113 patients from the U.S. during this time period, which represents slightly more than 90% of the total number of operated adrenal grafted patients. Followup data were obtained on 110 cases (97%). Only 64 (58.2%) of these patients did not have some type of complication. Serious abdominal complications occurred in 9 (8.2%) patients; medical complications occurred in 21 (19.1%) patients, and intracranial complications occurred in 11 (13.6%) patients. The complication rate decreased from 1987 (Table 1) to 1988 (Table 2) as surgical experience increased and patient selection criteria improved[6]. As with the UPF Registry, 6 patients died in the first year, but none in the perioperative period. Qualitative overall assessment of improvement occurred in 63% of patients with followup data at one year (n = 83) or 47% of the patients enrolled in this study. These data are similar to the UPF study despite being qualitative instead of quantitative.

A number of important lessons were learned about how to improve the safety and efficacy of the procedure. Adrenal glands obtained from a transabdominal approach had a much higher rate of abdominal (p < 0.001) and medical (p < 0.01) complications compared to the retroperitoneal approach. A

Table 1. *Incidence of Complications Reported to the AANS GRAFT Project for Patients Receiving Transplant in 1987*

	Abdominal		Medical		Intracranial		Total*	
	n	%	n	%	n	%	n	%
Nontransient	7	10.6	15	22.7	6	9.1		
Transient	3	4.5	8	12.1	11	16.7		
Any complications*	7	10.6	18	27.3	15	22.7	32	48.5
No complications*	59	89.4	48	72.7	51	77.3	34	51.5

* Complication per patient excluding multiple complications on the same patient.

Table 2. *Incidence of Complications Reported to the AANS GRAFT Project for Patients Receiving Transplant in 1988*

	Abdominal		Medical		Intracranial		Total*	
	n	%	n	%	n	%	n	%
Nontransient	2	5.0	6	15.0	3	7.5		
Transient	2	5.0	5	12.5	1	2.5		
Any complications*	2	5.0	7	17.5	2	7.5	12	30.0
No complications*	38	95.0	33	82.5	37	92.5	28	70.0

* Complication per patient excluding multiple complications on the same patient.

potentially more important finding for future studies relates to the intracranial technique. Centers that used stereotactic grafting techniques exclusively had fewer intracranial (p < 0.05) and total (p < 0.05) complications compared to those using craniotomy with or without stereotactic guidance. Although none of the patient demographics correlated with overall improvement at one year, there was a trend for younger patients and less severely disabled patients, as measured by Hoehn and Yahr stage or Schwab and England scores to do better, most likely because of the lower complication rate. The AANS GRAFT Project is unaware of anyone in the U.S. who is continuing to use the Madrazo technique.

Future of Adrenal Medullary Grafting

Several lines of investigation have been pursued based on a better understanding of the underlying mechanisms of CNS grafting. Autopsy results from patients with adrenal medullary grafts demonstrated a paucity of cell survival[3,20,40,44,46,52,79,94], but potential for dopaminergic axonal sprouting from the host[40,52]. Chromaffin cells are known to need NGF for morphologic conversion from an endocrine phenotype to a neuronal phenotype, and recent evidence suggests that there is little NGF in the caudate of nonhuman primates[53]. One logical extension that has been found effective in adult rats is the infusion of NGF after grafting[76,91]. Backlund and colleagues have initiated a study with this technique and have reported some degree of success[75].

The sprouting response and lack of surviving cells suggest to some that the lesion alone may be important to the observed improvement[13,24]. Clinical studies with lesions in this area performed by Meyer are frequently misinterpreted. None of his patients had a lesion of the caudate alone[14,70]. Most of his patients had removal of the anterior limb of the internal capsule as far back as the genu. Although autopsy data

unfortunately are not available, he undoubtedly damaged the globus pallidus and putamen. The occasional improvement he did observe was immediate and transient. He did not demonstrate late improvement as with CNS grafting. This area was revisited by Cooper using stereotactic lesioning techniques, but again the overall results were felt to be a failure[19]. Nonhuman primate studies did show that sham lesions could produce modest behavioral effects[8]. The question remained whether cells were important. Cografting of adrenal medullary tissue with peripheral nerve tissue to supply Schwann cells, and thereby a ready source of NGF, dramatically improved chromaffin cell survival[53] and demonstrated marked behavioral improvement[23]. Clinical studies[64,96] have been initiated using adrenal medullary peripheral

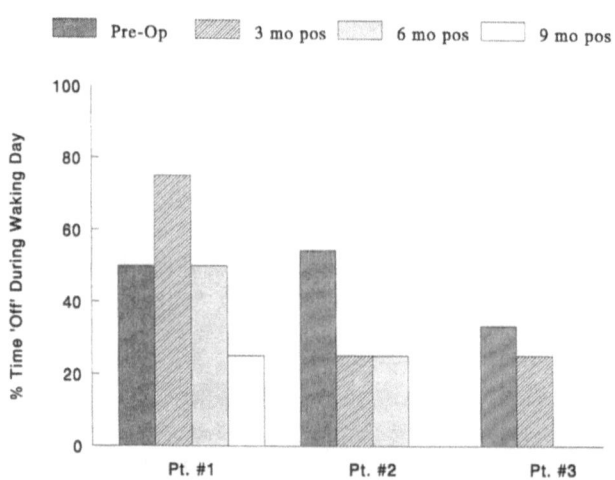

Fig. 1. The results of preliminary data on 3 patients with adrenal medullary and peripheral nerve cografts are illustrated. Using the United Parkinson Disease Rating Scale, the percent OFF time was recorded from the mean of multiple preoperative evaluations and postoperative evaluations at 3, 6, and 9 months. These data were presented by Watts *et al.*[96] at the IVth International Neurotransplantations Meeting

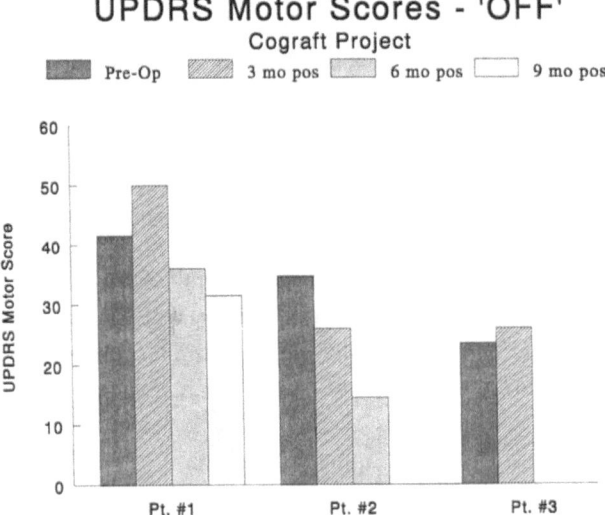

UPDRS Motor Scores - 'OFF'
Cograft Project

Pre-Op   3 mo pos   6 mo pos   9 mo pos

Fig. 2. This graft demonstrates the subset of motor scores from the United Parkinson Disease Rating Scale (*UPDRS*) during OFF times from the same patients as discussed in Fig. 1

nerve cografts stereotactically implanted into the caudate and putamen (Figs. 1 and 2).

While the adrenal gland continues to be an attractive source for dopamine because autografts avoid immunological problems. There remains a question of the quality of the adrenal medullary tissue in parkinsonian patients[90,93]. Other sources of dopamine autologous tissues are available[5,10,57,77]. Superior cervical ganglion cells have been demonstrated in nonhuman primates to improve parkinsonian behavior, and this technique has been used in clinical studies[38,43,45]. Further refinements in the quantity and quality of autologous tissue are anticipated from tissue culturing and cell separation techniques. One could envision in the future a stereotactic biopsy of the adrenal gland, culturing of chromaffin cells, and subsequent stereotactic CNS grafting.

Fetal Mesencephalic Grafting for Parkinson's Disease

The strongest basic science case can be made for fetal ventral mesencephalic grafting in the treatment of Parkinson's disease[5,11,57,77]. The initial fetal mesencephalic grafting in a Parkinson patient was performed using an open craniotomy approach to the caudate and tissue from a spontaneously aborted fetus at 14 weeks gestation[68]. To critics, this represented the wrong surgical technique, target, fetal tissue source, and fetal gestational age[21,25]. Despite this, persistent success has been reported[69]. They also grafted fetal adrenal medullary tissue, which has not produced as

good results,[69] probably for a number of reasons[21]. The database on these patients is still small and without a sufficiently long followup to analyze with confidence.

Almost all the other centers involved with fetal grafting are employing the stereotactic technique[18,22,27,28,39,41,42,58-62,65,71,81]. These centers have varied dramatically in the number and age of the fetus used as well as the target. Hitchcock *et al.*[39-41] use a stereotactic technique with fetal tissue of 11 to 19 weeks gestation. Although experimental data suggest that 6 to 9 weeks is optimal[15,16,32], they believe there are data to demonstrate that the older fetuses can be used[82]. They are also systematically varying the target but cannot recommend the best target or targets[42].

After an apparently unsuccessful set of patients, technical problems in the implantation technique have been resolved by Lindvall and colleagues[58-62]. Their technique continues to vary with the most recent use of multiple fetuses into bilateral and multiple targets of MPTP patients[61]. It remains to be determined whether the greater improvement seen in these patients is due to the multiple targeting or the fact that the MPTP patients may represent a more pure dopaminergic lesion.

The most rigorous attempt at behavioral quantification has been performed by Freed *et al.*[27,28]. Their effort has also been one of gradually increasing the number of fetuses and multiple targets. While some centers use immunosuppressants on all their patients and others never do, Freed's group has elected to alternate between using and not using immunosuppressants. The rationale for not using immunosuppressants is based on nonhuman primate data which evidence long-term fetal survival without immunosuppressant, as well as the added cost and risk from the immunosuppressive agents[26]. They have developed techniques to allow delayed implantation while maintaining cellular viability with tissue cultures[28]. The other American study at Yale also allows delayed implantation by using cryopreservation[84,85]. Both of these techniques allow systematic analysis of the fetal tissue to confirm the presence of dopaminergic cells and to insure the absence of infection.

A number of serious problems remain with these clinical studies. In order to facilitate interpretation of these results, the use of the Core Assessment Program for Intracerebral Transplantations (CAPIT) evaluations are recommended[98]. Despite encouraging changes on positron emission tomography suggesting increase in dopaminergic containing tissue, there remains the need for definitive anatomical confirmation that these changes have some relevance. Anatomical proof of successful fetal grafting and reinnervation of the host striatum is lacking. A single autopsy study revealed very few dopaminergic cells surviving[83]. It turns out, however, that this patient had progressive supranuclear palsy which had not been particularly responsive to CNS grafting[51]. Since it will not be possible to reinnervate the entire striatum, the question remains as to how the dopamine can be restored in other areas since dopamine penetrates brain substance and CSF very poorly[9,54,93]. Animal data suggest that pharmacological therapy for the Parkinson patients may, in fact, inhibit graft survival[89]. There are, however, contradictory data which suggest that this is not the case[12].

Future of Fetal Grafting for Parkinson's Disease

All of the major fetal grafting centers worldwide are claiming some degree of success. It is interesting to note that none of these centers uses the exact same technique they advocated even one year ago. Before full acceptance can be given, larger numbers of patients with more consistent results and use of established evaluation systems will be required. The problem in evaluating the reports from these centers is the lack of an internationally recognized movement disorders specialist to ensure reliable clinical examinations. Seminal questions still need to be answered:

Who should be included in the admission criteria?
What surgical technique should be used?
When in the course of the disease should the patient be grafted?
Where should the graft be placed?
Why should this technique be used as opposed to evolving medical therapy or alternative ablative surgery?

The latter, of course, will require a randomized multicenter trial, which requires the establishment of a widely accepted technique.

Some of the answers may be obtained through animal research. The MPTP nonhuman primate model is an excellent means to this end. Unfortunately, this area remains underutilized and underfunded. It also should be the primary testing ground for other grafting techniques that may replace fetal grafts. Techniques of encapsulated cells[1] or genetically engineered cells[33] are currently receiving most of the attention, but there is also an increasing awareness of the utility of neurotrophic factors which may allow the host to self-repair or even prevent disease from occurring[74]. The combination of some type of cellular replacement and neurotrophic factor(s) appears to be a powerful tool for the future.

One largely overlooked problem still remains: the immunology of these allografts[5,57,77,97]. Rejection or autoimmune disorders have not been reported. Fetal mesencephalic allografts in monkeys and humans do not appear to stimulate the afferent limb of the immune system (unpublished data). But rejection remains possible for the remainder of the host's life. Nonhuman primate data suggest that this can occur. Further work is necessary to determine the degree of risk and how to prevent the development of rejection.

Conclusion

A great deal of work is required before CNS transplantation will be accepted as therapy for Parkinson's disease or any other disorder. Some of this involves clinical research and basic science research. Unquestionably, there will be a series of transitions as we increase our knowledge of CNS repair. We must be astute as to the point when enough research is enough. There are some reasonable standards to apply. First, however, the technique should be evaluated thoroughly in animal models. Then must be matched the degree of risk with the severity of the disease.

Diseases such as Alzheimer's and Huntington's, which are fatal and for which there is no effective therapy, would merit more aggressive and earlier application of clinical trials. The problem, however, is that the animal models may not be reflective of these clinical disorders. In this situation, further evaluation of the pathophysiology is required. Clinical trials for Huntington's disease have been criticized because the scientific rationale for these clinical trials has not been well established[87]. In other diseases such as epilepsy, in which the patient is expected to live a long time and for which there is good treatment, extreme caution should be exercised before entering precipitously into CNS transplantation trials based on a small number of experiments in models that may or may not reflect the pathophysiology of the disease. Thus, there is a three-way balancing act between the weight of the basic science data, the risk of the surgery, and the gravity of the disease being treated.

Almost every animal model of disease is being investigated for potential therapeutic intervention with CNS transplantation. Hopefully, at the appropriate time, well-performed studies will be able to test their efficacy. The CNS is too complicated to anticipate simple replacement to "cure" any disease. Nonetheless, therapy through biopharmacological replacement, compensation repair, and preventive intervention should be possible. It is an exciting time for neuroscience and neurotransplantation. The scientific leaders in the field are concerned[49,88] and caution in clinical studies needs to be exercised lest the field lose its credibility and delay introduction of effective reparative therapy for the CNS. It is indeed time to "benefit from more patience rather than more patients"[88].

Acknowledgements

We wish to extend our appreciation to our collaborators, Drs, Alan Freeman, Allen Mandir, P. Michael Iuvone, Ray L. Watts, Nelson B. Watts, Sam Graham, Aftab A. Ansari, Connie Hill, and Carl Herring. We would also like to thank Joyce Klemm, Nickie Skinner, and Emile Worthy for their technical assistance, and Dr. Anne LeMoine for typing the manuscript. This work was supported in part by grants from the American Parkinson Disease Association (R.L.W., R.A.E.B.), United Parkinson Foundation (R.L.W.), NIH R01-NS24312 (R.L.W.), NIH R01-NS24340 (R.A.E.B., R.L.W.), VA Merit Award (R.A.E.B.), NIH RR-00165 (Yerkes Primate Research Center), and the Emory University Parkinson Research Fund.

References

1. Aebischer P, Tresco PA, Winn SR, Greene LA, Jaeger CB (1991) Long-term crossspecies transplantation of a polymer-encapsulated dopamine-secreting cell line. Exp Neurol 111: 269–275
2. Allen GS, Burns RS, Tullipan NB (1989) Adrenal medullary transplantation to the caudate nucleus in Parkinson's disease: Initial clinical results in 18 patients. Arch Neurol 46: 487–491
3. Apuzzo MLJ, Neal JH, Waters CH, Appley AJ, Boyd SD, Couldwell WT, Wheelock VH, Weiner LP (1990) Utilization of unilateral and bilateral stereotactically placed adrenomedullary-striatal autografts in parkinsonian humans: Rationale, techniques, and observations. Neurosurgery 26: 746–757
4. Backlund EO, Granberg PO, Hamberger B, Knutsson E, Martensson A, Sedvall G, Seiger A, Olson L (1985) Transplantation of adrenal medullary tissue to striatum in parkinsonism: First clinical trials. J Neurosurg 62: 169–173
5. Bakay RAE (1990) Transplantation into the central nervous system: A therapy of the future. Stereotact Neurosurg 1: 881–895
6. Bakay RAE (1991) Selection criteria for CNS grafting into Parkinson's disease patients. In: Lindvall A, Björklund A, Widner H (eds) Intracerebral transplantation in movement disorders, Chapter 12. Elsevier, Amsterdam, pp 137–148
7. Bakay RAE, Allen GS, Apuzzo M, Borges LF, Bullard DE, Ojemann GA, Oldfield EH, Penn R, Purvis JT, Tindall GT (1990) Preliminary report on adrenal medullary grafting from the American Association of Neurological Surgeons GRAFT Project. In: Dunnet SB, Richards SR (eds) Neural transplantation: From molecular basis to clinical application. Elsevier, Amsterdam, Chapter 67, pp 603–617
8. Bankiewicz KS, Plunkett RJ, Kopin IJ, Jacobowitz DM, London WT, Oldfield EH (1988) Transient behavioral recovery in hemiparkinsonian primates after adrenal medullary allografts. Prog Brain Res 78: 543–749
9. Bankiewicz KS, Plunkett RJ, Mefford I, et al (1990) Behavioral recovery from MPTP induced parkinsonism in monkeys after intracerebral tissue implants is not related to CSF concentrations of dopamine metabolites. In: Dunnet SB, Richards SR (eds) Neural transplantation: From molecular basis to clinical application. Elsevier, Amsterdam, pp 561–571
10. Bing G, Notter MFD, Hansen JT, Gash DM (1988) Comparison of adrenal medullary, carotid body, and PC12 cell grafts in 6-OHDA lesioned rats. Brain Res Bull 20: 399–406
11. Björklund A, Brundin P, Isacson O (1988) Neuronal replacement by intracerebral neural implants in animal models of neurodegenerative disease. In: Waxman SG (ed) Physiological basis for functional recovery in neurological disease: Advances in neurology, Vol 47. Raven, New York, pp 455–492
12. Blunt S, Jenner P, Marsden CD (1991) The effect of chronic L-dopa treatment on the recovery of motor function in 6-hydroxydopamine lesioned rats receiving ventral mesencephalic grafts. Neuroscience 40: 453–464
13. Bohn MC, Cupit F, Marciano F, Gash DM (1987) Adrenal medulla grafts enhance recovery of striatal dopaminergic fibers. Science 237: 913–916
14. Browder EJ, Meyers R (1940) A surgical procedure for post-encephalitic tremors. Trans Am Neurol Assoc 66: 176–177
15. Brundin P, Strecker RE, Widner H, et al (1988a) Human fetal dopamine neurons grafted in a rat model of Parkinson's disease: Immunological aspects, spontaneous and drug-induced behaviour, and dopamine release. Exp Brain Res 70: 192–208
16. Brundin P, Strecker RE, Widner H, et al (1988b) Survival and function of dissociated dopamine neurones grafted at different development stages or after being cultured in vitro. Dev Brain Res 39: 233–243
17. Cahill DW, Olanow CW (1990) Autologous adrenal medulla to caudate nucleus neurons grafted in advanced Parkinson's disease: 18 month results. In: Dunnet SB, Richards SR (eds) Neural transplantation: From molecular basis to clinical application. Elsevier, Amsterdam, pp 637–642
18. Cesaro P, Degos JD, Keravel Y, Peschanski M, INSERM Clinical Research Network for Intracerebral Transplantation (1992) Transplantation of fetal ventral mesencephalic neurons in patients with Parkinson's disease: The French experience. Restor Neurol Neurosci 4: 222
19. Cooper IS (1965) Surgical treatment of parkinsonism. Ann Rev Med 16: 309–330
20. Dohan FC, Robertson JT, Feler C, Schweitzer J, Hall C, Robertson JH (1988) Autopsy findings in a Parkinson's disease patient treated with adrenal medullary to caudate nucleus transplant. Soc Neurosci Abstr 7: 4
21. Dwork AJ, Pezzoli G, Fahn S, Hill R (1988) Transplantation of fetal substantia nigra and adrenal medulla to the caudate nucleus in two patients with Parkinson's disease. N Engl J Med 319: 370–371
22. Dymecki J, Zabek M, Mazurowski W, Stelmachow J, Zawada E (1992) 30-month results of foetal dopamine cell transplantation into the brain of parkinsonian patients. Restor Neurol Neurosci 4: 223
23. Ellis JE, Byrd LD, Bakay RAE (1992) A method for quantitating motor deficits in a nonhuman primate following MPTP-induced hemiparkinsonism and cografting. Exper Neurol 115: 376–387
24. Fiandaca MS, Kordower JH, Hansen JT, Jiao SS, Gash DM (1988) Adrenal medullary autografts into the basal ganglia of cebus monkeys: Injury-induced regeneration. Exp Neurol 102: 76–91

25. Freed CR (1988) Transplantation of fetal substantia nigra and adrenal medulla to the caudate nucleus in two patients with Parkinson's disease. N Engl J Med 319: 370–371

26. Freed CR (1991) Comments on brain tissue transplantation without immunosuppression. Arch Neurol 48: 259–260

27. Freed CR, Breeze RE, Rosenberg NL, Kriek E, Lone T, Wells T, Grafton S, Huang H, Mazziotta J, Sawle G, Brooks D (1992) Implants of human embryonic mesencephalic dopamine cells improve motor performance and reduce drug requirements in patients with severe Parkinson's disease 5 to 45 months after transplant. Restor Neurol Neurosci 4: 230

28. Freed CR, Breeze RE, Rosenberg NL, Schneck SA, Wells TH, Barrett JN, Grafton ST, Huang SC, Eidelberg D, Rottenberg DA (1990) Transplantation of human fetal dopamine cells for Parkinson's disease: Results at 1 year. Arch Neurol 47: 505–512

29. Freed WJ (1983) Functional brain tissue transplantation: Reversal of lesion-induced rotation by intraventricular substantia nigra and adrenal medulla grafts, with a note on intracranial retinal grafts. Biol Psychiatry 18: 1205–1267

30. Freed WJ, Cannon-Spoor HE, Krauthamer E (1986) Intrastriatal adrenal medulla grafts in rats: Long-term survival and behavioral effects. J Neurosurg 65: 664–670

31. Freed WJ, Morihisa JM, Spoor E, et al (1981) Transplanted adrenal chromaffin cells in rat brain reduce lesion-induced rotational behaviour. Nature 292: 351–352

32. Freeman TB, Spence MS, Boss BD, Spector DH, Strecker RE, Olanow CW, Kordower JH (1991) Development of dopaminergic neurons in the human substantia nigra. Exp Neurol 113: 344–353

33. Gage FH, Fisher LJ, Jinnah HA, et al (1990) Grafting genetically modified cells to the brain: Conceptual and technical issues. In: Dunnet SB, Richards SR (eds) Neural transplantation: From molecular basis to clinical application. Elsevier, Amsterdam, pp 5–8

34. Gildenberg PL, Pettigrew LC, Merrell R, Butler I, Conklin R, Katz J, DeFrance J (1990) Transplantation of adrenal medullary tissue to caudate nucleus using stereotactic techniques. Stereotact Funct Neurosurg 54 + 55: 268–271

35. Goetz CG, Olanow CW, Koller WC, Penn RD, Cahill D, Morantz R, Stebbins G, Tanner CM, Klawans HL, Shannon KM, Comella CL, Witt T, Cox C, Waxman M, Gauger L (1989) Multicenter study of autologous adrenal medullary transplantation to the corpus striatum in patients with advanced Parkinson's disease. N Engl J Med 320: 337–341

36. Goetz CG, Stebbins GT, Klawans HL, Koller WC, Grossman RG, Bakay RAE, Penn RD, and the United Parkinson Foundation Neural Transplantation Registry (1991) United Parkinson Foundation Neurotransplantation Registry on adrenal medullary transplants: Presurgical, and 1- and 2-year followup. Neurology 41: 1719–1722

37. Goetz CG, Tanner CM, Penn RD, Stebbins GT, Gilley DW, Shannon KM, Klawans HL, Comella CL, Wilson RS, Witt T (1990) Adrenal medullary transplant to the striatum of patients with advanced Parkinson's disease: 1-motor and psychomotor data. Neurology 40: 273–276

38. Hakura T, Kamei I, Nakai KY, Nakakita K, Imai T, Komai N (1988) Autotransplantation of the superior cervical ganglion into the brain. J Neurosurg 68: 955–959

39. Henderson BTH, Clough CG, Hughes RC, Hitchcock ER, Kenny BG (1991) Implantation of human fetal ventral mesencephalon to the right caudate nucleus in advanced Parkinson's disease. Arch Neurol 48: 822–827

40. Hirsch EC, Duyckaerts C, Javoy-Agid F, Hauw JJ, Agid Y (1990) Does adrenal graft enhance recovery of dopaminergic neurons in Parkinson's disease? Ann Neurol 27: 676–82

41. Hitchcock ER, Clough CG (1990) Stereotactic implantation of adrenal medulla and fetal mesencephalon (STIM)—the UK experience. In: Dunnet SB, Richards SR (eds) Neural transplantation: From molecular basis to clinical application. Elsevier, Amsterdam, pp 723–728

42. Hitchcock E, Henderson B, Hughes R, Clough C, Kenny B, Detta A (1992) United Kingdom experience with neural transplantation for advanced Parkinson's disease. Restor Neurol Neurosci 4: 230–231

43. Horvath M, Pasztor E, Palkovits M, Solyom A, Tarczy M, Lekka N, Csanda E (1990) Autotransplantation of superior cervical ganglion to the caudate nucleus in three patients with Parkinson's disease (preliminary report). Neurosurg Rev 13: 119–122

44. Hurtig H, Joyce J, Sladek JR, Trojanowski JQ (1989) Postmortem analysis of adrenal-medulla-to-caudate autograft in a patient with Parkinson's disease. Ann Neurol 25: 607–614

45. Itakura T, Kamei I, Nakai K, Naka Y, Nakakita K, Imai H, Komai N (1988) Autotransplantation of superior cervical ganglion into the brain: A possible therapy for Parkinson's disease. J Neurosurg 68: 955–959

46. Jankovic J, Grossman R, Goodman C, Pirozzolo F, Schneider L, Zhu Z, Scardino P, Garber AJ, Jhingran SG, Martin S (1989) Clinical, biochemical, and neuropathologic findings following transplantation of adrenal medulla to the caudate nucleus for treatment of Parkinson's disease. Neurology 39: 1227–1234

47. Jiao S, Zhang W, Cao J, et al (1988) Study of adrenal medullary tissue transplantation to striatum in parkinsonism. Progr Brain Res 78: 575–580

48. Jiao S, Ding Y, Zhang W, Cao J, Zhang Z, Zhang Y, Ding M, Zhang Z, Meng J (1989) Adrenal medullary autografts in patients with Parkinson's disease. N Engl J Med 321: 324–325

49. Joynt RJ, Gash DM (1987) Neural transplants: Are we ready? Ann Neurol 22: 455

50. Kelly PJ, Ahlskog JE, van Heerden JA, Carmichael SW, Stoddard SL, Bell GN (1989) Adrenal medullary autograft transplantation into the striatum of patients with Parkinson's disease. Mayo Clin Proc 64: 282–290

51. Koller WC, Morantz R, Vetere-Overfield B, Waxman M (1989) Autologous adrenal medullary transplant in progressive supranuclear palsy. Neurology 39: 1066–1068

52. Kordower JH, Cochran E, Penn RD, Goetz CG (1991) Putative chromaffin cell survival and enhanced host-derived TH-fiber innervation following a functional adrenal medulla autograft for Parkinson's disease. Ann Neurol 29: 405–412

53. Kordower JH, Fiandaca MS, Notter MFD, Hansen JT, Gash DM (1990) NGF-like trophic support from peripheral nerve for grafted rhesus adrenal chromaffin cells. J Neurosurg 73: 418–428

54. Kroin JS, Kao LC, Zhang TJ, Penn RD, Klawans HL, Carvey PM (1991) Dopamine distribution and behavioral alterations resulting from dopamine infusion into the brain of the lesioned rat. J Neurosurg 74: 105–111

55. Lewin R (1988) Cloud over Parkinson's therapy. Science 240: 390–392

56. Lieberman A, Ransohoff J, Berczeller P, Goldstein M (1990) Adrenal medullary transplants as treatment for advanced Parkinson's disease. In: Dunnet SB, Richards SR (eds) Neural transplantation: From molecular basis to clinical application. Elsevier, Amsterdam, pp 665–670

57. Lindvall O (1989) Transplantation into the human brain: Present status and future possibilities. J Neurol Neurosurg Psychiatry [Suppl]: 39–54

58. Lindvall O, Backlund EO, Farde L, Sedvall G, Freedman R, Hoffer B, Nobin A, Seiger A, Olson L (1987) Transplantation in Parkinson's disease: Two cases of adrenal medullary grafts to the putamen. Ann Neurol 22: 457–468

59. Lindvall O, Brundin P, Widner H, Rehncrona S, Gustavii B, Frackowiak R, Leenders KL, Sawle G, Rothwell JC, Marsden CD, Björklund A (1990) Grafts of fetal dopamine neurons survive and improve motor function in Parkinson's disease. Science 247: 574–577

60. Lindvall O, Rehncrona S, Brundin P, Gustavii B, Astedt B, Widner H, Lindholm T, Björklund A, Leenders KL, Rothwell JC, Frackowiak R, Marsden CD, Johnels B, Steg G, Freedman R, Hoffer BJ, Seiger A, Bygdeman M, Stromberg I, Olson L (1989) Human fetal dopamine neurons grafted into the striatum in two patients with severe Parkinson's disease: A detailed account of methodology and a 6-month follow-up. Arch Neurol 46: 615–631

61. Lindvall O, Widner H, Rehncrona S, Brundin P, Odin P, Gustavii B, Frackowiak R, Leenders KL, Sawle G, Rothwell JC, Björklund A, Marsden CD (1992a) Long-term survival and function of fetal dopaminergic grafts in patients with Parkinson's disease. Restor Neurol Neurosci 4: 230

62. Lindvall O, Widner H, Rehncrona S, Brundin P, Odin P, Gustavii B, Frackowiak R, Leenders KL, Sawle G, Rothwell JC, Björklund A, Marsden CD (1992b) Transplantation of fetal dopamine neurons in Parkinson's disease: One-year clinical and neurophysiological observations in two patients with putaminal implants. Ann Neurol 31: 155–165

63. Lopez-Lozano JJ, Bravo G, Abascal J, Clinical Puerta de Hierro Neural Transplantation Group (1991) Grafting of perfused adrenal medullary tissue into the caudate nucleus of patients with Parkinson's disease. J Neurosurg 75: 234–243

64. Lopez-Lozano JJ, Bravo G, Abascal J, CPH Neural Transplantation Group (1992) First clinical trial of cografting of autologous adrenal medulla and peripheral nerve in Parkinson's disease. Restor Neurol Neurosci 4: 207

65. Lopez-Lozano JJ, Bravo G, Abascal J, Dargallo J, Salmean J, CPH Neural Transplantation Group (1992) Madrid experience in implants of adrenal medulla and neural tissue in Parkinson's disease. Restor Neurol Neurosci 4: 194

66. Madrazo I, Drucker-Colin R, Diaz V, Martinez-Mata J, Torres C, Becerril JJ (1987a) Open microsurgical autograft of adrenal medulla to the right caudate nucleus in two patients with intractable Parkinson's disease. N Engl J Med 316: 831–834

67. Madrazo I, Drucker-Colin R, Leon V, Torres C (1987b) Adrenal medulla transplanted to caudate nucleus for treatment of Parkinson's disease: Report of 10 cases. Surg Forum 38: 510–512

68. Madrazo I, Leon V, Torres C, Aguilera MC, Varela G, Alvarez F, Fraga A, Drucker-Colin R, Ostrosky F, Skurovich M, Franco R (1988) Transplantation of fetal substantia nigra and adrenal medulla to the caudate nucleus in two patients with Parkinson's disease. N Engl J Med 318: 51 (letter)

69. Madrazo I, Franco-Bourland R, Ostrosky-Solis P, Aguilera M, Cuevas C, Zamorano C, Morelos A, Magallon E, Guizar-Sahagun G (1990) Fetal homotransplants (ventral mesencephalon and adrenal tissue) to the striatum of parkinsonian subjects. Arch Neurol 47: 1281–1285

70. Meyers R (1942) The present status of surgical procedures directed against the extrapyramidal diseases. NY State J Med 42: 535–543

71. Molina H, Quiñones R, Alvarez L, Ortega I, Gonzalez C, Muñoz J, Castellanos O, Cuétara K, Piedra J, Garcia JC, Macias R, Pavón N, Araujo F, Torres O, León M, Suárez C, Córdova F, Hernández O, Lastra O, Rojas MJ, Muchuli F, Dávila JD (1992) Stereotactic transplantation of foetal ventral mesencephalic cells: Cuban experiences from four patients with idiopathic Parkinson's disease. Restor Neurol Neurosci 4: 230

72. Molina H, Quinones R, Suarez O (1987) Open microneurosurgical autograft of adrenal medulla to the caudate nucleus in a patient with Parkinson's disease. J Neurol Sci 5 [Suppl 7]: 64

73. Olanow CW, Koller W, Goetz CG, Stebbins GT, Cahill DW, Gauger LL, Morantz R, Penn RD, Tanner CM, Klanwans HL, Shannon KM, Comella CL, Witt T (1990) Autologous transplantation of adrenal medulla in Parkinson's disease: 18-month results. Arch Neurol 47: 1286–1289

74. Olson L (1990) Grafts and growth factors in CNS: Basic science with clinical promise. In: Ohye C, Gildenberg P, Franklin PO (eds) Proceedings of Xth Meeting of the World Society for Stereotactic and Functional Neurosurgery. Karger, Basel

75. Olson L, Backlund EO, Ebendal T, Freedman R, Hamberger B, Hansson P, Hoffer B, Lindblom U, Meyerson B, Strömberg I, Sydow O, Seiger A (1991) Intraputaminal infusion of nerve growth factor to support adrenal medullary autografts in Parkinson's disease: One-follow-up of first clinical trial. Arch Neurol 48: 373–381

76. Olson L, Stromberg I, Herrera-Marschitz M, Ungerstedt V, Ebendal T (1985) Adrenal medullary tissue grafted to the dopamine-denervated rat striatum: Histochemical and functional effects of additions of nerve growth factor. In: Björklund A, Stenevi U (eds) Neural grafting in the mammalian CNS. Elsevier, Amsterdam, pp 503–518

77. Oyesiku NM, Bakay RAE (1992) Central nervous system transplantation. In: Crockard A, Hayard R, Hoff JT (eds) Neurosurgery: Scientific basis of clinical practice. Blackwell, Cambridge, MA, pp 448–469 (in press)

78. Penn RD, Goetz CG, Tanner CM, Klawans HL, Shannon KM, Comella CL, Witt TR (1988) The adrenal medullary transplant operation for Parkinson's disease: Clinical observations in five patients. Neurosurgery 22: 999–1004

79. Peterson DI, Price ML, Small CS (1989) Autopsy findings in a patient who had an adrenal-to-brain transplant for Parkinson's disease. Neurology 39: 235–238

80. Petruk KC, Wilson AF, Schnidel DR, et al (1990) Treatment of refractory Parkinson's disease with adrenal medullary autografts utilizing two-stage surgery. In: Dunnet SB, Richards SR (eds) Neural transplantation: From molecular basis to clinical application. Elsevier, Amsterdam, pp 671–676

81. Pezzoli G, Motti E, Zecchinelli A, et al (1990) Adrenal medullary autografts in 3 parkinsonian patients: Results using two different approaches. In: Dunnet SB, Richards SR (eds) Neural transplantation: From molecular basis to clinical application. Elsevier, Amsterdam, pp 677–682

82. Postans R, Detta A, Hitchcock ER (1992) The proliferative capacity of second trimester human mesencephalon. Restor Neurol Neurosci 4: 221

83. Redmond DE, Leranth C, Spencer DD, Robbins R, Vollmer T, Kim JH, Roth RH (1990) Fetal neural graft survival. Lancet 336: 820–822

84. Redmond DE, Marek KL, Robbins RJ, Naftolin F, Vollmer T, Leranth C, Roth RH, Price LH, Gjedde A, Bunney BS, Sass KJ, Elsworth JD, Makuch R, Gulanski BI, Serrano C, Spencer DD (1992) Human fetal substantia nigra grafts in 11 patients with Parkinson's disease: Preliminary clinical results. Restor Neurol Neurosci 4: 231

85. Robbins RJ, Torres-Aleman J, Leranth C, Bradberry CW, Deutch AY, Welsh S, Roth RH, Spencer D, Redmond DE, Naftolin F (1990) Cryopreservation of human brain tissue. Exp Neurol 107: 208–213

86. Rosenstein JM, Brightman MW (1978) Intact cerebral ventricle as a site for tissue transplantation. Nature 276: 83–85

87. Sanberg PR, Norman AB (1988) Adrenal transplants for Huntington's disease? Nature 335: 122 (letter)

88. Sladek JR Jr, Shoulson S (1988) Neural transplantation: A call for patience rather than patients. Science 240: 1386–1388

89. Steece-Collier K, Collier TJ, Sladek CD, Sladek JR Jr (1989) Chronic L-DOPA treatment decreases the viability of grafted

and cultured embryonic rat mesencephalic dopamine neurons. Abstr Soc Neurosci 15: 533–536

90. Stoddard SL, Ahlskog JE, Kelly PJ, Tyce GM, van Heerden JA, Zinsmeister AR, Carmichael SW (1989) Decreased adrenal medullary catecholamines in adrenal transplanted parkinsonian patients compared to nephrectomy patients. Exp Neurol 104: 218–222

91. Strömberg I, Herrera-Marschitz M, Ungerstedt U, Ebendal T, Olson L (1985) Chronic implants of chromaffin tissue into the dopamine-denervated striatum: Effects of NGF on survival, fiber growth, and rotational behavior. Exp Brain Res 60: 335–349

92. Thompson L (1992) Fetal transplants show promise. Science 257: 868–870

93. Tyce GM, Ahlskog JE, Carmichael SW, Chritton SL, Stoddard SL, van Heerden JA, Yaksh TL, Kelly PJ (1989) Catecholamines in CSF, plasma, and tissue after autologous transplantation of adrenal medulla to the brain in patients with Parkinson's disease. J Lab Clin Med 114: 185–192

94. Waters C, Itabashi HH, Apuzzo MLJ, Weiner LP (1990) Adrenal caudate transplantation: Postmortem study. Mov Disorders 5: 248–250

95. Watts RL, Bakay RAE (1991) Autologous adrenal medulla-to-caudate grafting for parkinsonism in humans and in nonhuman primates. Front Neuroendocrin 12: 357–378

96. Watts RL, Freeman A, Graham S, Bakay RAE (1992) Early experience with intrastriatal adrenal medulla/nerve cografting in Parkinson's disease. Restor Neurol Neurosci 4: 194

97. Widner H (ed) (1990) Immunological basis for intracerebral reconstructive transplantation. Doctoral dissertation. University of Lund, Lund, Sweden

98. Widner H, Langston JW, Brooks D, Fahn S, Freeman T, Goetz C, Watts R (1992) CAPIT—Core Assessment Program for Intracerebral Transplantations. Restor Neurol Neurosci 4: 223

99. Wuerthele SM, Freed WJ, Olson L, et al (1981) Effect of dopamine agonists and antagonists on the electrical activity of substantia nigra neurons transplanted into the lateral ventricle of the rat. Exp Brain Res 44: 1–10

Correspondence: Roy A. E. Bakay, M.D., Associate Professor of Neurological Surgery, Department of Neurological Surgery, The Emory Clinic, 1327 Clifton Road, N.E., Atlanta, GA 30322, U.S.A.

Acta Neurochir (1993) [Suppl] 58: 17–19

Computer Assisted CT-Guided Stereotactic Transplantation of Foetal Ventral Mesencephalon to the Caudate Nucleus and Putamen in Parkinson's Disease

H. Molina, R. Quiñones, I. Ortega, L. Alvarez, J. Muñoz, C. Gonzalez, and **C. Suárez**

Centro Iberolatinoamericano de Trasplante y Regeneración del Sistema Nervioso, Ciudad de la Habana, Cuba

Summary

We report our preliminary results related to CT-guided stereotactic transplantation of foetal ventral mesencephalic cell suspension into the striatum of five patients with idiopathic Parkinson's disease. The mean age was 51 years, the evolution time of the disease ranged from 7 to 14 years, and all of them had motor complications associated with chronic L-dopa therapy. The patients were evaluated according to the Core Assessment Program for Intracerebral Transplantations (CAPIT) for one year before and three months after surgery. The postoperative clinical assessment demonstrated significant improvement of neurological symtoms and reduction of daily L-dopa dosage.

Keywords: Parkinson's disease; transplantation; foetal tissue; ventral mesencephalon.

Introduction

On the basis of experimental work with functional grafting of embryonic nervous tissue to brain in animal models of neurological disorders[1,3] and in view of the need for new therapeutic alternatives for many patients with severe Parkinson's disease, a Neurotransplantation Program was initiated in Cuba in 1984. In 1987 we performed adrenal medulla autotransplantation on three Parkinsonian patients[9]; from January, 1988 to April, 1990, thirty patients underwent foetal ventral mesencephalic tissue transplantation into the caudate nucleus using an open micro-surgical approach[10]. The post-operative assessment carried out over 52 months demonstrated reduction of the time spent in the "off" phase, stabilization of the motor state, and statistically significant improvement of neurological symtoms from 3 to 36 months after operation[10].

A stereotactic neurotransplantation technique was first performed in Cuba in 1992. To date, stereotactic implantations of foetal mesencephalic tissue has been undertaken in five Parkinsonian patients. The preliminary results and conclusions after a follow up of three months will be briefly reported.

Patients and Methods

Five patients with idiopathic Parkinson's disease (Hoehn and Yahr Stage 4) were selected according the CAPIT criteria[5]. Mean age at time of surgery was 51 years and the evolution time of the disease ranged from 7 to 13 years. They had been on L-dopa treatment for a mean period of 9 years, and all had motor complications associated with this therapy. The patients underwent clinical assessment based on international rating scales and on CAPIT recommendations[5] for one year before and three months after operation. The pre- and post-surgical protocol also included neurophysiological recordings, CT, MRI, SPECT and a battery of biochemical, immunological and neuropsychological studies.

Using the CT-guided Leksell Stereotactic System, foetal ventral mesencephalic cell suspension was transplanted into the right caudate and putamen. Tissue from one ventral mesencephalon was implanted per patient, the postconception ages (according to crown-rump length and Carnegie stages) of the foetuses were 8–13 weeks; the mean time between abortion (legal and routine suction abortions) and implantation was 2 hours 42 minutes (range from 2 h 20 min to 3 h).

Transplantation was performed under local anaesthesia and without sedation in order to assess the neurological condition of the patients during surgery. After application of the Leksell Stereotactic frame model G, we selected the images from CT-scanner, and converted them to IBM compatible format for their further analysis in the Montreal-Leksell CT Surgical Planning System (SPS). The distal targets were chosen both directly from the display of the SPS and by reference to Stereotactic Atlases (Schaltenbrand and Wahren and Tallairach); the proximal targets were selected by taking the longest distance to the previously defined distal targets in the caudate and putamen. Based on the SPS we obtained and visualized the tracts between the entry point and the distal targets; coronal and parasagital reconstructions of the images were performed to improve three-dimensional visualization of the trajectories. The graft tissue was injected using Rehncrona's cannula (outer diameter 1.0 mm, inner diameter 0.8 mm) along three linear tracts in four patients and two linear tracts in the fourth case; the length of these trajectories ranged from 5 to 20 mm. The amount of mesencephalic cell suspension was 2.5 ul per implantation and 20–50 ul per patient; between each injection there was a 2-minute delay; the cannula was then retracted 1.4–2.8 mm, and after the final injection of each trajectory we waited 10 minutes before withdrawing the cannula from the brain. The grafted cell viability ranged from 52 to 80%.

Immunosuppression therapy was started on the day of transplantation using a combination of Cyclosporine A, 2 mg/kg daily and Prednisona, 1 mg/kg/24 h.

Results and Discussion

The clinical assessment of the patients, carried out during the twelve months before surgery and three months after transplantation, revealed a progressive reduction of both the daily time spent in "off" conditions from 58% before surgery to 13% three months after transplantation. The number of daily "off" periods per day decreased from 4.4 to 1.5 (Fig. 1). In the 3rd postoperative month, the patients showed a significant bilateral improvement of their neurological performance (from 108.8 to 59.5 in UPDRS) (Fig. 2), more marked contralaterally to the transplantation site; a less severe disability during "off" periods and increase of "on" time, with a reduction of dyskinesias from 31 to 89%.

It has been suggested that antiparkinsonian drugs, especially L-dopa, may inhibit the development of the graft[10]. The patients' sensitivity to L-dopa increased after transplantation allowing a gradual reduction of the mean dose from 1000 mg per day before surgery to 409.5 mg/daily, three months after implantation (Fig. 3).

These results, which are in agreement with those reported by other authors[2,4,7,8], suggest that the

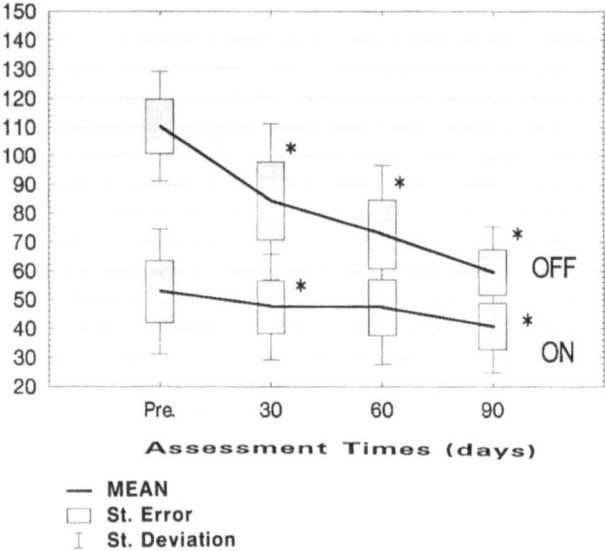

Fig. 2. Clinical evolution (UPDRS). Progressive clinical improvement from 1 to 3 months after grafting. P ≤ 0.05 Wilcoxon matched pairs test

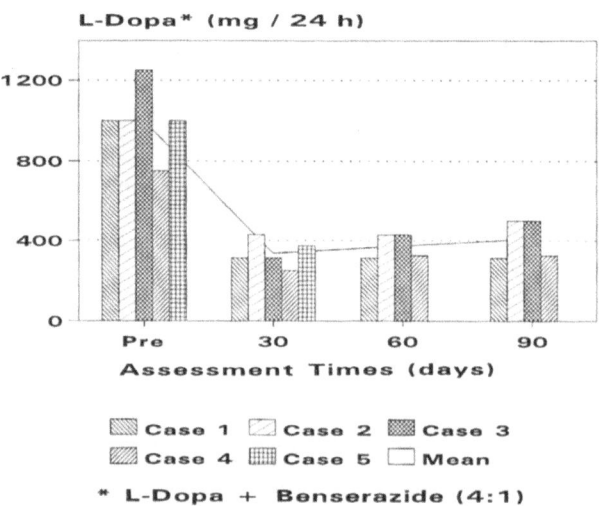

Fig. 3. Comparative daily L-dopa dose. Post-operative progressive reduction of daily L-dopa requirement (P ≤ 0.01). *250 mg L-dopa + benserazide (4:1)

probable initial improvement after grafting may be due to early release from the host striatum of dopamine and/or trophic factors[6]; the foetal tissue itself or the inflammatory cells are attracted to the graft; reinervation of the host striatum by the foetal mesencephalic tissue occurs. Increased sprouting from the remaining nigrostriatal neurons might be the major mechanism operating later in the evolution.

Dopaminergic drug responses were assessed before and after surgery; the results of the pharmacological

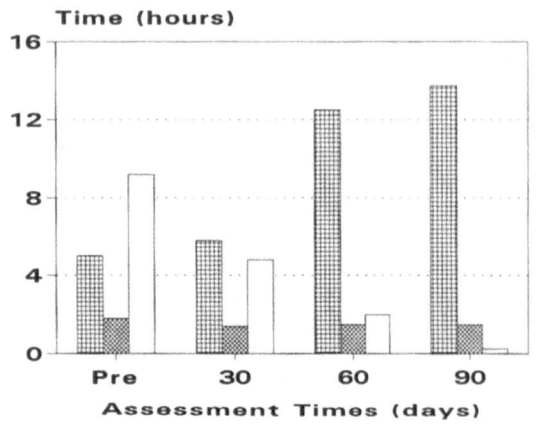

Fig. 1. Evolution of clinical fluctuations. Significant improvement of the clinical fluctuations after grafting (P ≤ 0.01)

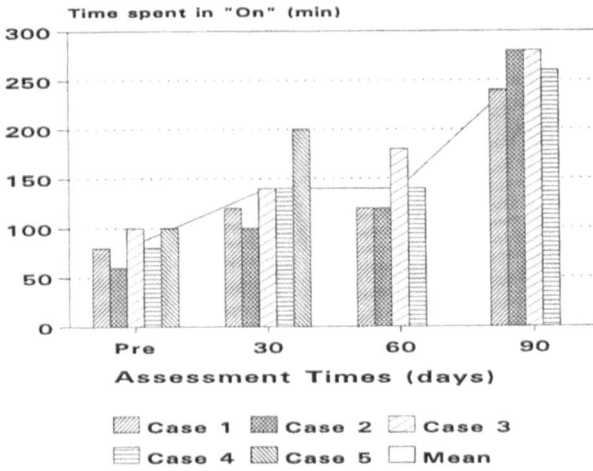

Fig. 4. Drug-induced "on" phase. The beneficial time-effect of L-dopa significantly increased after transplantation (P ≤ 0.03)

tests proved postoperative changes in the pattern of the single dose L-dopa response: the motor performance improved significantly, the "on" latency shortened from 76 minutes pre-surgery to 25 minutes three months after grafting, and the beneficial effect of L-dopa increased from 84 minutes before to 265 minutes in the 3rd postoperative month (Fig. 4). This improvement after transplantation is best explained by a dopaminergic reinnervation of the striatum. We have previously reported similar results in 30 patients who received foetal implants by an open microsurgical approach[10]. The post-surgical improvement in our patients can be summarized as follows: neurological performance, response to L-dopa treatment and quality of life improved; parkinsonian symptoms, daily requirements of L-dopa, and motor complications associated with L-dopa therapy were reduced. Based on research conducted during more than four years in the field of neural transplantation and on the results of our most recent clinical experiences, we conclude that stereotactic transplantation of dopaminergic cells may improve the neurological condition of Parkinsonian patients suffering from motor side effects of L-dopa therapy.

References

1. Bjorklund A, *et al* (1982) Cross-species neural grafting in a rat model of Parkinson's disease. Nature 298: 652–654
2. Freed C, *et al* (1991) Foetal neural implants for Parkinson's disease: results at 15 months. In: Lindvall O, *et al* (eds) Intracerebral transplantation in movement disorders, Vol 4. Elsevier, Amsterdam, pp 69–77
3. Freed W, *et al* (1984) Embryonic brain grafts in an animal model of Parkinson's disease. Criteria for human application. Appl Neurophysiol 47: 16–22
4. Hitchcock E, *et al* (1991) Stereotactic implantation of foetal mesencephalon. In: Lindvall O, *et al* (eds) Intracerebral transplantation in movement disorders, Vol 4. Elsevier, Amsterdam, pp 79–86
5. Langston J, *et al* (1992) Core assessment program for intracerebral transplantations (CAPIT). Mov Disord 7: 2–13
6. Lieberman A, *et al* (1990) Adrenal medullary transplants as treatment for advanced Parkinson's disease. In: Dunnett S, *et al* (eds) Neural transplantation: From molecular basis to clinical applications, Vol 82. Elsevier, Amsterdam, pp 665–669
7. Lindvall O, *et al* (1990) Grafts of foetal dopamine neurons survive and improve motor function in Parkinson's disease. Science 247: 574–577
8. Lindvall O, *et al* (1992) Transplantation of foetal dopamine neurons in Parkinson's disease: One-year clinical and neurophysiological observations in two patients with putaminal implants. Ann Neurol 31: 155–165
9. Molina H, *et al* (1987) Open microneurosurgical autograft of adrenal medulla to the caudate nucleus in a patient with Parkinson's disease. Ital J Neurol Sci 5: 64–68
10. Molina H, *et al* (1991) Transplantation of human foetal mesencephalic tissue in caudate nucleus as treatment for Parkinson's disease: the Cuban experience. In: Lindvall O, *et al* (eds) Intracerebral transplantation in movement disorders, Vol 4. Elsevier, Amsterdam, pp 99–110

Correspondence: Prof. Hilda Molina, M.D., General Director, Centro Iberolatinoamericano de Trasplante y Regeneración del Sistema Nervioso, Avenida 25 15805e/158y 160, Playa, Ciudad de la Habana, Cuba.

Acta Neurochir (1993) [Suppl] 58: 20–23

Regeneration of Adult Dorsal Root Axons into Transplants of Dorsal or Ventral Half of Foetal Spinal Cord

Y. Itoh[1,2], M. Kowada[2], and A. Tessler[1]

[1] Philadelphia Veterans Administration Hospital, and Department of Anatomy and Neurobiology, The Medical College of Pennsylvania, Philadelphia, PA 19129, U.S.A., and [2] Department of Neurosurgery, Akita University School of Medicine, Akita, Akita 010, Japan

Summary

Severed dorsal root axons regenerate into the transplants of foetal spinal cord (FSC) and form synapses there. It is unknown whether the growth is specific to transplants of dorsal half FSC, a normal target of most dorsal root axons, or whether it is due to properties shared by transplants of ventral half FSC. We used calcitonin gene-related peptide immunohistochemistry to label subsets of regenerated host dorsal root axons, and morphometric analysis to compare neuronal populations within both transplants.

Adult Sprague-Dawley rats received intraspinal grafts of dorsal or ventral half FSC (E14), and the L4 or L5 dorsal root was cut and juxtaposed to the grafts. Three months later sagittal sections were prepared for immunohistochemistry and Nissl-Myelin stain. Histograms of the perikaryal area showed that the transplants of dorsal half FSC consisted of small neurons predominantly, whereas transplants of ventral half FSC consisted of neurons of variable sizes. Dorsal root axons regenerated into both transplants, but growth into dorsal half FSC was more robust.

These results indicate that both transplants provide an environment that supports dorsal root regeneration, but that the environment provided by dorsal half FSC is more favorable. Transplants of dorsal half FSC may offer advantages for the long-term goal of repairing of damaged spinal cord circuits.

Keywords: Embryonic spinal cord transplant; dorsal root ganglion neuron; regeneration; dorsal half of foetal spinal cord.

Introduction

The cut central axons of adult rat dorsal root ganglion (DRG) neurons do not penetrate the spinal cord[7]. When transplants of foetal spinal cord (FSC) are substituted for adult spinal cord, however, cut DRG axons regrow into the transplants and establish synapses there[2,3,9]. The extent to which dorsal root axons labelled for calcitonin gene-related peptide (CGRP) regenerate into the transplants persists unchanged for over 60 weeks after grafting[4]. It is therefore suggestive of the proposal that FSC trans-plants can be used to repair damaged spinal cord circuits. It is unknown whether the growth is specific to dorsal half FSC transplants, a normal target of most DRG axons, or whether it is due to properties shared by ventral half FSC transplants.

In the present study we used CGRP immunohisto-chemistry to label subsets of regenerated dorsal root axons because many DRG neurons are immunoreactive for CGRP and because we have found it to be a more sensitive indicator of regenerated axons than methods that rely on axon transport or diffusion of HRP[2,9]. We also used morphometric analysis to compare neuronal populations within both transplants.

Materials and Methods

Surgery

Twenty seven female adult (200–300 g) Sprague-Dawley rats were used as graft recipients. The rats were anaesthetized with ketamine (76 mg/kg), xylazine (7.6 mg/kg), and acepromazine (0.6 mg/kg), and the lumbar enlargement was exposed. After transection of the left L4/L5 dorsal root, the distal portion of the root was reflected caudally. A dorsal quadrant cavity was aspirated from the lumbar enlargement. Dorsal half (N = 15) and ventral half (N = 12) of spinal cord were dissected from homologous rat pups (E14) and intro-duced into the cavity. The cut dorsal root stump was juxtaposed to the dorsal surface of the transplants. The superficial wound were closed in layers.

CGRP Immunohistochemistry

Three months later the animals were deeply anaesthetized with Nembutal (50 mg/kg, i.p.) and perfused with 4% paraformaldehyde in 0.1 M phosphate buffer pH 7.4. To identify regenerated axons, every fifth cryostat section (14 μm) in the sagittal plane was processed for CGRP immunohistochemistry. Sections were reacted with primary antiserum against CGRP, and immersed in biotinylated goat antirabbit IgG and avidin-biotinylated horseradish peroxidase complex. The chromagen was DAB.

To evaluate transplant morphology and the dorsal root-transplant and transplant-host spinal cord interfaces, the adjacent sections were stained with cyanine R for myelin and counterstained with cresyl violet. This procedure has been described in detail[2,3,4,9].

Quantitative Analysis

The extent to which CGRP-labelled axons regenerated into transplants was measured for dorsal half FSC (N = 5) and ventral half FSC (N = 5) transplants. We used a point-counting stereological analysis to measure the area occupied by CGRP-labelled axons. Sagittal sections that contained labelled axons were examined under a light microscope. A micrometer 10 mm × 10 mm in size ($10^4 \mu m^2$) composed of 1-mm grid squares (Olympus, Tokyo) that was fitted in an ocular lens was used as a sampling lattice, and the number of times that CGRP-labelled axons intersected the corners of the grid squares was counted. To evaluate the arborization of CGRP-immunoreactive axons, we also measured the distribution of labelled axons in 3 sagittal sections by making composite montages that consisted of all the individual sampling lattices examined and the results were averaged. These procedures have been described in detail[3].

To compare the neuronal populations in both transplants, sagittal sections stained with cresyl violet and cyanine R were examined. Perikaryal area of cells contained in a sampling rectangle (7,450 μm^2) was measured using a Bioquant System IV (R&M Biometrics, TN). Three consecutive sections were examined per transplant. A histogram was made from these data to show the differences of cellular composition between both transplants.

The significance of the differences between both transplants was determined by Mann-Whitney two sample test (p < 0.05).

Results

Transplants of dorsal and ventral half FSC survive in the adult host spinal cord and differentiate into patterns that are characteristic for each region. Dorsal half FSC transplants contain regions that resemble substantia gelatinosa based on the presence of numerous small neurons and relative paucity of myelination. Ventral half FSC transplants composed of neurons of variable sizes showed abundant myelination. Transplants are generally well-integrated with host dorsal root and spinal cord.

CGRP Immunohistochemistry

Dorsal root axons immunoreactive for CGRP regenerate into every transplant examined. CGRP-labelled axons show distinctive patterns of distribution within both transplants. In dorsal half FSC, CGRP-labelled axons arborize extensively near the surface of transplants and in some portions the axons form dense bundles (Fig. 1). In ventral half FSC, CGRP-immunoreactive axons extend sparsely but

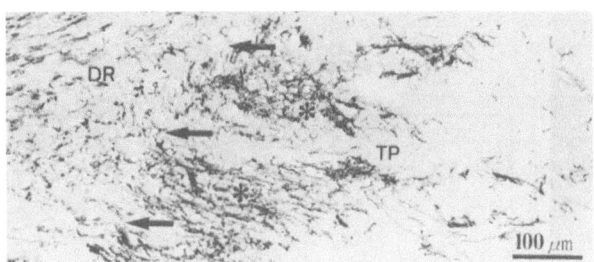

Fig. 1. CGRP-immunoreactive axons in the transplants of dorsal half of E14 spinal cord 3 months after transplantation. Sagittal section. Regenerated axons cross the interface (arrows) between host dorsal root (*DR*) and transplant (*TP*) and form a dense plexus (*) near the interface. Interface was identified in the adjacent Nissl-stained section. Bar = 100 μm

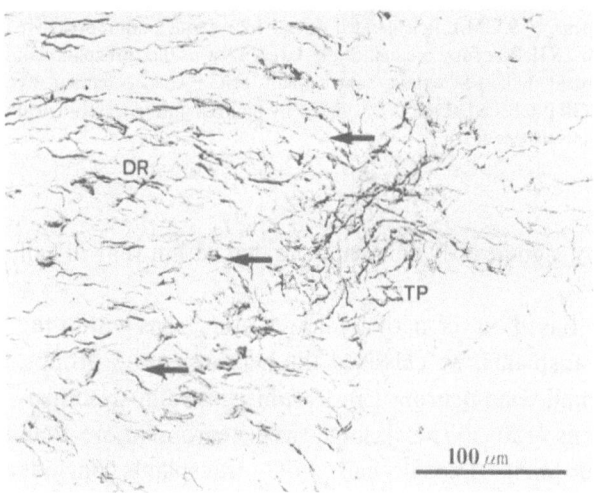

Fig. 2. CGRP-immunoreactive axons in the transplants of ventral half of E14 spinal cord 3 months after transplantation. Sagittal section. Regenerated axons cross the dorsal root (*DR*)-transplant (*TP*) interface (arrows) and grow extensively within the transplant without the formation of obvious plexuses

diffusely through the transplants and individual axons but not bundles of axons can be recognized (Fig. 2).

Quantitative Analysis

The point-counting stereological analysis shows the area fraction of both transplants occupied by regenerated CGRP-labelled axons. In dorsal half FSC transplants these axons occupy a mean area of $5.42 \times 10^4 \mu m^2$. In ventral half FSC the area occupied is approximately 40% of that in dorsal half FSC. Regenerated CGRP-labelled axons therefore occupy a significantly larger area in dorsal half FSC than in ventral half FSC (Fig. 3). The mean area of CGRP-innervated regions within dorsal half FSC is

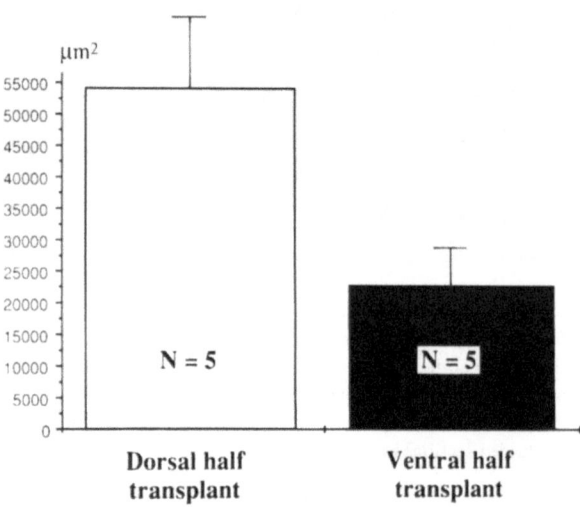

Fig. 3. Comparison of area occupied by CGRP-labelled axons (mean ± S.E.M.). Regenerated dorsal root axons immunoreactive for CGRP occupy a significantly larger area in the transplants of dorsal half FSC than ventral half FSC. Area occupied by CGRP-labelled axons is calculated by point-counting stereological analysis (see text)

not significantly different from that within ventral half FSC.

Based on the perikaryal area of neurons within the transplants, we classified the neurons into 3 groups: small-sized neurons (50–150 μm^2); medium-sized neurons (150–300 μm^2); large-sized neurons (more than 300 μm^2). Dorsal half FSC transplants include small-sized neurons significantly greater than ventral half FSC transplants, whereas ventral half FSC contain medium- and large-sized neurons greater than dorsal half FSC. These results confirm our qualitative observations.

Discussion

The pattern and extent of dorsal root ingrowth differed between transplants of dorsal half and ventral half of FSC. In dorsal half FSC regenerated axons stayed relatively close to the host dorsal root transplant interface, arborized extensively, and were often tangled together in plexuses. In ventral half FSC, regenerated axons were distributed widely and sparsely, and grew as individual axons rather than in bundles or plexuses. These qualitative morphological observations were confirmed by our quantitative studies. Although cut dorsal roots labelled for CGRP regenerate into both transplants, the area occupied by regenerated CGRP-labelled axons within dorsal half FSC is more robust than ventral half FSC. Both

transplants therefore provide an environment conductive to dorsal root regeneration, but dorsal half FSC, a normal target of most primary afferent fibers, provide additional more specific cues for growth. It is unlikely that inappropriate targets taken from ventral half FSC reproduce precisely the conditions found in dorsal half FSC. Growth into inappropriate targets is therefore consistent with the concept that the early stages of axon extension depend on molecules that are expressed generally throughout the developing nervous system[5].

Our observations that regenerated DRG axons grew more densely within dorsal half FSC than ventral half FSC suggest the presence of target-specific cues for pathfinding and target recognition that are not provided by ventral half FSC. These results are similar to those of *in vitro* studies showing that neurites of explanted foetal DRG axons grew and arborized more abundantly within co-cultured explants of spinal cord than of tectum[8]. Surface macromolecules likely to mediate the formation of specific pathways include glycoproteins that are expressed transiently by discrete populations of neurons[1].

Since there is a ventral-to-dorsal gradient of proliferation in the development of the spinal cord, the ventral motor system develops earlier than does the dorsal sensory system. As development proceeds, proliferation diminishes in the ventral cord, and by E15 only the most dorsal portion of the ventricular zone remains active[6]. Since adult dorsal roots begin to regrow into the FSC transplants soon after grafting[4], at the early stage of the axon elongation regeneration of adult rat dorsal roots is likely to be additionally enhanced by the precusor neurons remained in the dorsal cord.

Our results suggest that ventral half FSC as well as dorsal half FSC provide the conditions under which cut dorsal root axons can grow and survive. The conditions that constitute a permissive environment for regenerating axons are therefore relatively nonspecific, but dorsal half FSC transplants nevertheless differ in the extent to which they satisfy the requirements for growth of dorsal roots. The results suggest that dorsal half FSC transplant supply additional more specific cues for pathfinding and target recognition.

References

1. Dodd J, Jessell TM (1988) Axon guidance and the patterning of neuronal projections in vertebrates. Science 242: 692–699

2. Itoh Y, Tessler A (1990a) Ultrastructural organization of regenerated adult dorsal root axons within transplants of fetal spinal cord. J Comp Neurol 292: 396–411

3. Itoh Y, Tessler A (1990b) Regeneration of adult dorsal root axons into transplants of fetal spinal cord and brain: A comparison of growth and synapse formation in appropriate and inappropriate targets. J Comp Neurol 302: 272–293

4. Itoh Y, Sugawara T, Kowada M, Tessler A (1992) Time course of dorsal root axon regeneration into transplants of fetal spinal cord: I. A light microscopic study. J Comp Neurol 323: 198–208

5. Jessell TM (1988) Adhesion molecules and the hierarchy of neural development. Review. Neuron 1: 3–13

6. Nornes HO, Das GD (1974) Temporal pattern of neurogenesis in spinal cord of rat. I. An autoradiographic study—time and sites of origin and migration and settling patterns of neuroblasts. Brain Res 73: 121–138

7. Reier PJ, Stensaas LJ, Guth L (1983) The astrocytic scar as an impediment to regeneration in the central nervous system. In: Kao CC, Bunge RP, Reier PJ (eds) Spinal cord reconstruction. Raven, New York, pp 163–196

8. Smalheiser NR, Peterson ER, Crain SM (1981) Neurites from mouse retina and dorsal root ganglion explants show specific behavior within co-cultured tectum or spinal cord. Brain Res 208: 499–505

9. Tessler A, Himes BT, Houle J, Reier PJ (1988) Regeneration of adult dorsal root axons into transplants of embryonic spinal cord. J Comp Neurol 270: 537–548

Correspondence: Yasunobu Itoh, M.D., Department of Neurosurgery, Akita University School of Medicine, 1-1-1 Hondo, Akita City, Akita 010, Japan.

Acta Neurochir (1993) [Suppl] 58: 24–26
© Springer-Verlag 1993

Electrophysiological Responses in Foetal Spinal Cord Transplants Evoked by Regenerated Dorsal Root Axons

Y. Itoh[1,2], A. Tessler[1], M. Kowada[2], and M. Pinter[1]

[1]Philadelphia Veterans Administration Hospital, and Department of Anatomy and Neurobiology, The Medical College of Pennsylvania, Philadelphia, PA 19129, U.S.A. and [2]Department of Neurosurgery, Akita University School of Medicine, Akita, Akita 010, Japan

Summary

Cut dorsal root axons regenerate into intraspinal transplants of foetal spinal cord (FSC) and establish synaptic connections there. The aim of the present study was to determine whether transplant neurons are driven synaptically in response to electrical stimulation of regenerated dorsal root axons.

Adult Sprague-Dawley rats received FSC transplants (E14) into dorsal quadrant cavities at the lumbar enlargement. The cut L4 or L5 dorsal root stump was placed at the bottom of the lesion cavity and secured between the transplant and host spinal cord. Four to ten weeks later the animals were prepared for electrical stimulation and recording. We stimulated regenerated dorsal roots and recorded extracellular signle unit post-synaptic activities which were evoked close to the dorsal root-transplant interface. We used intracellular recording to observe several examples of monosynaptic EPSPs in transplant neurons evoked by dorsal root stimulation.

These results indicate that the regenerated dorsal root axons establish functional connections with neurons within the transplants and suggest that FSC transplants can be used to reconstruct functional connections between neurons that have been interrupted by spinal cord injury.

Keywords: Embryonic spinal cord transplant; dorsal root ganglion neuron; regeneration; electrophysiology.

Introduction

The cut central axons of adult rat dorsal root ganglion (DRG) neurons do not regenerate into the spinal cord[5]. When normal spinal cord is replaced by a transplant of foetal spinal cord (FSC), cut DRG axons cross the dorsal root-transplant interface, grow within the transplant[7], and establish synapses with transplant neurons[2]. These synapses persist for over 60 weeks after grafting and resemble morphologically those formed in normal spinal cord[4]. Similarities between regenerated and normal synapses have important implications for the therapeutic potential of the transplantation technique, but only if the regenerated synapses are functional and if the regenerated axons can activate transplanted neurons. We therefore studied whether FSC transplant neurons are driven synaptically in response to electrical stimulation of regenerated DRG axons.

Materials and Methods

Surgical Procedures

Seventeen female adult (200–300 g) Sprague-Dawley rats were used as graft recipients. The rats were anaesthetized with ketamine (76 mg/kg), xylazine (7.6 mg/kg), and acepromazine (0.6 mg/kg), and the lumbar enlargement was exposed by a laminectomy of the T13 or L1 vertebra. After transection of the left L4 or L5 dorsal root, the distal portion of the root was reflected caudally. A dorsal quadrant cavity 3 mm in length was aspirated from the left side of the lumbar enlargement. FSC was then dissected from Sprague-Dawley rat pups (E14) and introduced into the cavity[6]. The cut dorsal root stump encircled with a 10-0 suture for later identification was placed at the bottom of the lesion cavity and secured between the transplant and host spinal cord. The dural opening and the superficial wound were closed in layers.

Electrophysiology

Four to 10 weeks later transplanted animals were anaesthetized with a mixture of ketamine and xylazine. All animals were ventilated with a respirator. End-tidal CO_2 and rectal temperature were continuously monitored and body temperature was maintained at 37 °C.

The animals were then placed in a mechanical system designed to immobilize the vertebral column. The dorsal surface of the transplant was exposed and identified by the presence of the suture looped around the dorsal root during initial surgery. The dorsal root was exposed and prepared for stimulation with a bipolar hook electrode. This electrode was used routinely for dorsal root stimulation with constant current pulses of 0.2 m/sec duration. All exposed tissues were covered with warmed mineral oil.

All recordings of unitary activity in the transplants were accomplished with glass micropipette electrodes. A silver-ball monopolar recording electrode was placed at the dorsal root-transplant interface to monitor the size of incoming volleys evoked by dorsal root stimulation. Nerve stimulus strength was expressed as a multiple of the threshold intensity for the most excitable fibers in the dorsal root. This threshold was determined by slowly increasing stimulus strength to a level producing the first sign of electrical activity detectable by the silver-ball electrode.

Results

Analysis of the potentials recorded at the dorsal root-transplant interface indicated that the maximal conduction velocities of the dorsal roots were 50–60 m/sec. These values are similar to those of normal dorsal roots. The maximal conduction velocities of regenerated dorsal root axons within the transplants were 2–4 m/sec, suggesting that most of these axons are unmyelinated at the survival periods of 4 to 10 weeks.

Extracellular records were obtained from a total of 40 single units activated by electrical stimulation of dorsal roots. Consistent with previous morphological findings[2,3,7], most single unit activity was encountered within approximately 1 mm of the dorsal root-transplant interface. A train of at least 2 shocks was needed to secure reliable firing. The electrical thresholds for orthodromic firing were 5–10 times that of the most excitable fibers in the dorsal roots. Orthodromic firing latencies ranged from 4 to 6 m/sec (Fig. 1). Based on the variance in the latency between the dorsal root stimulus and the unit spike, units were classified into two types. A histogram of the latencies

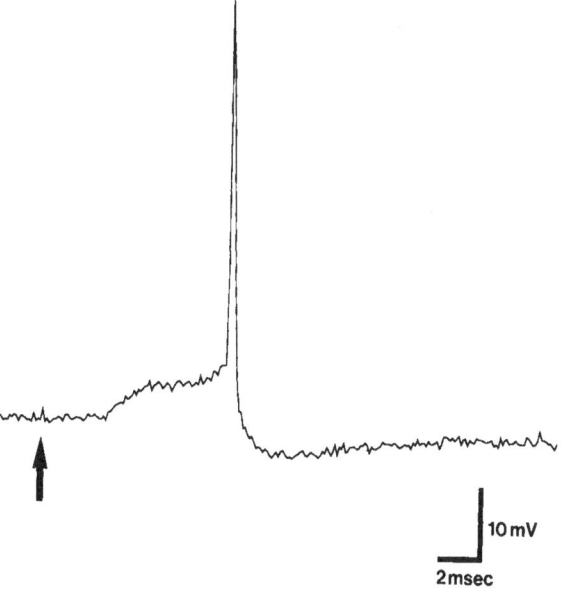

Fig. 2. The record shows an example of an intracellular-recorded EPSP and action potential taken from foetal spinal cord transplants following dorsal root stimulation

of type I units showed a relatively narrow band, whereas type II showed a broad band. Units in which the response to high frequency stimulation (100 Hz) of the dorsal root was studied exhibited failures of spike initiation.

In several units, the dorsal root was stimulated with a train of high frequency stimuli (1,000 Hz, 5 m/sec train). Following these trains, the number of spikes evoked by single or paired dorsal root stimuli was increased and remained elevated for several minutes. These results indicate the presence of post-tetanic potentiation of primary afferent EPSPs in transplant neurons.

We obtained intracellular records from 8 units. EPSPs with monosynaptic latencies were evoked by single shock stimulation of the dorsal root (Fig. 2). In one unit in which recording was exceptionally stable, the EPSP amplitude was observed to fluctuate considerably.

Discussion

The results of this study demonstrate that neurons within FSC transplants can be driven synaptically in response to electrical stimulation of regenerated DRG axons. The features of the neuronal firing observed extracellularly in response to stimulation of DRG axons support this condition. Synaptically-driven units display a variance in the latency between the onset of the dorsal root stimulus and the resulting unit

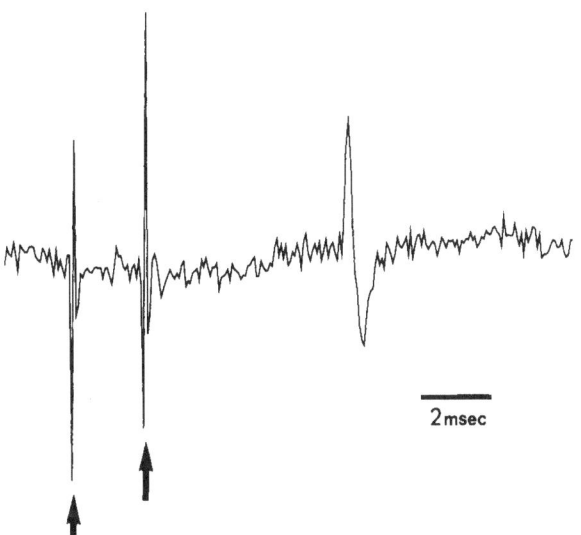

Fig. 1. The record illustrates an example of synaptically-driven neuronal activity in foetal spinal cord transplants following electrical stimulation of regenerated dorsal root fibers

spike that is greater than latencies observed for antidromic activation. This variance arises in part because of random fluctuations in the amplitude of the evoked EPSPs[4]. The latency variations that we observed in single unit activity are compatible with those expected of synaptic activation. Our data indicate the presence of 2 types of functional synaptic connections characterized by appreciable differences in the latency variance. We do not yet understand the basis for this difference, but one possibility is that these connections differ in strength.

Another feature of the unit activity we observed that reflects synaptic activation is an inability to follow high frequency stimulation of regenerated dorsal root axons. Some of the units we encountered could not follow stimulus frequencies of 100 Hz. Such failure can be related to depression of EPSPs that occurs during high frequency activation of primary afferent fibers[1]. In the case of antidromic firing or conduction along primary afferent fibers, such failure does not occur until much higher stimulation frequencies are employed. Our finding of enhanced unit firing following a train of high frequency stimuli to the implanted dorsal root is also compatible with synaptic activity. In other primary afferent synapses, such stimuli patterns evoke post-tetanic potentiation of EPSP amplitude that can increase the probability of post-synaptic action potential firing. These extracellular indications of synaptic activity were confirmed by our finding of monosynaptic EPSPs during intracellular recording from several neurons located within the transplants. These results therefore demonstrate that the regenerated dorsal root axon synapses that we have previously observed within transplants are functional by electrophysiological criteria.

One way in which transplants can contribute to recovery of function following spinal cord injury is by serving as a substrate for the reformation of reflex arcs. A necessary component in this process is the type of functional synaptic connection we have now demonstrated. Such connections complete the afferent limb of the requisite reflex arc. Less information is available about the efferent limb of such arcs, although several possibilities are suggested by available anatomical data. One possibility is that the axons of transplant neurons project to host muscle and in this way reconstitute an interrupted segmental reflex.

References

1. Curtis DR, Eccles JC (1960) Synaptic action during and after repetitive stimulation. J Physiol (Lond) 150: 374–398
2. Itoh Y, Tessler A (1990) Ultrastructural organization of regenerated adult dorsal root axons within transplants of fetal spinal cord. J Comp Neurol 292: 396–411
3. Itoh Y, Sugawara T, Kowada M, Tessler A (1992) Time course of dorsal root axon regeneration into transplants of fetal spinal cord: I. A light microscopic study. J Comp Neurol 323: 198–208
4. Jack JJB, Miller S, Porter R, Redman SJ (1971) The time course of minimal excitatory postsynaptic potentials evoked in spinal motoneurones by group Ia afferent fibres. J Physiol (Lond) 215: 353–380
5. Reier PJ, Stensaas LJ, Guth L (1983) The astrocytic scar as an impediment to regeneration in the central nervous system. In: Kao CC, Bunge RP, Reier PJ (eds) Spinal cord reconstruction. Raven, New York, pp 163–196
6. Reier PJ, Bregman BS, Wujek JR (1986) Intraspinal transplantation of embryonic spinal cord tissue in neonatal and adult rats. J Comp Neurol 247: 275–296
7. Tessler A, Himes BT, Houle J, Reier PJ (1988) Regeneration of adult dorsal root axons into transplants of embryonic spinal cord. J Comp Neurol 270: 537–548

Correspondence: Yasunobu Itoh, M.D., Department of Neurosurgery, Akita University School of Medicine, 1-1-1 Hondo, Akita City, Akita 010, Japan.

Psychosurgery

Acta Neurochir (1993) [Suppl] 58: 29–33

Present-Day Indications for Capsulotomy

P. Mindus

Department of Psychiatry, Neuropsychiatric Unit, Karolinska Hospital, Stockholm, Sweden

Summary

Although the majority of patients with anxiety disorders respond well to behavioural techniques of exposure and response prevention, to pharmacotherapy, or, more commonly, to combinations of the two approaches, a small percentage of patients remain refractory and are severely disabled by their symptoms. Some of these individuals constitute candidates for neurosurgical intervention, e.g. cingulotomy and capsulotomy. Therefore, such operations are performed, if to a very limited extent, both in the United States and in Europe to-day.

At the Karolinska, patients are accepted for capsulotomy who suffer from chronic, severe, incapacitating, and otherwise intractable anxiety disorders, i.e. obsessive-compulsive disorder, generalized anxiety disorder and phobias. The present-day inclusion and exclusion criteria are described, and the safety and the efficacy of capsulotomy in these extreme forms of anxiety disorders are discussed.

Keywords: Anxiety disorders; capsulotomy; indications; results.

Present-Day Indications for Capsulotomy

Capsulotomy and other so-called psychosurgical procedures were originally developed for the treatment of severe psychotic illness. With the advent of antipsychotic drugs, this indication was soon abandoned. In the intervening years it has become evident, however, that patients with severe, chronic, and otherwise intractable anxiety disorders may benefit from neurosurgical intervention such as capsulotomy. The present-day selection practices, indications, and contraindications for capsulotomy used at Karolinska Hospital, Stockholm, may be summarized as follows:

General Guidelines

General guidelines have been developed conjointly by workers in the field over several years and are, consequently, quite similar across the procedures. In brief, the guidelines include the following:

The patient has been able to function adequately or almost adequately in private and professional lives at least for some years, before falling ill; the illness is causing considerable suffering, as evidenced by ratings; because of the symptoms, the patient's psychosocial functioning is clearly reduced, as evidenced by ratings; the patient himself clearly acknowledges that he is ill and requests, in no vague terms, neurosurgical treatment; the illness has been subject to intensive, up-to-date psychiatric treatment for such a long time that the conclusion has been reached that it is, indeed, refractory to such treatment. In practice, this means a minimum of five years.

Current Inclusion Criteria for Capsulotomy

1. The patient fulfills the current diagnostic criteria for generalized anxiety disorder, obsessive-compulsive disorder, panic disorder with agoraphobia, or social phobia, either alone or coexistant.
2. The duration of illness exceeds five years.
3. The disorder is causing substantial suffering.
4. The disorder is causing substantial reduction in the patient's psychosocial functioning.
5. Current treatment options tried systematically for at least five years have either been without appreciable effect on the symptoms, or must be discontinued due to intolerable side effects.
6. The prognosis, without neurosurgical intervention, is considered poor.
7. The patient gives informed consent.
8. The patient accepts to participate in the pre-operative evaluation programme.
9. The patient accepts to participate in the post-operative rehabilitation programme.
10. The referring physician is willing to acknowledge

responsibility for the postoperative long term management of the patient.

Current Exclusion Criteria for Capsulotomy

1. Age below 20 or over 65 years.
2. The patient has a complicating other, current or lifetime, Axis I diagnosis, for example organic brain syndrome, delusional disorder, or manifest abuse of alcohol, or sedative, or illicit drugs. For a condition to qualify as "complicating", it must substantially complicate function, or treatment, or the patient's ability to comply with treatment, or lead to serious adverse events such as overdosage, paradoxical reactions etc.
3. A complicating current DSM-III-R Axis II diagnosis from clusters A (e.g., paranoid personality disorder) or B (e.g., borderline, antisocial, or histrionic personality disorder) may constitute a relative contraindication. A current cluster C personality disorder (e.g., avoidant or obsessivecompulsive personality disorder) is not to be considered a contraindication since it may, in fact, disappear with successful treatment of the coexistent anxiety disorder (for a review, see[8]).
4. The patient has a complicating, current Axis III diagnosis with brain pathology, e.g. atrophy or tumour.

Original Indications for Capsulotomy

For an understanding of how these indications were developed, a brief historical review may be of interest. The French neurosurgeon Talairach and co-workers[1] were the first to report on the use of selective lesions in the anterior limb of the internal capsule. They were unimpressed by the results in schizophrenia, but reported satisfactory results in patients with "névroses anxieuses". A few years later, the Swedish neurosurgeon Lars Leksell used his stereotactic system to produce radiofrequency (RF) thermolesions in the internal capsule, i.e. capsulotomy, in patients with various psychiatric disorders. Although schizophrenia was intended as the main indication, the results in schizophrenic patients were disappointing[2], with a satisfactory response in only 14%. The best results were obtained, however, in patients suffering from obsessional neurosis, 50% of whom were rated as having a satisfactory response to capsulotomy. Somewhat less favourable results were recorded in patients with anxiety neurosis, with a satisfactory response in 20%.

These results led Bingley and associates[3] to select OCD patients only for capsulotomy. A satisfactory result was rated in 70% of their severely disabled patients. Further support for the efficacy of capsulotomy in patients with severe OCD was derived from the findings of a Spanish group[4,5], who published combined data of their own and those of other authors comprising a total of 149 OCD patients undergoing capsulotomy, and found that satisfactory results had been reported in 72% of the cases.

The Relative Efficacy of Capsulotomy in OCD and in Non-OCD

The efficacy of capsulotomy in OCD being well established, the group at Karolinska[6,7,8] extended the indication to include also non-OCD anxiety disorders such as phobias, generalized anxiety disorder, and panic disorder, the rationale being, interal, the considerable phenomenological similarites and overlap between these disorders, and the wellknown high rates of co-existence among the anxiety disorders. The high rate of concurrent anxiety disorders observed in the capsulotomy cases (see[8]) may be one contributing factor in the intractability of their illnesses.

In search of any differential effect of capsulotomy in OCD and non-OCD anxiety disorders, Mindus[8] compared prospectively the CPRS (a widely used psychiatric rating scale) scores of 10 consecutive OCD patients with those of 14 consecutive patients suffering from non-OCD anxiety disorders such as generalized anxiety disorder, panic disorder with agoraphobia, and social phobia. Each subgroup was rated before RF capsulotomy, and at two, six and twelve months after it. The main reduction in scores was apparent already at the two month follow-up ($t = 9.74$, $p < 0.001$). From that rating session and later, there were neither significant additional reductions in scores, nor significant inter-group differences. Thus, at the 12 month follow-up, the reductions in scores were of similar magnitude in both diagnostic subgroups. With regard to the scores on the Global Scale included in the CPRS, both subgroups improved on this measure (the OCD subgroup: $t = 2.75$, $p < 0.05$; the non-OCD subgroup: $t = 12.45$, $p < 0.001$). Significant correlations were obtained between the scores of these modern scales and those on the modified Pippard postoperative rating scale, widely used in the field. The Pippard scores indicated satisfactory response in 60% of the OCD patients and in 93% of the non-OCD patients.

Table 1. *Summary of Capsulotomy Studies in OCD and Non-OCD Patients by Outcome*

Reference	n	OCD		n	Non-OCD	
		A + B (%)	D + E (%9)		A + B (%)	D + E (%9)
Herner, 1961	18	50	28	15	20	40
Bingley et al., 1977	35	70	0	—	—	—
Kullberg, 1977	8	38	15	5	60	0
Rylander, 1979	7	57	42	7	86	0
Burzaco, 1980	85	73	22	—	—	—
Fodstad et al., 1982	2	100	0	—	—	—
Mindus et al., 1987	7	71	0	—	—	—
Mindus, 1991	10	60	0	14	93	0
Total	172	67	8	41	51	15

A Free of symptoms; *B* much improved; *D* unchanged, *E* worse.

Four of the five poor responders were OCD patients. It appeared that the results in the non-OCD group were more homogeneous with regard to their outcome on the Pippard scale than in the OCD group of patients, who tended to be more evenly distributed across outcome classifications. The proportion of patients rated as free of symptoms was, however, similar in both diagnostic subgroups (OCD: 5/10 or 50%; non-OCD: 7/14 or 50%).

In order to obtain an estimate of the relative efficacy of capsulotomy in the two diagnostic subgroups, the above data were combined with those from other capsulotomy studies reporting diagnosis and outcome in sufficient detail to permit a meta-analysis[8]. Eight studies[2,3,5,6,8-11] comprising 172 patients met these criteria. Using the scores on the Pippard postoperative rating scales as the dependent measurement and contrasting the two best outcome categories to the two worst outcome categories, the following results were found. Whereas 67% of the OCD patients were reported to have benefited from capsulotomy, 8% did not. In the non-OCD patient group, 51% were reported to have satisfactory results, whereas 15% did not (see Table 1).

Although the data base for OCD patients is considerably larger than that for non-OCD patients, the results appear comparable, suggesting that both diagnostic subgroups may benefit. What is more, the odds are good even for these severely disabled patients to benefit from capsulotomy. Furthermore, the findings may be taken as evidence of some common brain basis and/or pathophysiology between these two types of anxiety disorders.

There are four procedures in current use in the treatment of otherwise intractable anxiety disorders, i.e. stereotactic subcaudate tractotomy, limbic leucotomy, cingulotomy and capsulotomy. As to the relative efficacy of these operations, a meta-analysis by Waziri[14] suggested capsulotomy to produce the best results. Kullberg[9] performed a head to head comparison of cingulotomy and capsulotomy in OCD patients, and arrived at the same conclusion.

What is the Price for Capsulotomy?

The above, salutary effects have not, of course, been achieved without risk. A detailed analysis of the risk-benefit issue has been published elsewhere[8]. In brief, capsulotomy carries little risk of somatic morbidity, and no deaths have been reported in conjunction with the procedure. But what about adverse personality changes and negative effects on cognitive functions? These important issues have been studied prospectively by several independent authors[3,4,12,13], who have all arrived at the same conclusion: negative effects are not to be expected following these modern procedures. In a recent, prospective study on 19 capsulotomy patients examined with an extensive battery of neuropsychological tests before, at one, and at seven years after capsulotomy, the main finding was that of no significant differences in neuropsychological functions over time (Nyman and Mindus, unpublished data).

These findings may not, of course, be interpreted to mean that the risk is negligible in a given case; only that it must be weighed against the risk of non-intervention, i.e. high risk of somatic, social and mental complications, including suicide (for a review, see[8]). Deferring the decision to operate on a given patient may not spare him complications.

How Useful Are the Selection Criteria?

The following clinical vignettes may be used to illustrate the clinical usefulness of the above selection criteria. (With the obvious exception of case 3, the cases are reported here with the consent of the patients. No case may be reproduced in lay publications).

Case 1: Bingley et al.[3] provide an example of the operation's effect on creativity, an ability which is difficult to evaluate by psychometric methods: "An archeologist, holding a highly qualified academic position, had for 15 years suffered from increasing obsessive-compulsive neurosis. He had been given all sorts of psychiatric treatment, including several years of psycho-analysis. A few months after the operation, he resumed his duties at an institution. He is now very productive in his work, and is working on his doctor's thesis" (page 296).

The present author has been able to retrospectively follow-up this patient; he reports his OCD to have significantly reduced his professional and private performance in the years prior to surgery, despite intensive, systematic trials with non-surgical methods. After capsulotomy (1972), his OC symptoms disappeared, and he resumed work on his research project. His doctoral thesis was so well received both nationally and internationally, he reports, that it was soon sold out. He is now professor and chairman of an archeological research center. Inspection of his list of publications shows it to include more than 160 articles and books, the vast majority of which were published after his operation.

Comment: Contrary to conventional wisdom regarding the alleged detrimental effects of "psychosurgery" (a misnomer which should be expunged from modern texts) on creativity, it would appear that it was this patient's OCD that impeded his creativity—not his operation.

Case 2: A male with intractable Generalized Anxiety Disorder underwent gamma capsulotomy using one isocenter and a 4 mm collimator with a target maximum dose of 160 Gy. Both the patient and the assessor were kept ignorant of the radiosurgical details and the neuroradiological findings. At the one year follow-up, no clinical effects were noted. It was then revealed by the surgeon that MRI displayed minimal signal changes in the left internal capsule only, whereas no signal change was detected in the right hemisphere (unilateral capsulotomy appears to be unproductive, see[8]). The patient, who was now desperate and demoralized, requested a second intervention which he underwent 1.5 years after the first one. At this time, the target maximum dose of 160 Gy was delivered with a 201 Cobalt-60 source Gamma Knife to a larger volume using three isocenters with 4 mm collimators, the idea being to tailor the field of irradiation according to the individual anatomy of the capsules. The clinical results of this second intervention were encouraging: at the follow-up at one year after the second intervention, this patient, for the first time in 15 years, was now free of symptoms and of drugs. He later relocated to another city, married and took up a new job. Serial postoperative follow-ups up to 4 years have demonstrated no undesired side effects. He is now leading a normal life.

Comment: At least in this accidental placebo-controlled single case study, it would appear that the placebo response, if any, was negligible. This is in line with previous findings[11].

Case 3: A 31 year old male with an almost life long history of a severe but atypical anxiety disorder, recurrent major depression, and borderline personality traits. He had made several serious suicide attempts. The evaluator was reluctant to accept the patient for capsulotomy, for several reasons, including the many atypical features of patient's illness. When he sensed this, he summoned a friend journalist who broadcasted in the media the patient's intention to commit suicide, should this last chance be denied him. Following this, the patient underwent capsulotomy, experienced a minimal, transient improvement, soon relapsed, and took his life.

Comment: His case is a tragic illustration of the following dilemma: if less suitable cases are accepted for surgery for humanitarian reasons, the homogeneity of the material may be compromised, and the overall results less favourable. This, in turn, may deter physicians from referring more suitable cases, initiating a vicious circle.

The Future

Given the remarkable progress in the diagnosis and the treatment of anxiety disorders in recent years, it is probable that, in the future, an even more refractory patient population will be referred for neurosurgical treatment. The clinical efficacy of capsulotomy and related operations may merit study in individuals with severe, chronic, otherwise intractable, self-mutilative or even life threatening behavioural problems encountered in some rare cases of Bipolar Disorder, Anorexia Nervosa, and Gilles de la Tourette's syndrome.

In the heyday of lobotomy, it would appear that too many were operated on too soon[15]. To-day, as it would seem, too few are operated on too late[6,8]. In both situations, the patients pay a price. It is therefore hoped that in this Decade of the Brain, physicians may overcome outdated attitudinal barriers and more often consider neurosurgical intervention in their desperately ill patients.

References

1. Talairach J, Hecaen H, David M (1949) Lobotomie préfrontal limitée par électrocoagulation des fibres thalamo-frontales a leur émergence du bras antérieur de la capsule interne. In: Congress Neurologique International. Masson, Paris 1949, p 1412
2. Herner T (1961) Treatment of mental disorders with frontal stereotactic thermo-lesions. A follow-up of 116 cases. Thesis. Acta Psychiatr Scand [Suppl] 158: 36
3. Bingley T, Leksell L, Meyerson BA, Rylander G (1977) Long term results of stereotactic capsulotomy in chronic obsessive-compulsive neurosis. In: Sweet WH, Obrador S, Martin-Rodriguez JG (eds) Neurosurgical treatment in psychiatry, pain and epilepsy. University Park Press, Baltimore, pp 287–299
4. Lopez-Ibor JJ, Lopez-Ibor Alino JJ (1977) Selection criteria for patients who should undergo psychiatric surgery. In: Sweet WH, Obrador S, Martin-Rodriguez JG (eds) Neurosurgical treatment in psychiatry, pain and epilepsy. University Park Press, Baltimore, pp 151–162
5. Burzaco J (1981) Stereotactic surgery in the treatment of obsessive-compulsive neurosis. In: Perris C, Struwe G, Jansson B (eds) Biological psychiatry. Elsevier, Amsterdam, pp 1103–1109
6. Rylander G (1979) Stereotactic radiosurgery in anxiety and obsessive-compulsive states: Psychiatric aspects. In: Hitchcock

ER, Ballantine HT Jr, Meyerson BA (eds) Modern concepts in psychiatric surgery. Elsevier, Amsterdam, pp 235–240

7. Meyerson BA, Mindus P (1988) Capsulotomy as treatment of anxiety disorders. In: Lunsford LD (ed) Modern stereotactic neurosurgery. Martinus Nijhoff, Boston, pp 353–364

8. Mindus P (1991) Capsulotomy in anxiety disorders. A multidisciplinary study. Thesis, Karolinska Institute, Stockholm

9. Kullberg G (1977) Differences in effect of capsulotomy and cingulotomy. In: Sweet WH, Obrador S, Martin-Rodriguez JG (eds) Neurosurgical treatment in psychiatry, pain and epilepsy. University Park Press, Baltimore, pp 301–308

10. Fodstad H, Strandman E, Karlsson B, West KA (1982) Treatment of chronic obsessive compulsive states with anterior capsulotomy or cingulotomy. Acta Neurochir (Wien) 62: 1–23

11. Mindus P, Bergström K, Levander SE, Norén G, Hindmarsh T, Thuomas K-Å (1987) Magnetic resonance images related to clinical outcome after psychosurgical intervention in severe anxiety disorder. J Neurol Neurosurg Psychiatry 50: 1288–1293

12. Mindus P, Nyman H, Rosenquist A, Rydin E, Meyerson BA (1988) Aspects of personality in patients with anxiety disorders undergoing capsulotomy. Acta Neurochir (Wien) [Suppl] 44: 138–144

13. Mindus P, Nyman H (1991) Normalization of personality characteristics in patients with incapacitating anxiety disorders after capsulotomy. Acta Psychiatr Scand 83: 283–291

14. Waziri R (1990) Psychosurgery for anxiety and obsessive-compulsive disorders. In: Noyes RJr, Roth M, Burrows GD (eds) Handbook of anxiety. Treatment of anxiety. Elsevier, Amsterdam, pp 519–535

15. Valenstein ES (1980) Historical perspective. In: Valenstein ES (ed) The psychosurgery debate. Freeman, San Francisco, pp 11–54

16. Mindus P (1979) Some thoughts on the anti-psychosurgery attitude in Sweden. In: Hitchcock ER, Ballantine HT Jr, Meyerson BA (eds) Modern concepts in psychiatric surgery. Elsevier, Amsterdam, pp 359–365

Correspondence: Prof. Per Mindus, M.D., Ph.D., Neuropsychiatric Unit, Department of Psychiatry, Karolinska Hospital, S-104 01 Stockholm, Sweden.

Acta Neurochir (1993) [Suppl] 58: 34–35
© Springer-Verlag 1993

Dorsomedial Thalamotomy as a Treatment for Terminal Anorexia: a Report of Two Cases

R. Zamboni, V. Larach, M. Poblete, R. Mancini, H. Mancini, V. Charlin[1], F. Parr, C. Carvajal[1], and R. Gallardo[1]

Servicio de Neurocirugía Estereotáxica, Instituto Psiquiátrico de Santiago, [1]Universidad de Chile, Chile

Summary

Anorexia nervosa has been considered a compulsive obsessive disorder (OCDS). We present a case report of two patients, male and female, suffering from an extremely severe, chronic and refractory anorectic syndrome. Both patients underwent bilateral stereotactic thalamotomy with involvement of the lamella medialis. Follow-up was 4 year and 2 year, respectively. Both patients have regained weight and improved significantly in terms of their obsessive-compulsive symptoms and in their quality of life. The results suggest that anorexia nervosa may be considered in the OCDS.

Keywords: Anorexia nervosa; psychosurgery; thalamotomy; obsessive compulsive disorder.

Introduction

Anorexia nervosa has been considered part of a group of pathologies that resemble so called obsessive-compulsive disorders spectrum (OCDS) (Rasmussen and Eisen 1992).

Our experience in psychosurgery since 1962 has been with a group of 51 patients whose clinical symptoms indicate OCDS. A bilateral dorsomedialis thalamotomy was performed in almost all of them and better results were obtained by combining thalamotomy with lesioning the lamella medialis, due to the significance of the thalamo-cortical system (Hassler and Dieckmann 1973).

We performed this procedure in two patients suffering from severe and terminal anorectic syndrome refractory to medical and psychiatric treatment who were in life threatening condition.

Methods

The patients were assessed pre- and postoperatively by psychiatric tests (BPRS, Hamilton Depression, Hamilton Anxiety) and a psychological test in case 2 (Wechsler, Bender-Bip, Rorschach, TAT).

The Riechert and Mundinger stereotactic system was used for the operation. The target point was marked in a ventriculogram using the base line foramen of Monroe-posterior commissure. The dorso medialis nucleus was located 14 mm behind the foramen and 4 mm above the base line. Lamella medialis was identified as 7 mm and 1 mm from the same base line. In the frontal plane the distance from the wall of the third ventricle was 4 mm for dorso medialis and 2.5 mm for lamella medialis. Two lesions of 3 mm in diameter were made in each dorsomedialis nucleus and only one in the lamella medialis nucleus on each side.

Material

Case 1

17-year-old male with a chronic schizophrenic psychosis with prominent obsessive compulsive features (compulsive spitting on people and objects, avoidance of direct contact with anyone, accompanied by rituals) with a severe anorectic syndrome. The patient required repeated hospitalization and suffered cardiac arrest due to his cachectic syndrome on two occasions.

Prior to surgery, the Corporal Mass Index (CMI-Quetellet) was 6.1 (normal 20–25) corresponding to 13 kg (height 146 cm). He was operated on with a CMI of 9.1, corresponding to 19 kg.

Postoperative follow up of 54 months. At present, the patient shows major improvement of his compulsive behavior. He spits only seldom and allows physical contact. Brief Psychiatric Rating Scale (BPRS) shows a decrease from 67 to 44. Hamilton Depression shows a decrease from 34 to 10. Hamilton Anxiety decreased from 18 to 2.

The Corporal Mass Index was 6.1 and increased to 19.4, weight gain from 13 kg to 42 kg.

Case 2

21-year-old female, 5 years of unsuccessful treatment for anorexia nervosa with deviant behavior (occasional psychotropic drug abuse and robbery) and several suicide attempts. Repeated hospitalization for severe cachexia (including coma) and clinical anorectic syndrome considered irreversible. Prior to surgery the CMI was 9.7, corresponding to 21 kg (height 147 cm). She was operated on with a CMI of 12.5, corresponding to 27 kg. Postoperative follow up of 23 months.

The deviant behavior has ceased. The patient is in full-time regular employment. BPRS decreased from 36 to 20. Hamilton Depression shows a reduction from 38 to 5. The Hamilton Anxiety decreased from 14 to 4. Wechsler preoperative full IQ: 96 and postop 101, and Bender-Bip shows normal range (I–A) in pre- and postoperative evaluations.

Rorschach and TAT show the following pre- to postoperative changes: a decrease of impulsive responses as well as improvement in empathic contact and an increase in cognitive functions with an overall decrease of depression and anxiety.

The CMI was 9.7 and increased to 17.8, weight gain from 21 kg to 38.5 kg.

Discussion

Reports indicate that anorectic patients share symptoms with OCD or OCDS patients (Rothenberg 1990; Pigott et al. 1991; Rasmusen and Eisen 1992). Our two patients suffered such obsessive-compulsive symptoms as rumination, ritual-performing, and compulsions that were related to severely disturbed eating habits. As these patients showed extremely severe, chronic, refractory eating disorders, which kept them at high risk of death, we decided to employ the stereotactic technique used in OCD patients, considering the positive results reported from using the combined thalamotomy-lamella medialis method to treat such disorders (Hassler and Dieckmann 1973; Zamboni 1981).

The results obtained in both patients show significant sustained improvement of the anorectic and OCD symptoms for 4 and 2 years of follow up, respectively. Both patients are no longer at risk of death, as they have regained weight and improved their CMI. Although they have not reached normal weight values, their behavior has improved, as is shown by their psychiatric and psychological assessment, as has their quality of life and that of their families. The results show that considering terminal anorexia as belonging to OCDS, just like Tourette's

syndrome and others mentioned previously, may contribute to the understanding of functional and structural relationships between these entities.

Conclusions

This case report shows that a combined medial thalamotomy-lamella medialis was a successful treatment for two patients suffering from severe, refractory, and life-threatening anorectic syndrome. These results using the same target applied for OCD patients suggest that in these severe refractory anorectic syndromes there could be a link between a neuroanatomical and neurophysio-pathological substrate.

Although our findings are based on only two cases, we suggest that a bilateral, combined medial thalamotomy-lamella medialis could be offered to those patients suffering terminal anorectic syndromes.

References

1. Hassler R, Dieckmann G (1973) Relief of obsessive-compulsive disorders, phobias and tics by stereotactic coagulation of the rostral intralaminar and medial-thalamic nuclei. In: Laitinen LV, Livingston KV (eds) Surgical approaches in psychiatry. MTP, pp 206–212
2. Pigott A, Altemus M, Rubenstein C, Hill J, Bihari K, L'Heureux F, Bernstein S, Murphy D (1991) Symptoms of eating disorders in patients with obsessive-compulsive disorder. Am J Psychiatr 148: 1552–1557
3. Rasmussen S, Eisen J (1992) The epidemiology and differential diagnosis of obsessive-compulsive disorder. J Clin Psychiatry 53: 4–10
4. Rothenberg A (1990) Adolescence and eating disorder: The obsessive-compulsive syndrome. Psychiatr Clin North Am 13: 469–488
5. Zamboni R (1981) Algunos alcances sobre psicocirugía. Rev Chil Neuropsiq XIX: 70–76

Correspondence: R. Zamboni, M.D., Instituto Psiquiátrico, Box 2677, Santiago, Chile.

Movement Disorders

Acta Neurochir (1993) [Suppl] 58: 39–44

Chronic VIM Thalamic Stimulation in Parkinson's Disease, Essential Tremor and Extra-Pyramidal Dyskinesias

A. L. Benabid, P. Pollak, E. Seigneuret, D. Hoffmann, E. Gay, and **J. Perret**

Department of Neurosciences, INSERM 318, and University Hospital of Grenoble, France

Summary

Stereotactic thalamotomy of the VIM (ventral intermediate) nucleus is considered as the best neurosurgical treatment for Parkinsonian and essential tremors. However, this surgery, especially when bilateral, still presents a risk of recurrence and neurological complications.

We observed that acute VIM stimulation at frequencies higher than 60 Hz during the mapping phase of the target suppressed the tremor of Parkinson's disease (PD) and essential tremor (ET). This effect was immediately reversible at the end of the stimulation. This was initially proposed as an additional treatment for patients already thalamotomized on the contralateral side, and then extended as a regular procedure for extra-pyramidal dyskinesias.

Since January 1987, we implanted 126 thalami in 87 patients (61 PD, 13 ET, 13 dyskinesias of various origins). Deep brain stimulation electrodes were stereotactically implanted under local anaesthesia, using stimulation and micro-recording to delineate the best site of stimulation. Electrodes were subsequently connected to implantable programmable stimulators. The optimal frequency was around 130 to 185 Hz.

The results (evaluated by a neurologist from $0 =$ no effect to $4 =$ perfect relief) are related to the type of tremor. Altogether, 71% of the 80 patients benefited from the procedure with grade 3 and 4 results. In 88% of the PD cases, the results were good (grade 3) or excellent (grade 4) and stable with time. Rigidity was moderately for a long improved but akinesia was not. The same level of improvement was observed in 68% of the ET patients and only in 18% of the other types of dyskinesias. A rebound effect was observed in 30% of the ET patients in whom the long term results decreased. In all patients, adverse effects were mild and always reversible. There was no operative morbidity.

The mechanism of action of electrical VIM stimulation is still unknown but could involve a jamming-based effect. However, the high rate of success, the extremely low morbidity, the reversibility and adaptability of chronic VIM stimulation makes this procedure safer than traditional thalamotomy, especially when bilateral surgery is indicated.

Keywords: Parkinson's disease; essential tremor; extra-pyramidal dyskinesias; thalamic stimulation; results.

Introduction

Parkinson's disease has always been targeted by either surgical or medical therapeutics. Based on the anecdotal observation of Cooper[10], thalamotomy has been for decades the major weapon used against rest and essential tremors[11-16,18,19,27,28], but with much less success on the two other Parkinsonian symptoms, rigidity and bradykinesia. The discovery of the dopamine deafferentation from substantia nigra neuronal death, its replacement by the dopamine precursor L-Dopa, opened an unprecedented successful therapeutic period. In the meantime, stereotaxy almost disappeared and had to find new applications to survive. At the beginning of the eighties, side effects of L-Dopa as well as its limits appeared and thalamotomy was again proposed as a possible solution when medical treatment failed. However, patients and neurologists had become accustomed to reversible side effects and were not prepared anymore to accept post-operative deficits. The quest for less invasive procedures, for safer approaches and even for reversible treatments was strengthened. The Ventral intermediate nucleus (Vim) of the thalamus had already been recognized as the best target for thalamotomy[4,11,14-16,19,21-23] and several authors had already reported the observation that stimulation during stereotactic surgery would influence tremor amplitude, most often in the sense of a reduction[1-4,20,22,29]. We observed that reduction of tremor was actually consistently associated with high frequency stimulation while low frequency (lower than 100 Hz) would be ineffective or even would increase tremor amplitude. We proposed the method as an alternative for thalamotomy in patients already operated on the controlateral side and then extended it to all surgical indications in the treatment of movement disorders[5-7]. We report now our experience over five years and the differential effects of this method on various types of tremor.

Methods and Material

Stereotactic Procedure

Patients are attached to a Talairach stereotactic frame under light general anaesthesia, ventriculography is performed by direct puncture of the frontal horn of the lateral ventricle, injection of iodine contrast medium is made and X-rays are taken in prone and supine positions. Vim target is drawn from the third ventricle landmarks according to the Guiot's scheme[28]: average values of the electrode lower tip coordinates are as follows: 25% of AC-PC line in front of PC, zero mm above the level of AC-PC line, 11.5 mm lateral to the wall of the third ventricle at the level of AC-PC line. Aiming at this theoretical target, electrode tracks are used to further define the limits of the effective area by electrophysiology. Bipolar concentric semi-microelectrodes and monopolar microelectrodes are used to stimulate (frequency: 130–185 Hz, pulse width: 60 microseconds) and to record neuronal activity and single cell units. Precise location of the target is achieved by determination of the site of best response to stimulation and eventually the record of cells synchroneous to tremor. Tremor is assessed by accelerometry and clinical examination under rest and stress conditions.

When the best target has been determined, a chronic deep brain stimulation electrode (Medtronic monopolar model SP, 5535 in the first 108 thalami and Medtronic quadripolar model 3387 in the last 18 thalami) is inserted, fixed to the bone by a nylon suture embedded in a drop of methyl metacrylate cement and connected to a percutaneous transitory extension. The 123 first electrodes have been inserted in a parallel plane to the mid-sagittal

plane. Figure 1 shows the global distribution of these electrodes. In the last two patients, a double oblique trajectory, in a 10° plane relative to the mid-sagittal plane has been used to implant 3 electrodes into Vim along its main axis.

After a week of test stimulation the electrodes are connected under general anaesthesia to programmable stimulators (Medtronic Itrel II) placed into a subcutaneous pouch in the subclavicular region.

Patient Selection

Eighty-seven patients suffering from disabling and drug-resistant tremor gave their informed consent and underwent this procedure. 61 patients had Parkinson's disease (PD), 13 had essential tremor (ET) and 13 patients had action tremor mainly related to mesencephalic lesions, including 4 with multiple sclerosis. Eleven patients had previously undergone contralateral thalamotomy and 39 (45%) had bilateral Vim stimulation at the same time, making a total of 50 out of 87 patients (57%) with bilateral thalamic surgery. The effect on tremor was scored independently by the neurologist on a 5-point scale (4 = complete disappearance of tremor in all circumstances. 3 = reappearance of a slight tremor on rare occasions, for instance under stress. 2 = moderate benefit. 1 = slight benefit without real improvement in daily life. 0 = no benefit at all or worsening of tremor).

Results

Acute Stimulation During Surgery

Suppression of tremor by Vim stimulation can be easily seen during the stereotactic procedure on a patient under local anaesthesia. Progressive increase in current intensity induces a progressive decrease in tremor amplitude when the stimulating electrode is located in a responsive area. In a given location, the delay between onset of stimulation and total arrest of tremor is shorter when the current intensity is higher. When the stimulator is turned off, tremor arrest may persist for a short time, never more than 30 seconds. In the Vim region, this effect is not accompanied by side effects except for higher intensities when the stimulator is abruptly turned on: paraesthesiae may be felt by the patient and are usually transient, which is not the case when the electrode is situated in the main somatosensory ventroposterolateral nucleus (VPL). As the stimulating electrode progresses from anterior to posterior and crosses the Vim limits, the threshold decreases to values about 1 mA, sometimes to 0.2 mA. As the electrode approaches the limit between Vim and VPL, paraesthesiae are more easily induced, and they last longer than those in Vim. As shown by Guiot the topography of paraesthesiae evoked from VPL may help to correct the laterality of the trajectory[14,28]. In VPL, paraesthesiae are elicited by very low intensities, far beyond the appearance of any effect on tremor.

Lateral view of Monopolar Electrodes

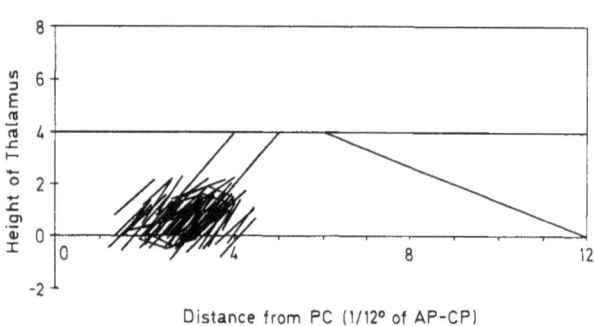

Frontal view of Monopolar Electrodes

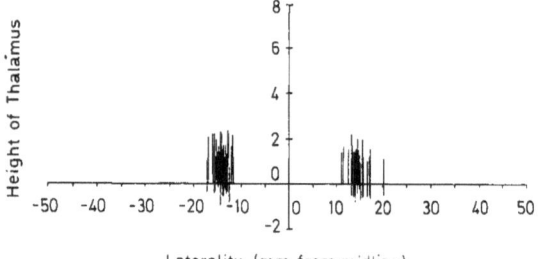

Fig. 1. Computer drawing of the 108 monopolar electrodes presented in relation to normalized anatomical landmarks. Lateral view: the AC-PC line is divided in 12 parts, the height of the thalamus in 8. The schematic representation of Vim extends from 2/12 to 3/12 on the AC-PC line and from 4/12 to 5/12 on the line corresponding to approx. half the height (dorsoventral) of the thalamus. Frontal view: verticality is represented in 1/8 of the height of the thalamus. Laterality is expressed in mm from the midline

Neuronal Activity During Surgery

Microrecording along the different trajectories used to locate Vim may demonstrate neurons expressing spontaneous activity. The recorded spikes fire randomly, or in bursts, and in the lower part of Vim, close to the plane of the AC-PC line, these bursts are synchronous with the tremor oscillations. We did not succeed so far in recording cells responding to proprioceptive inputs into the Vim, but in the VPL cells may be found which have receptive fields to superficial somatosensory inputs, the topography of which may be helpful in correcting the laterality of the final target.

There is a good correspondence between the threshold current to induce tremor arrest, the presence of synchroneously burst firing cells and the immediate vicinity of persistent somatosensory induced paraesthesiae in the VPL.

The ideal implantation site appears to be the anterior border of Vim, where the tremor suppression effect is well obtained and which is far enough from VPL to avoid annoying paraesthesiae which would prevent reaching the adequate level of stimulation to suppress tremor.

Electrode Placement

The statistical distribution of the final position of the electrodes covers the Vim diagram which is therefore partly validated (Fig. 1). Coordinates of this series of electrodes are given in Table 1. The scattering of the electrode positions is due in part to the imprecision of the surgical positioning but also to the choice of the final target according to the results of the electrophysiological investigation.

Table 1. *Average and Standard Deviation of the Co-ordinates of the Monopolar Electrodes in the two Series of patients (PD + ET and Dyskinesias of Various origins) and in the Series with Tetrapolar Electrodes (DBS4). The antero-posterior distance is from the anterior limit of the posterior commissure, as seen on the ventriculogram, expressed in 1/12° of the AC-PC length. Height and laterality are given in millimeters. The differences between the PD + ET group and the Dyskinesias group are not statistically significant. They cannot be compared to the DBS4 group as the shape and size of this tetrapolar electrode led to different co-ordinates of the aimed target*

Group	Antero-Posterior 1/12 of AC-PC line	Height/AC-PC mm	Laterality/Midline mm
PD + ET	2.6 ± 0.6	0 ± 1	13.7 ± 1.2
Dyskinesias	2.7 ± 0.7	− 0.2 ± 2	13.2 ± 1.4
DBS 4	2.1 ± 0.5	− 1.2 ± 0.8	14.4 ± 1.2

Morbidity and Mortality

There was no mortality in this series. Two patients had secondary skin ulceration of the scalp in front of the electrode-to-extension connection. In one patient, this led 6 months after implant to temporary removal of the extension which was replaced 3 months later. In the second patient, the electrode and the extension had to be removed 12 months after implant and were successfully re-implanted 4 months later. One patient with multiple sclerosis presented 10 days after implant a small haemorrhage around the tip of the electrode. He developed a moderate aphasia and slight right hemiparesis which improved within 2 months. At the time of bleeding, tremor which was totally controlled by stimulation disappeared even when the stimulator was turned off and it never recurred. Two other patients had asymptomatic intracranial microhaematomas detected on routine postoperative CT scan.

Side Effects

The adverse effects were mild and accepted by the patients because of their relief from tremor and because they immediately disappeared when stimulation was decreased or stopped. They consisted of contralateral paraesthesiae (9%), limb dystonia (9%), dysequilibrium (7.6%) and dysarthria (15% on the whole: 6% with bilateral stimulation, 7.5% with previous unilateral thalamotomy and contralateral stimulation, and 1.5% with unilateral stimulation). Dysarthria was therefore observed in 14% of the bilaterally stimulated patients and in 50% of the patients previously thalamotomized. No spontaneous psychological disturbance was reported. Suddenly switching on the stimulator could induce transient (a few seconds) and not disabling contralateral paraesthesiae. Switching off the stimulator induced a transient rebound tremor in about half of the patients which made them use the stimulator at night. However, continuous stimulation induced in some patients, and mainly in those with action tremor, a kind of "tolerance" with decreasing efficacy of stimulation.

Tremor Suppression

Immediately after surgery, a thalamotomy-like effect was responsible for a transitory tremor suppression for a few days. During the test period, various combinations of stimulation parameters were evaluated. The best effect/side effects ratio was observed for a pulse width of about 60 microseconds, the lowest possible with the Medtronic Itrel I and II stimulators. The threshold intensity versus frequency necessary to suppress totally the tremor was assessed: the minimum was a plateau from about 100 to 2000 Hz. The stimulators are therefore set to 130 and 185 Hz, the two highest available frequencies on the ITREL I or II. The voltage value was actually set according to the patient choice based on his compromise between benefit and side effects. This voltage increased during the first 6 weeks. Part of this was due to a progressive increase of impedance (measurable on the ITREL II) from about 750 to 1000 ohms during the first 4 weeks. The average voltage at the latest follow-up for each patient was 2.7 volts (range 0.4–5.5 V). A good result, such as permanent total suppression or slight reappearance on rare occasions (scores $3 + 4$), was obtained in 71% of the operated sides. More precisely, this major benefit was obtained in 88% of cases with PD, 68% of cases with ET, and 18% of cases related to other causes. In these latter cases, the effect was often complete during the first postoperative month but an action component recurred later. Rest tremor was better controlled than action tremor, distal limb tremor better than proximal or axial tremor, upper limb better than lower limb tremor. In all cases the effect was strictly simultaneous with stimulation, without significant delay of onset or posteffect at arrest.

Tremor was the only Parkinsonian sign influenced by Vim stimulation. In one-third of the patients with PD, L-Dopa doses could be decreased by more than 30%.

Discussion

Technical Aspects

Correlation between the electrode position and the stimulation effect shows that the most efficient area is generally situated at the level of the AC-PC line. Electrodes placed on the anterior border of Vim provided equally good results as those situated on the posterior border of Vim but with less paraesthetic side effects. Laterality may be difficult to determine. It appears that the responsive part of Vim is very close to the internal capsule and that the main part of the nucleus is parallel to it. Stereotactic MRI with cuts in the plane of the trajectory depicts the position of the pyramidal tract and helps positioning of the electrode in an oblique trajectory parallel to the internal capsule. This was done in our two last patients. However, as

stressed already by Guiot[15,28] and other[17,22,25,26,30], electrophysiology is the ultimate method to optimize the position of an electrode for a thalamotomic lesion: stimulation may identify VPL as well as the pyramidal tract and it is essential for showing during surgery to what extent the tremor may be suppressed by stimulation. Recording of neuronal activity may provide additional information of scientific value. Moreover, it may help to define the optimal site for a stimulating electrode corresponding to where neurons exhibit "burst" activity synchroneously with tremor[14,21-24]. Beneath this region, neuronal silence is characteristic of the internal capsule.

Therapeutic Benefit

Indications for Vim stimulation seem to be similar to those for thalamotomy. Obviously, thalamotomy[11-13,19], is often more cost-effective, since the patient may be quickly discharged from the hospital, there is no foreign implanted material, and there is no need for a future replacement of an energy source. However, thalamotomy has some well known drawbacks: if tremor reappears, a new surgical procedure is needed, and adverse effects may be present in the form of motor neglect, sensory loss or even hemiplegia. Bilateral procedures are often associated with neuropsychological deficits[18]. Vim stimulation has no severe complications, adverse effects are reversible when stimulation intensity is decreased, bilateral procedures do not induce neuropsychological deficits, and recurrence may often be controlled by increasing the stimulation intensity.

As for thalamotomy, rest tremor and essential tremor are the best indications for stimulation. The transient effect on the action component of tremor as well as the development of tolerance are matters of concern. In multiple sclerosis the postural component of the tremor is usually well controlled but there is little effect on the action component and the cerebellar dysmetria.

Mechanism of Action

There are currently no data which could provide an explanation of the effect of Vim stimulation on tremor. The concepts of neural interactions between the cortex and the nigro-striatal system do not account for the possible role of dopamine deafferentation. The diagram of Alexander provides a theoretical synthesis mainly valid to explain bradykinesia and the subthalamic nucleus is not included for the control of tremor. The frequency response curve demonstrates the key importance of this parameter.

However, the range of frequency is far higher than the range of excitability of cells and corresponds better to that of fibers. Vim stimulation does not suppress tremor by means of an excitatory effect since Vim destruction by thalamotomy or by inactivation with lidocaine injection produces the same effect. Inhibitory mechanisms are also unlikely in view of the relatively high frequency of stimulation required to obtain the effect. A jamming of the neuronal network processing proprioceptive inputs to the thalamus is a more likely background. Feeding the system with an artificial neural noise could interfere with the positive feedback loop comprising muscular proprioception—thalamus—cortex and spinal motor neurons.

Conclusion

Vim stimulation provides a new therapeutic approach of Parkinsonian and essential tremors. A five-year experience has permitted one to define the best indications and to observe the long term effects among which the problem of "tolerance" in some cases needs to be solved. This treatment is favoured by the fact that the risk of side-effects is minimal, there are no neuropsychological effects during bilateral stimulation, and it can be combined with a contralateral thalamotomy. A better understanding of the mechanism, finding the optimal stimulus parameters and the electrode site, and even new targets[8,9], will further improve the results and probably also widen the indications.

Acknowledgments

This work was supported by grants from CNAMTS-INSERM, Region Rhône-Alpes, Association France Parkinson and Grenoble Joseph Fourier University.

References

1. Albe-Fessard D (1988) Interactions entre recherches fondamentale et clinique. Deux exemples tirés d'une experience personnelle. Can J Neurol Sci 15: 324–332
2. Albe-Fessard D, Arfel G, Guiot G (1963) Activités électriques caractéristiques de quelques structures cérébrales chez l'homme. Ann Chir 17: 1185–1214
3. Albe-Fessard D, Arfel G, Guiot G, Hardy J, Vourc'h G, Hertzog E, Aléonard P (1962) Dérivations d'activités spontanées et évoquées dans les structures cérébrales profondes de l'homme. Rev Neurol (Paris) 106: 89–105
4. Albe-Fessard D Arfel G, Guiot G, Hardy J, Vourc'h G, Hertzog E, Aléonard P (1961) Identification et délimitation précise de certaines structures sous-corticales de l'homme par l'électrophysiologie. Son interêt dans la chirurgie stéreotaxique des dyskinesies. C R Acad Sci Paris 253: 2412–2414
5. Benabid AL, Pollak P, Louveau A, Henry S, de Rougemont J (1987) Combined (thalamotomy and stimulation) stereotactic surgery of the VIM thalamus nucleus for bilateral Parkinson's disease. Appl Neurophysiol 50: 344–346
6. Benabid AL, Pollak P, Gervason C, Hoffmann D, Gao DM, Hommel M, Perret JE, de Rougemont J (1991) Long-term suppression of tremor by chronic stimulation of the ventral intermediate thalamic nucleus. Lancet 337: 403–406
7. Benabid AL, Pollak P, Hommel M, Gaio JM, De Rougemont J, Perret J (1989) Traitement du tremblenment parkinsonien par stimulation chronique du noyau ventral intermédiaire du Thalamus. Rev Neurol (Paris) 145: 320–323
8. Bergmann H, Wickmann T, DeLong MR (1990) Reversal of experimental Parkinsonism by lesions of the subthalamic nucleus. Science 249: 1436–1438
9. Brice J, McLellan L (1980) Suppression of intention tremor by contingent deep-brain stimulation. Lancet i: 1221–1222
10. Cooper IS (1953) Ligation of anterior choroidal artery for involuntary movement parkinsonism. Psychiatr Q 27: 317–319
11. Derôme PJ, Jedynak CP, Visot A, Delalande O (1986) Traitement des mouvements anormaux par lésions thalamiques. Rev Neurol (Paris) 142: 391–397
12. Fox MW, Ahlskog JE, Kelly PJ (1991) Stereotactic ventrolateralis thalamotomy for medically refractory tremor in postlevodopa era Parkinson's disease patients. J Neurosurg 75: 723–730
13. Goldman MS, Ahlskog JE, Kelly PJ (1992) The symptomatic and functional outcome of stereotactic thalamotomy for medically intractable essential tremor. J Neurosurg 76: 924–928
14. Guiot G, Arfel G, Derôme P (1968) La chirurgie stéreotaxique des tremblements de repos et d'attitude. Gazette Médicale de France 75: 4029–4056
15. Guiot G, Derome P, Arfel G, Walter S (1973) Electrophysiological recordings in stereotaxic thalamotomy for parkinsonism. Prog Neurol Surg 5: 189–221
16. Guiot G, Derome P, Trigo JC (1967) Le tremblement d'attitude. Indication la meilleure de la chirurgie stéréotaxique. Presse Méd 75: 2513–2518
17. Jasper HH, Bertrand G (1966) Recording from microelectrode in stereotaxic surgery for Parkinson's disease. J Neurosurg 24: 219–221
18. Matsumoto K, Shichijo F, Fukami T (1984) Long-term followup review of cases of Parkinson's disease after unilateral or bilateral thalamotomy. J Neurosurg 60: 1033–1044
19. Narabayashi H (1989) Stereotaxic Vim thalamotomy for treatment of tremor. Eur Neurol 29 [Suppl 1]: 29–32
20. Nashold BS, Slaughter DG (1969) Some observations on tremor. In: Gillingham FJ (eds) Third symposium on Parkinson's disease. Linvingstone London, pp 241–246
21. Ohye C, Maeda T, Narabayashi H (1977) Physiologically defined VIM nucleus. Its special reference to control of tremor. Appl Neurophysiol 39: 285–295
22. Ohye C, Nakamura R, Fukamachi A, Narabayashi H (1975) Recording and stimulation of the ventralis intermedius nucleus of the human thalamus. Confin Neurol 37: 258
23. Ohye C, Narabayashi H (1979) Physiological study of presumed ventralis intermedius neurons in the human thalamus. J Neurosurg 50: 290–297
24. Ohye C, Shibazaki T, Hirai T, Wada H, Hirato M, Kawashima

Y (1989) Further physiological observations on the ventralis intermedius neurons in the human thalamus. J Neurophysiol 61: 488–500

25. Schaltenbrand G, Spuler H, Wahren W, Rümler B (1971) Electroanatomy of the thalamic ventro-oral nucleus based on stereotaxic stimulation in man. Z Neurol 199: 259–276

26. Sem-Jacobsen CW (1966) Depth-electrographic observations related to Parkinson's disease. J Neurosurg 24 [Suppl 10]: 388–402

27. Stellar S, Coope IS (1968) Mortality and morbidity in cryothalamectomy for Parkinson's disease. A statistical study of 2868 consecutive operations. J Neurosurg 28: 459–467

28. Taren J, Guiot G, Derome P, Trigo JC (1968) Hazards of stéréotaxic thalamotomy. Added safety factors in corroborating X-ray target localization with neurophysiological methods. J Neurosurg 29 173–182

29. Tasker RR (1986) Effets sensitifs et moteurs de la stimulation thalamique chez l'Homme. Applications cliniques. Rev Neurol (Paris) 142: 316–326

30. Tasker RR, Organ LW, Hawrylyshyn PA (1982) The thalamus and midbrain of man. Thomas, Springfield, Illinois 505 pp

Correspondence: A. L. Benabid, M. D., Department of Neurosciences, INSERM 318, and University Hospital of Grenoble, BP217X-38043, Grenoble, France.

Acta Neurochir (1993) [Suppl] 58: 45–47

Tremor Reduction by Microinjection of Lidocaine During Stereotactic Surgery

A. G. Parrent, R. R. Tasker[1], and J. O. Dostrovsky[2]

Division of Neurosurgery, University Hospital, London, Ontario, [1]Division of Neurosurgery, The Toronto Hospital, Toronto, and [2]Department of Physiology, University of Toronto, Toronto, Ontario, Canada

Summary

We report our experience with lidocaine microinjection into the thalamus in 10 patients undergoing stereotactic thalamotomy for the treatment of Parkinsonian or non-Parkinsonian tremor. 18 injection sites in 4 patients with Parkinson's disease and 22 sites in 6 patients with other forms of tremor have been compared with respect to the effect of microstimulation. In over two thirds of cases the test microinjection replicated the effects of microstimulation.

Long term follow up will be required to determine whether lesions made on the basis of lidocaine induced tremor suppression will result in a lower rate of tremor recurrence than those based on stimulation induced tremor suppression.

In those patients in whom stimulation induced tremor suppression occurs but tremor arrest cannot be produced with lidocaine microinjection, chronic thalamic stimulation may be an alternative for the long term control of tremor in these patients.

Keywords: Tremor; movement disorders; stereotactic surgery; lidocaine; thalamotomy.

Introduction

Various anatomical and physiological criteria have been proposed to identify sites at which stereotactic lesions should stop tremor in Parkinson's disease and other conditions. Lesions are usually made at sites where microelectrode recording identifies neurons that fire in relation to tremor (tremor cells) and where microstimulation induces tremor arrest or substantial tremor reduction, both features considered reliable indicators of an adequate lesion site. However, we have occasionally seen tremor recur after making lesions at sites that meet both of these criteria.

The mechanism by which electrical stimulation arrests tremor is not known. It may act by the blockade of neural circuits required for the expression of tremor or by activation of inhibitory neural circuits. If neural activation is the mechanism of stimulation induced tremor arrest, then it may not be predictive of lesion efficacy.

We have recently added the strategy of microinjection of small aliquots of lidocaine into the thalamus at potential lesion sites, assuming that this drug would produce its effects only by neural blockade, and better predict the effects of an ablative lesion. Experimental work has shown that microinjections of 1–2 μl of lidocaine into the brain will diffuse into a spherical volume 1.9–2.4 mm diameter[1,2] producing a reversible increase in the threshold to electrical stimulation[3] and decrease in glucose utilization[2].

Methods

This report is based on our experience with 10 patients undergoing stereotactic thalamotomy for the treatment of tremor associated with: Parkinson's disease—4, multiple sclerosis—2, essential tremor —1, post stroke tremor—1, post traumatic tremor—1, cerebellar tremor of unknown etiology—1.

The technique of microelectrode recording and microstimulation has been described in detail previously and will only be briefly reviewed. Computer generated sagittal thalamic maps were produced based on the coordinates of the anterior and posterior commissures identified on axial CT scans carried out after the application of a stereotactic frame to the patient's head. The tactile relay nucleus (ventrocaudal nucleus) and the more rostral kinesthetic nucleus (ventrointermedius nucleus) were explored using microelectrode recording and microstimulation (tungsten microelectrode, impedance 0.5–2.0 MΩ).

Injections of lidocaine were made at potential lesion sites as determined by the presence of kinesthetic or tremor cells related to the affected limb and/or tremor improvement with microstimulation (300 Hz, 0.1 ms pulse width, 1–5 s trains, up to 100 μA). Prior to an injection, baseline assessment of tremor, voluntary movements and sensation was made and recorded on videotape. A 25 μl Hamilton syringe was used to inject 1–2 μl of 2% lidocaine hydrochloride at each site through a 25 gauge stainless steel tube inserted through the same cannula used to guide the microelectrode. Follow up assessments of tremor, voluntary movement, sensation and speech were recorded at intervals after the injection and compared to the baseline preinjection assessment.

Results

Fifty lidocaine injections were made in 10 patients. An accurate determination of the time course of observable effect on tremor was available for 19 injections in 5 patients. The mean time to the onset of an observable effect was 69 s (30–360 s), mean time to peak effect was 91 s (30–180 s) and mean duration of effect was 171 s (90–300 s).

The effect of microstimulation and lidocaine on tremor was compared in 2 subgroups of patients, those with parkinsonian tremor[4], and those with other types of tremor[6].

Eighteen injection sites in 4 patients with Parkinson's disease were assessed (Table 1). There was concurrence between the effect of microstimulation and lidocaine microinjection at 12 sites (67%). At 11 of these sites tremor was suppressed by both, and at 1 site tremor was not affected by either. At 6 sites there was non concurrence between the effects of microstimulation and lidocaine microinjection. At 2 of these sites tremor was not affected by microstimulation but was suppressed after lidocaine microinjection, and at 4 sites tremor was suppressed by microstimulation but not affected by lidocaine. At one of the latter sites lidocaine produced subjective numbness in the contralateral

fingers, prompting us to explore further rostrally, away from the thalamic tactile area.

In patients with other forms of tremor, 22 sites in 6 patients were compared with respect to the effect of microstimulation and lidocaine microinjection (Table 2). There was concurrence at 15 sites (68%), 7 of which produced tremor alteration (tremor suppression in 6, and tremor drive in 1), and 8 producing no effect on tremor. There was non concurrence at 7 sites. Lidocaine microinjection produced tremor suppression at 6 sites where microstimulation had no effect, and microstimulation arrested tremor a 1 site where lidocaine produced no effect.

Discussion

Our experience with lidocaine microinjection into the thalamus in movement disorder patients shows that it is safe, and in over two thirds of cases replicates the effects of microstimulation.

Long term follow up will be required to determine whether lesions made on the basis of lidocaine induced tremor suppression result in a lower rate of tremor recurrence than those based on stimulation induced tremor suppression. Such determination will likely be hampered by the fact that in any given patient several lesions may be made, some of which are based on the effect of lidocaine microinjection and others on microstimulation.

In patients with non parkinsonian tremor, modification of tremor with macro- and microstimulation is much less common than in Parkinson's disease[4]. Lidocaine microinjection may allow identification of potential lesion sites in such cases. In our group of patients with non parkinsonian tremor the most frequent pattern of mismatch between the effect of lidocaine microinjection and microstimulation was the suppression of tremor with lidocaine in the absence of response to stimulation. In one patient with post stroke tremor no sites could be found at which microstimulation had any effect on tremor, but two areas were found where lidocaine substantially reduced the tremor. Two thalamic lesions were made and, 16 months later tremor was still reduced.

In those patients in whom stimulation induced tremor suppression occurs but tremor arrest cannot be produced with lidocaine microinjection, thalamic stimulation may be suppressing tremor by the activation of inhibitory neural circuits. Chronic thalamic stimulation may be an alternative for the long term control of tremor in these patients[5,6].

Table 1. *Comparison of the Effects of Microstimulation and Lidocaine Microinjection in 4 Patients with Parkinsonian Tremor (16 Stimulation and Injection Sites)*

Lidocaine	Stimulation	
	No effect	TA/TR
No effect	1	4
TA/TR	2	11

TA tremor arrest; *TR* tremor reduction.

Table 2. *Comparison of the Effects of Microstimulation and Lidocaine Microinjection in 6 Patients with Non Parkinsonian Tremor (22 Stimulation and Injection Sites)*

Lidocaine	Stimulation	
	No effect	TA/TR/TD
No effect	8	1
TA/TR	6	7

TA tremor arrest; *TR* tremor reduction; *TD* tremor drive.

References

1. Myers RD (1966) Injection of solutions into cerebral tissue: Relation between volume and diffusion. Physiol Behav 1: 171–174
2. Martin JH (1991) Autoradiographic estimation of the extent of reversible inactivation produced by microinjection of lidocaine and muscimol in the rat. Neurosci Lett 127: 160–164
3. Sandküler J, Maisch, Zimmerman M (1987) The use of local anaesthetic microinjections to identify central pathways: A quantitative evaluation of the time course and extent of the neuronal block. Exp Brain Res 68: 168–178
4. Tasker RR, Organ LW, Hawrylyshyn PA (1982) The thalamus and midbrain of man. Thomas, Baltimore, pp 304–308
5. Blond S, Siegfried J (1991) Thalamic stimulation for the treatment of tremor and other movement disorders. Acta Neurochir (Wien) [Suppl] 52: 109–111
6. Benabid AL, Pollak P, Gervason C, Hoffman D, Gao DM, Hommel M, Perret JE, deRougement J (1991) Long-term suppression of tremor by chronic stimulation of the ventral intermediate thalamic nucleus. Lancet 337: 403–406

Correspondence: Andrew Parrent, M.D., Division of Neurosurgery, University Hospital London, Ontario, N6A 5AS, Canada.

Acta Neurochir (1993) [Suppl] 58: 48–52
© Springer-Verlag 1993

External and Implanted Pumps for Apomorphine Infusion in Parkinsonism

P. Pollak, A. L. Benabid, P. Limousin, C. L. Gervason, and E. Jeanneau-Nicolle[1]

Department of Clinical and Biological Neurosciences, INSERM U-318, Joseph Fourier University of Grenoble, and [1]Laboratory of Pharmacology, Centre Hospitalier Universitaire de Grenoble, France

Summary

Continuous delivery of dopaminergic agents to the striatum is a major challenge to improve the treatment of Parkinson's disease. Apomorphine is one of the best candidates because of its solubility and its D1 and D2 receptor agonist properties.

Seventeen Parkinsonian patients suffering from severe L-dopa-induced on-off effects were treated by continuous subcutaneous (SC) infusion with a portable minipump. Administration of intracerebroventricular (ICV) apomorphine was carried out in 7 macaca fascicularis monkeys using implanted programmable pumps. Four of the monkeys were made Parkinsonian by MPTP injections.

In patients receiving apomorphine, the mean duration of daily off periods was reduced by 61%. Psychiatric side effects were rare but SC nodules occured in all patients and the external infusion method was therefore difficult to implement. In monkeys, the implanted system was well tolerated. ICV apomorphine infusion led to CSF apomorphine concentrations higher than the same apomorphine dose infused i.m. Motor function was considerably improved in two MPTP monkeys during the time of ICV infusion and 30 min after its arrest. Long-term ICV administration could not be carried out because of catheter blockage and/or apomorphine toxicity.

SC and ICV apomorphine infusions are efficient for controlling motor activity in Parkinsonism but long-term toxicity remains to be studied further.

Keywords: Parkinsonism; apomorphine; intracerebroventricular infusion; MPTP monkey.

Introduction

Initially, the response to doses of L-dopa in patients with Parkinson's disease (PD) is smooth, stable and long-lasting. After several years of dopatherapy most patients develop fluctuations in motor response and abnormal involuntary movements[11]. Many patients develop a "threshold effect", when the sinuous fluctuations are replaced by more severe and abrupt "all-or-none" switches from "on" (mobile with dyskinesias) to "off" (akinetic). The frequent, sudden and more or less unpredictable occurrence of severe akinesia constitutes the on-off phenomenon. For the majority of patients, treatment becomes difficult.

Continuous intravenous infusions of L-Dopa have proved to be efficient in the maintenance of mobility in such patients, suggesting a role for the peripheral pharmacokinetics[15]. Central pharmacodynamic factors should also be taken into account. In rats, studies using dopaminergic drugs have shown that striatal dopaminergic receptors can rapidly become tolerant and, during the course of some hours, their sensitivity to dopaminergic agonism is restored. Intermittent administration induces hypersensitivity, whereas continuous administration induces hyposensitivity. It is believed that only the D2 system is important for the motor effects of dopaminergic drugs, but more recent evidence suggests that the full expression of dopaminergic pharmacological effects requires activation of both types of receptors[18]. Thus, the most suitable drug for PD should activate D1 and D2 dopamine receptors, and have a long-lasting activity allowing smooth fluctuations in concentrations at the dopamine receptor sites. Hitherto, the drug of choice has been L-dopa which has a short-lasting non-selective action upon dopamine receptors. The maintenance of continuous delivery of dopamine to the striatum is difficult because it has a short plasma half-life, an erratic gastro-intestinal absorption, it competes with large neutral amino acids and its metabolite 3-O-methyldopa, and it is dependent upon its enzymatic conversion to dopamine within the striatum.

Among direct dopaminergic agonist drugs, apomorphine is one of the best candidates because of its solubility and its powerful D1 and D2 receptor agonist properties[10]. Its large first-pass hepatic effect prevents its oral route of administration. Owing to its short half-life, a continuous delivery of apomorphine with variable flow rates necessitates the use of programmable pumps. Sites of infusion could be the subcutaneous (SC), vascular and CSF spaces. Intracerebroventricular (ICV) infusions are not mandatory because apomorphine easily crosses the brain-blood-barrier, but it may have some theoretical advantages. Delivery close to the site of action avoids transport problems, averts peripheral side effects, and long-term infusion is known to be technically feasible in humans.

In Parkinsonian patients, apomorphine has been infused in the SC space using an external portable minipump[4,14,16]. We will shortly report here our experience of this technique. During the last five years apomorphine ICV infusion has not yet been performed. ICV infusion of dopamine has been carried out in a single patient using a non programmable implanted pump with encouraging results[7]. In order to test the feasibility of ICV infusion of apomorphine, we performed experiments in a primate model of PD.

Methods

External SC Apomorphine Infusion in PD Patients

Seventeen Parkinsonian patients suffering from severe L-dopa-induced on-off effects resistant to other therapeutical strategies (including intermittent apomorphine injections) were treated by continuous SC infusion with a portable minipump. Their mean age was 55 (range 34–73) years and disease duration 16 years. To avoid nausea, vomiting and postural hypotension, domperidone, a peripheral dopamine receptor antagonist (20 mg t.i.d.) was administered 2 days before apomorphine onset and continued at least for 2 weeks. Apomorphine solution (Apokinon® 1%, Aguettant Laboratory, Lyon, France) was infused through a 27 G needle inserted subcutaneously in the abdomen wall and the site was changed at least daily. The Graseby MS 26 syringe driver (Graseby DMS, Montpellier, France) was used, allowing continuous infusion at a rate up to 12 mg/h plus additional bolus doses. Attempts were made to decrease L-dopa intake compensating to an increase of the apomorphine flow rate.

Implanted Programmable Pump in Monkeys

Animal model: Seven male adult cynomolgus monkeys (macaca fascicularis), weighing 3–4 kg were used in this study. The neurotoxic MPTP hydrochloride (RBI, Natick, USA) was dissolved freshly in NaCl 0.9%, to a final concentration of 1 mg/ml, and injected via a lower leg vein in an animal anaesthetized with intramuscular (IM) 10 mg/kg ketamine (Imalgène® 1000, Rhône Mérieux, France). The MPTP dose was 0.33 mg/kg in the first injections, administered twice a week until the animal exhibited bradykinetic symptoms. The dose was doubled if the first six injections were ineffective. Five monkeys were made Parkinsonian by MPTP injections.

Implantation of the delivery device: All surgery was performed under sterile conditions with animals anaesthetized with ketamine. A synchroMed™ DAD Medtronic pump (Medtronic, Minneapolis, USA) was implanted intraperitoneally, after the completion of a stable MPTP induced Parkinsonism in two monkeys and before the

administration of MPTP in three. This pump is programmable by telemetry for infusion mode and flow rate (0.006 to 0.9 ml/h). It weighs 185 g, and its reservoir volume is 18 ml. It can be filled by transcutaneous puncture of a rubber membrane. An outflow catheter (Spinal catheter dowcorning silastic®, Tokyo, Japan) was placed in the right lateral ventricle using stereotactic technique, 30 mm posterior to the orbital arch, 2 mm off the midline, 15–18 mm in depth. A radiographic control using a contrast injection (Iopamiron 200®, Schering, Germany) was made. The catheter was fixed to the skull with dental cement. It was connected to the pump by another Medtronic silicone rubber catheter tunnelled subcutaneously. Two access ports (Clinical Plastic Products, LaChaux-de-Fonds, Switzerland) were placed subcutaneously in the occipital region of the skull. One was joined to the intraventricular catheter for bolus injections and the other was connected to the cisterna magna for CSF samples.

Control of delivery: The pumps were initially filled with 0.9% NaCl and programmed at the minimal rate of flow for maintaining permeability. The material function was checked by X-rays using injections of iodine contrast via the access port, and by scanning after isotope injection into the pump.

Stability of apomorphine and concentrations used: Stability of the apomorphine solution (concentration from 1 to 10 mg/ml) in the pump was evaluated up to one month, in vivo and in vitro in a non-implanted Medtronic pump maintained at 37 °C. Apomorphine concentration was measured with high performance liquid chromatography using electrochemical detection[2]. For short-term drug administration, apomorphine was diluted in 0.9% NaCl to obtain a final concentration of 1 mg/ml.

Motor evaluation: Motor activity was assessed at regular intervals during the process of MPTP induced Parkinsonism and after apomorphine administrations, using the subjective scales of the University of Laval (Quebec, Canada) and of Kurlan et al. (1991), with a video control. During apomorphine ICV infusions, we have also evaluated the number of to-ings, fro-ings, jumps, and the time of motionlessness.

Results

SC Apomorphine Infusion in PD Patients

Only 7 out of 17 patients continued SC apomorphine infusion for more than a year, up to 55 (mean 47 ± 7) months. Eleven patients were available for evaluation. Mean infusion rates were 97 ± 44 µg/kg/h in daytime, and 6 patients were infused at night with a lower mean flow rate of 39 µg/kg/h. Patients used a mean of 4 additional boluses per day (approximately 3 mg each). The total daily dose was 95 ± 28 mg or 1.7 ± 0.8 mg/kg, with little change as time passed. The mean duration of daily off periods was reduced by 61% (from a mean of 6.8 to 2.6 hours). L-dopa dosage was reduced by 53% on average (from 790 to 370 mg/d), and one patient could stop L-dopa. During the residual off motor periods, akinesia was improved in 4/11 patients. The severity of hyperkinesia was improved in 5/11 patients in parallel with the decrease in L-dopa dose.

SC nodules occured in all patients. It was the only adverse effect in the 7 patients treated on a long-term basis. The reasons why 8 patients gave up the infusions

Table 1. *Characteristics of MPTP Injections and Parkinsonian Scores in Three Monkeys*

	Monkey 1	Monkey 2	Monkey 3
MPTP			
Total dose (mg)	24.8	14.8	4.9
Injection number	12	11	4
Time from 1st to last injection (Mo)	14	5	4
Parkinsonian score			
University of Laval scale (/10)	7	5	4
Kurlan's scale (/29)	16	10	8

were the following: toxic cutaneous reactions and/or difficulty to cope with the technique, 3 cases; prolongation of diphasic dyskinesias, 3 cases; behavioural side-effects, 2 cases. Five of these 8 patients changed from continuous to intermittent apomorphine injections. Two patients died for reasons unrelated to apomorphine administration. All biological tests remained within their normal ranges.

Implanted Programmable Pump in Monkeys

Animal model: MPTP induced a long-lasting marked Parkinsonian syndrome in 5 primates. Only 3 of them (Table 1) were available for the apomorphine administration test. Tremor was slight and present during action contrary to the classical rest tremor of Parkinsonian patients.

Implantation and control of delivery device: Despite its relative large volume compared to the monkey size, the pumps were well tolerated up to 18 months when implanted before the administration of MPTP. After completion of the MPTP induced Parkinsonian motor state, two monkeys died in the postoperative period following pump implantation and one monkey after pump change. Some surgical revisions were necessary for disconnections of the catheters from the access ports or moving from the ventricles or the cisterna magna.

Control of apomorphine stability: The concentration

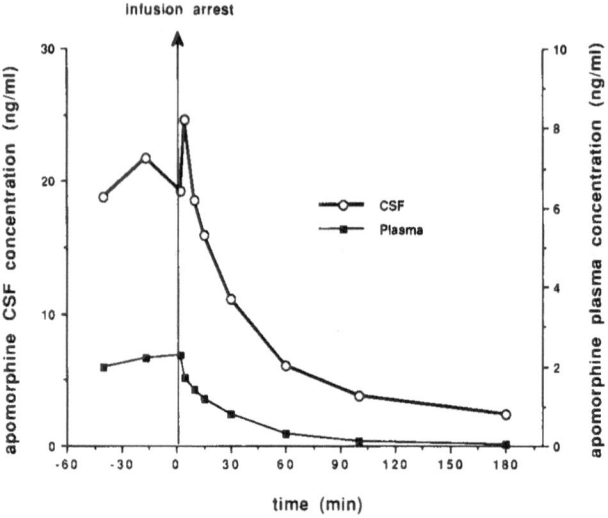

Fig. 1. ICV infusion of 50 µg/kg/h apomorphine for 8 hours followed by pump arrest

of apomorphine remained stable in the pump, *in vitro* at 37 °C and *in vivo* for up to one month.

Pharmacokinetic study: To characterize pharmacokinetic parameters, a total of 13 apomorphine administrations were carried out in 3 monkeys, with different i.m. and ICV bolus dose injections (50–100 µg/kg) and infusions (25–100 µg/kg/h).

The absolute bioavailability of the i.m. route was close to 1 because the area under curve was similar after IV or i.m. injection of the same dose. Plasma concentration-time profiles of apomorphine were biphasic. After i.m. bolus injections (3 tests), C_{max} was 8 to 29 times higher in plasma than in CSF, plasma T_{max}

Table 2. *Effective Doses of Apomorphine and Chronology of Apomorphine Induced Motor Activity in Three MPTP Monkeys*

	Monkey 1	Monkey 2	Monkey 3
IM Boluses			
Effective dose (µg/kg)	50	50	150
Latency for effect onset (min)	3 to 7	2	3
Duration of effect (min)	30 to 40	60	60
ICV Infusions			
Effective flow rate (µg/kg/h)	100	140	—
Duration of effect after pump arrest (mm)	30	30	—

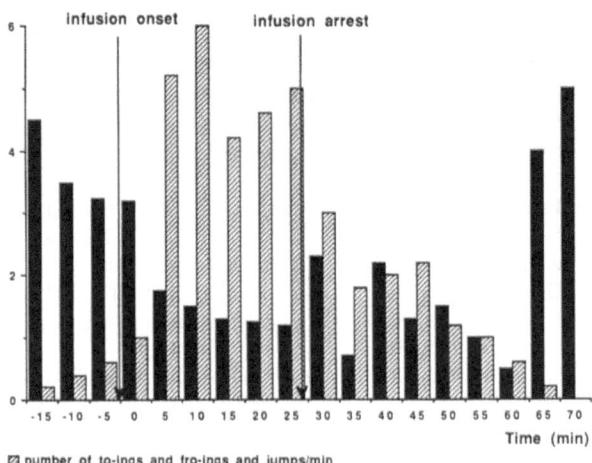

infusion onset   infusion arrest

Time (min)

▨ number of to-ings and fro-ings and jumps/min
■ duration of motionlessness (min)

Fig. 2. Motor assessment during consecutive 5 min periods following an ICV infusion of 50 µg/kg apomorphine for 28.5 min

was brief (3–8 min), as well as the elimination half-life (21–32 min). After ICV bolus injections (2 tests), the elimination CSF half-life was shorter (10–15 min) than plasma half-life. After at least 4 hours of continuous i.m. infusion, the apomorphine levels at the steady state were 7 times higher in the plasma than in the CSF (2 tests), whereas after ICV infusion the CSF levels were 22 times higher than in the plasma (Fig. 1 showing one out of 5 tests). Apomorphine plasma levels at the steady state after continuous infusion i.m. were double those following ICV infusion at the same flow rate.

Evaluation of motor activity: The reversal of Parkinsonian symptoms in MPTP primates with apomorphine was equally effective with i.m. bolus injections and ICV infusions. The dose required, the delay and duration of efficacy are shown in Table 2. In one monkey, i.m. injections induced motor improvement and dyskinesias at the same time whereas the anti-Parkinsonian effect was not accompanied by dyskinesias after ICV infusions. Figure 2 shows the number of to-ings, fro-ings, jumps, and time of motionlessness of a monkey before, during, and after an ICV 50 µg/kg apomorphine infusion for 28.5 min. This monkey had a tendency to rotate toward the contralateral side of the infused ventricle.

Another monkey experienced dyskinesias, from the first i.m. apomorphine injections at effective doses. The amount of dyskinesia was dose dependent. Dyskinesias affected the legs and the trunk, and were choreic and dystonic in nature. They had the same chronology, body topography and appearance whatever the type and route of apomorphine administration was employed.

The maximum duration of apomorphine ICV infusion at effective doses in one monkey was 36 h when the animal died. Pathological examination of the brain showed a temporal lobe herniation related to the expansion of the cerebral hemisphere infused, with a dark green mush in the ventricle suggesting apomorphine auto-oxidation. A further histological study is in progress.

Discussion

Continuous S.C. apomorphine infusion has led to a marked motor improvement in a selected group of severely disabled Parkinsonian patients. Yet, these patients must respond to L-dopa because sustained apomorphine administration induces the same motor effect as L-dopa, but with a longer lasting effect. This signifies that apomorphine has the same non-selective dopaminergic pharmacological profile as L-dopa. Continuous administration of apomorphine induces a more sustained dopaminergic striatal activity than the oral route of L-dopatherapy. Peak-dose dyskinesias are improved if L-dopa dosage is concomitantly reduced. This favours the S.C. continuous infusion mode of administration of apomorphine instead of giving repeated injections at the beginning of each off period[13]. It is notable that apomorphine induces few adverse psychiatric effects compared to lisuride, another dopaminergic agonist administered subcutaneously[17]. However, patients develop cutaneous nodules at the apomorphine infusion sites and experience difficulties in coping with the technique of external infusion. These are the main limitations for this mode of treatment. Implantation of a delivery programmable device would theoretically be useful.

In primates, susceptibility to MPTP varied from one animal to another[9]. Repetitive injections of small MPTP doses over a long period seem to induce a more stable Parkinsonian syndrome than the use of few injections with higher doses[8]. MPTP makes the animals too fragile to resist major surgery. Therefore, pumps must be implanted before rendering monkeys Parkinsonian.

Programmable pump implantation inside the abdominal peritoneum is feasible in monkeys. With this location skin necrosis can be avoided[19]. We encountered some difficulties with implanted catheters in the cisterna magna and the lateral ventricle because of the small size of the CSF spaces in these monkeys.

Apomorphine pharmacokinetic parameters following i.m. injections are similar to those obtained in humans after S.C. injections[5]: the T_{max} is brief, the drug is completely absorbed and rapidly cleared from the plasma. These pharmacokinetic data may explain the rapid onset and brief duration of the motor effects. Elimination of apomorphine is even faster in the CSF than in plasma suggesting that some of the drug infused into the CSF is transferred directly into the plasma via the usual bulk flow route of CSF absorption[6]. Comparison between i.m. and ICV continuous infusions shows that the plasma-CSF ratio is low, in keeping with the lipophilic nature of the apomorphine molecule which permits an easy brain-blood barrier entry. A CSF route of administration would imply a saving in dosage by about 2 only, because the blood in brain capillaries would substantially contribute to apomorphine tissue levels.

The ability of apomorphine to reverse Parkinsonian symptoms and to induce dyskinesias was markedly similar to what occurs in humans[4]. The same is also described with L-dopa[1]. In two monkeys, the motor effects were similar regardless of the mode or route of administration. In another monkey, only the ICV infusions, and not the i.m. injections were able to induce an improvement in Parkinsonism without dyskinesia. The results of this study are too preliminary, and obtained only with short-term ICV infusions, to permit any conclusion. One monkey exhibited dyskinesias from the first effective i.m. injection of apomorphine, suggesting that the duration of the dopaminergic treatment is not a mandatory factor in the genesis of dyskinesias[3,12].

In conclusion, apomorphine is a very potent dopamine agonist at the D1 and D2 dopaminergic receptors and it may induce a sustained motor benefit in Parkinsonian patients or monkeys when treated by a continuous infusion. The ICV route of infusion is interesting from a pathophysiological point of view and it would not enable a substantial reduction of the dose in comparison with a peripheral route of administration. Moreover, although apomorphine auto-oxidation in the pump may be prevented by the addition of anti-oxidants, this is not the case in tissue where it can have a toxic effect as dopamine[19]. Therefore, long-term toxicity remains to be evaluated before carrying out ICV apomorphine infusion in Parkinsonian patients.

References

1. Bédard PJ, Di Paolo T, Falardeau P, Boucher R (1986) Chronic treatment with L-DOPA, but not bromocriptine, induces dyskinesia in MPTP-parkinsonian monkeys. Correlation with (^{3}H) spiperone binding. Brain Res 379: 294–299

2. Bianchi G, Landi M (1985) Determination of apomorphine in rat plasma and brain by high-performance liquid chromatography with electrochemical detection. J Chromatogr 338: 230–235

3. Boyce S, Clarke CE, Luquin R, Peggs D, Robertson RG, Mitchell IJ, Sambrook MA, Crossman AR (1990) Induction of chorea and dystonia in Parkinsonian primates. Mov Disorders 5: 3–7

4. Frankel JP, Lees AJ, Kempster PA, Stern GM (1990) Subcutaneous apomorphine in the treatment of Parkinson's disease. J Neurol Neurosurg Psychiatry 53: 96–101

5. Gancher ST, Woodward WR, Boucher B, Nutt JG (1989) Peripheral pharmacokinetics of apomorphine in humans. Ann Neurol 26: 232–238

6. Harbaugh RE, Saunders RL, Reeder RF (1988) Use of implantable pumps for central nervous system drug infusions to treat neurological disease. Neurosurgery 23: 693–698

7. Horne MK, Butler EG, Gilligan BS, Wodak J (1989) Intraventricular infusion of dopamine in Parkinson's disease. Ann Neurol 26: 792–794

8. Kurlan R, Kim MH, Gash DM (1991) The time course and magnitude of spontaneous recovery of Parkinsonism produced by intracarotid administration of 1-methyl-4-phenyl-1,2,3,6-tetrahydropyridine to monkeys. Ann Neurol 29: 677–679

9. Kurlan R, Kim MH, Gash DM (1991) Oral levodopa dose-response study in MPTP-induced hemiparkinsonian monkeys: Assessment with a new rating scale for monkey parkinsonism. Mov Disord 6, 2: 111–118

10. Lal S (1988) Apomorphine in the evaluation of dopaminergic function in man. Prog Neuropsychopharmacol Biol Psychiatry 12: 117–164

11. Marsden CD, Parkes JD, Quinn N (1981) Fluctuations of disability in Parkinson's disease: Clinical aspects. In: Marsden CD, Fahn S (eds) Movement disorders. Butterworth, London, pp 96–122

12. Nutt JG (1990) Levodopa-induced dyskinesias: Review, observations, and speculations. Neurology 40: 340–345

13. Pollak P, Champay AS, Gaio JM, Hommel M, Benabid AL, Perret J (1990) Administration sous-cutanée d'apomorphine dans les fluctuations motrices de la maladie de Parkinson. Rev Neurol 146: 116–122

14. Pollak P, Champay AS, Hommel M, Perret JE, Benabid AL (1989) Subcutaneous apomorphine in Parkinson's disease. J Neurol Neurosurg Psychiatry 52: 544

15. Quinn N, Parkes JD, Marsden CD (1984) Control of on-off phenomenon by continuous intravenous infusion of levodopa. Neurology 34: 1131–1136

16. Stibe CMH, Lees AJ, Kempster PA, Stern GM (1988) Subcutaneous apomorphine in parkinsonian on-off oscillations. Lancet I: 403–406

17. Vaamonde J, Luquin MR, Obeso JA (1991) Subcutaneous lisuride infusion in Parkinson's disease. Response to chronic administration in 34 patients. Brain 114: 601–614

18. Walters JR, Bergstrom DA, Carlon JH, Chase TN, Braun AR (1987) D1 dopamine receptor activation required for post-synaptic expression of D2 agonist effects. Science 236: 719–722

19. Yébenes JG De, Fahn S, Lovelle S, Jackson-Lewis V, Jorge P, Mena MA, Reiziz J, Bustos JC, Magarinos C, Martinez A (1987) Continuous intracerebroventricular infusion of dopamine and dopamine agonists through a totally implanted drug delivery system in animal models of Parkinson's disease. Mov Disord 2: 143–158

Correspondence: Dr. P. Pollak, Department of Clinical and Biological Neurosciences, Clinique Neurologique, Centre Hospitalier Universitaire de Grenoble, BP 217, F-38043 Grenoble Cedex 9, France.

Acta Neurochir (1993) [Suppl] 58: 53–55

Clinical Evaluation of Computed Tomography-Guided versus Ventriculography-Guided Thalamotomy for Movement Disorders

M. I. Hariz and **A. T. Bergenheim**

Department of Neurosurgery, University Hospital, Umeå, Sweden

Summary

38 patients with Parkinsonian or essential tremor who underwent thalamotomy based on ventriculographic coordinates were compared to 23 patients whose thalamotomy was performed on the basis of computed tomography (CT)-derived target coordinates. The comparison between the two groups concerned the age, sex, duration of disease, target side, intra-operative target correction, duration of postoperative stay in hospital, transient side-effects, permanent complication, and tremor alleviation. The study showed that the surgical results in terms of tremor control were the same in the two groups. However, the percentage of post-operative transient side-effects was higher and the duration of stay in hospital was longer for the ventriculographic patients than for the CT patients.

Keywords: Thalamotomy; ventriculography; computed tomography; stereotactic surgery.

Introduction

Ventriculography is still the dominating method for calculation of target coordinates in stereotactic surgery for movement disorders[3,5,10,16,18,21]. Several computed tomography (CT)-guidance methods have been presented and dealt mainly with the accuracy of the stereotactic frame and of the CT study as well as other geometrical aspects[1,8,12,13,17,20,22]. As far as we know, there is no systematic report on the clinical outcome of CT-guided stereotaxis as compared to the ventriculography-guided stereotactic surgery for Parkinson's disease and essential tremor.

Since 1988, we have used exclusively a stereotactic CT study for calculation of target coordinates prior to thalamotomy[6,7,9]. The accuracy of the method has been assessed in previous publications[6-8]. Given that the ultimate goal of thalamotomy is to alleviate or "cure" the tremor in patients with movement disorders, we have compared the clinical results in 2 series of patients operated upon in the thalamus using either ventriculography or CT guidance during 1979–1982 and 1988–1991 respectively.

Patients and Methods

Sixty one patients with Parkinson's disease or essential tremor who underwent unilateral thalamotomy were studied. The target's coordinates were provided by air ventriculography (VG) in 38 patients treated between 1979 and 1982 and by a stereotactic CT study using the Laitinen Stereoadapter[12] in 23 patients treated between 1988 and 1991. The distribution of the patients between the VG and CT group, according to the diagnosis, sex, age, duration of disease and to the side of the target, are listed in Table 1. In both groups, the thalamic target was defined 7–9 mm anterior to the posterior commissure, 12–15 mm lateral to the midline of the third ventricle, and between zero and 2 mm above to the level of the intercommissural line. For surgery, the same stereotactic frame[11] was used during both periods. Radiofrequency (RF) electrodes with either 2 mm or 5 mm bare tip were used in the VG patients whereas only an electrode with 2 mm bare tip was used in the CT patients. In both groups, impedance monitoring and electrical stimulation were used to corroborate the anatomical target physiologically. The records of the patients were studied carefully to review the surgical parameters and the condition of the patients after surgery. The nurse's notes concerning the postoperative condition of the patients were also studied. All patients' follow-up examinations were reviewed. The surgical results were considered excellent if the tremor was completely abolished and the patient did not present a disabling complication of the surgery. The results were considered fair if the tremor was reduced or if the patient was tremor-free but with a slight complication. The results were poor when surgery had no effect on the tremor and/or when the complications of surgery were significantly affecting the daily life of the patient. For statistical analysis, Mann-Whitney U-test or chi-square test were used.

Results

Table 1 lists the patients' characteristics, the surgical parameters, and the results in both groups. In the VG group there were slightly more Parkinsonian patients than tremor patients whereas in the CT group it was the opposite. However, the difference was not signifi-

Table 1. *Characteristics, Surgical Parameters, and Outcome in Patients Operated upon by Thalamotomy Using Ventriculographic or CT-Determined Coordinates.* Mean, S.D., and range are given. Statistical analysis with Mann-Whitney U-test or chi-square test (*)

	VG	CT	p
No. of patients	38	23	
Diagnosis Parkinson/tremor	21/17	10/13	ns*
Sex male/female	27/11	16/7	ns*
Age (years)	54.4 (16.3; 15–77)	64.2 (11.3; 41–86)	<0.05
Duration of disease (years)	10.7 (9.7; 2–40)	16.6 (11.6; 4–40)	<0.05
Target side left/right	27/11	17/6	ns*
No. of probe introductions	1.4 (0.8; 1–4)	1.3 (0.6; 1–3)	ns
Intraop. target correction (mm)	0.5 (0.9; 0–3)	0.6 (0.9; 0–3)	ns
Duration of hospital stay (days)	5.0 (2.1; 2–9)	2.6 (1.3; 1–9)	<0.0001
Transient side effects (%)	73.5	26.0	<0.001*
Permanent complication (%)	13.1	8.7	ns*
Follow-up time (months)	11.0 (17.2; 1–96)	5.3 (3.7; 2–18)	ns
Excellent result (%)	57.9	78.3	ns*
Excellent or fair result (%)	86.8	86.9	ns*

VG ventriculography; *CT* computed tomography.

cant. The male/female ratio and the distribution of the patients according to the side operated upon were approximately the same in both groups. The age of the patients was significantly higher and the duration of the disease was significantly longer in the CT group than in the VG group. The number of introductions of the electrode before finding the physiological target as well as the distance between the anatomical and physiological targets were about the same in both groups. The duration of the postoperative hospitalization was significantly longer in the VG patients compared to the CT patients. The postoperative side-effects, the follow up time, the permanent complications of the surgery and the surgical results in terms of tremor control are listed in Table 1. In the VG group there were significantly more transient side-effects than in the CT group. The other parameters were not significantly different.

Discussion

The aim of this study was to find out whether the results of stereotactic surgery for tremor are different between patients operated on using ventriculography (VG)-derived target coordinates and patients operated on using computed tomography (CT)-derived coordinates. The two series, although including consecutive patients, do not include patients operated upon bilaterally in the thalamus nor patients who underwent re-operation after failed primary surgery. Furthermore, the reason for not including the patients operated on in the interval between the two reviewed periods, i.e. during 1983–1987, is that during this period a new CT-guided stereotactic system was under development

in our department and the stereotactic surgery performed during this period relied sometimes on both VG and CT in parallel, which has been analysed in previous works[6–8,12].

The present study is retrospective and concerns patients operated on in a routine fashion. We feel that the retrospectivity of this review should per se eliminate the eventual bias of trying *a priori* to embellish the VG method and denigrate the CT method or vice versa. The study disclosed a significant difference between the VG and CT groups concerning 4 of the 13 reviewed parameters (Table 1). The CT-operated patients were significantly older than the VG patients, the mean ages being 64.2 and 54.4 years, respectively. The duration of disease was also longer for the CT patients than for the VG patients (16.6 and 10.7 years respectively), which ought to be a logical consequence of the difference in patients' age. Despite this, the duration of postoperative stay in hospital was significantly longer and the occurrence of transient side-effects was significantly higher in the VG patients than in the CT patients, while no significant differences were found in the surgical results and permanent complications.

The transient side-effects included mainly symptoms generally attributed to air VG such as prolonged headache and fever, nausea, vomiting, and the so called psycho-organic syndrome[19], and sometimes symptoms attributed to the lesion itself such as dysarthria/dysphasia, dysphonia, confusion, numbness, paresis and disturbance of balance. These symptoms occurred alone or in different combinations in 28 of the 38 VG-patients and in 6 of the 23 CT-patients. Most of the symptoms were provoked by air-VG which affected the condition of the patients enough to prolong the duration of hospitalization. This in turn resulted in a more costly surgery for the VG-patients than for the CT-patients. It has been recommended by several authors that thalamotomy should not be performed on patients older than 65 years because of the higher surgical risks and the lower rate of successful surgery[2,4,15]. Our study demonstrates that by avoiding VG, elderly patients not only can be operated with satisfying results, but they can better tolerate the surgery. Therefore we believe that it is mainly the ventriculography and not the thalamotomy itself which may be harmful to the senior patients. Other workers have also come to the insight that VG can be potentially dangerous and may increase the morbidity[1,14].

At the follow-up examination, 5 of the 38 VG patients (13.15%) and 3 of the 23 CT patients (13.04%) had poor results. As for the permanent complications, 5 of the VG patients showed dysarthria/dysphasia, dysphonia, mental confusion, numbness, and disturbance of balance, respectively. Among the CT patients, 2 presented with dysphasia/dysarthria and disturbance of balance respectively.

In conclusion, this study showed that CT-guided thalamotomy equals the VG-based thalamotomy in terms of tremor control and permanent complications. However, the avoidance of ventriculography permitted surgery on elderly patients, with fewer side-effects and with shorter stay in hospital. Provided that the CT-guidance technique is accurate, CT-based thalamotomy can benefit older patients and at a lower cost than the ventriculography-guided procedure.

References

1. Aziz T, Torrens M (1989) CT-guided thalamotomy in the treatment of movement disorders. Br J Neurosurg 3: 333–336
2. Bravo G, Parera C, Seiquer G (1966) Neurological side-effects in a series of operations on the basal ganglia. J Neurosurg 24: 640–647
3. Fox MW, Ahlskog JE, Kelly PJ (1991) Stereotactic ventrolateralis thalamotomy for medically refractory tremor in post levodopa era Parkinson's disease patients. J Neurosurg 75: 723–730
4. Gildenberg PL (1985) Surgical therapy of movement disorders. In: Wilkins RH, Rengachary SS (eds) Neurosurgery, Vol 3. McGraw-Hill, New York, pp 2507–2516
5. Gildenberg PL (1988) General concepts of stereotactic surgery. In: Lunsford LD (ed) Modern stereotactic neurosurgery. Nijhoff, Boston, pp 3–11
6. Hariz MI (1990) A non-invasive adaptation system for computed tomography-guided stereotactic neurosurgery; Thesis. University of Umeå Medical Dissertation, No 269, Umeå, Sweden
7. Hariz MI (1991) Clinical study on the accuracy of the Laitinen's non-invasive CT-guidance system in functional stereotaxis. Stereotact Funct Neurosurg 56: 109–128
8. Hariz MI, Bergenheim AT (1990) A comparative study on ventriculographic and computed tomography-guided determinations of brain targets in functional stereotaxis. J Neurosurg 73: 565–571
9. Hariz MI, Bergenheim AT (1992) Stereotactic thalamotomy. J Neurosurg 76: 891
10. Kelly PJ (1988) Contemporary stereotactic ventralis lateral thalamotomy in the treatment of parkinsonian tremor and other movement disorders. In: Heilbrun MP (ed) Stereotactic neurosurgery, Vol 2. Williams and Wilkins, Baltimore, pp 133–148
11. Laitinen L (1971) A new stereoencephalotome. Zentralbl Neurochir 32: 67–73
12. Laitinen LV, Liliequist B, Fagerlund M, Eriksson AT (1985) An adapter for computed tomography-guided stereotaxis. Surg Neurol 23: 559–566
13. Latchaw RE, Lunsford LD, Kennedy WH (1985) Reformatted imaging to define the intercommissural line for CT-guided stereotaxic functional neurosurgery. AJNR 6: 429–433
14. Marks PV, Wild AM, Gleave JRW (1991) Long-term abolition of parkinsonian tremor following attempted ventriculography. Br J Neurosurg 5: 505–508
15. Matsumoto K, Shichijo F, Fukami T (1984) Long-term follow-up review of cases of Parkinson's disease after unilateral or bilateral thalamotomy. J Neurosurg 60: 1033–1044
16. Mundinger F, Birg W (1988) The imaging-compatible Riechert-Mundinger system. In: Lunsford LD (ed) Modern stereotactic neurosurgery. Nijhoff, Boston, pp 13–25
17. Rosenfeld JV, Barnett GH, Palmer J (1991) Computed tomography guided stereotactic thalamotomy using the B-R-W system for non-parkinsonian tremor disorders. Stereotact Funct Neurosurg 56: 184–192
18. Siegfried J, Rea GL (1988) Thalamotomy for Parkinson's disease. In: Lunsford LD (ed) Modern stereotactic neurosurgery. Nijhoff, Boston, pp 333–340
19. Siegfried J, Zumstein H (1976) Thalamotomies stéréotaxiques pour troubles fonctionnels chez les personnes âgées. Neurochirurgie 22: 536–539
20. Spiegelmann R, Friedman WA (1991) Rapid determination of thalamic CT-stereotactic coordinates. A method. Acta Neurochir (Wien) 110: 77–81
21. Tasker RR, Yamashiro K, Lenz F, Dostrovsky JO (1988) Thalamotomy for Parkinson's disease: Microelectrode technique. In: Lunsford LD (ed) Modern stereotactic neurosurgery. Nijhoff, Boston, pp 297–314
22. Zamorano L, Dujovny M, Malik G, Mehta B, Yakar D (1987) Factors affecting measurements in computed-tomography-guided stereotactic procedures. Appl Neurophysiol 50: 53–56

Correspondence: Marwan I. Hariz, M.D., Ph.D., Department of Neurosurgery, University Hospital, S-90185 Umeå, Sweden.

Stereotactic Techniques and Application

Acta Neurochir (1993) [Suppl] 58: 59–60

Stereotactic Localization Using Fast Spin-Echo Imaging in Functional Disorders

J. A. Taren[1], D. A. Ross[1], and **S. S. Gebarski[2]**

Departments of [1]Neurosurgery and of [2]Radiology, University of Michigan, Ann Arbor, Michigan, U.S.A.

Summary

Fast Spin-Echo Magnetic Resonance Imaging facilitates multiplanar target and trajectory selection in functional disorders by rapidly delineating gray matter, white matter, vascular, and ventricular structures. Errors due to anatomic variation or co-existing lesions can be avoided as are movement artifacts. Although not a substitute for physiologic target corroboration, the method facilitates safety and efficacy of "functional" stereotactic procedures.

Keywords: Magnetic resonance imaging; conventional spin-echo sequences; fast spin echo; target delineation.

Contemporary Magnetic Resonance Imaging facilitates stereotactic management of functional disorders by providing multiplanar CNS structure identification without ionizing radiation. Conventional Spin-Echo sequences (CSE) reduce magnetic field inhomogeneity and magnetic-susceptibility effects with respect to gradient MR scans, but suffer from relatively long acquisition times and the subsequent movement artifacts due to patient discomfort. Fast Spin-Echo (FSE) permits similar discrimination of gray matter, white matter, CSF, and blood vessels but in much less time than CSE imaging. Potential drawbacks include some flow artifacts which are not, however, a significant problem.

Methodology

In a consecutive series of patients suffering from Seizures, Dyskinesias, Obsessive Compulsive Disorders, or Chronic Intractable Pain, electrodes or lesions were placed stereotactically. All imaging was done with a GE 1.5 Tesla Signa unit and the Lexsell MRI-Compatible frame. Axial, Coronal, and sagittal images were obtained using Fast Spin-Echo technique employing a repetition time (TR) of 3000 msec and a pseudo echo time (pTE) of 17 msec. A matrix of 256 × 256, one excitation (NEX), a field of view (FOV) of 26 cm, and a slice thickness (THK) of 4.0 mm were used. Thirty-six to forty images were obtained in slightly over 3 minutes. Image processing was performed on a work station after transfer of the data via a network.

For Chronic Intractable Pain, targets included the internal capsule (IC), the Ventral Postero Lateral Nucleus (VPL) the Ventral Postero Medial Nucleus (VPM), the Para Ventricular Gray (PVG), or Ventral Para Aqueductal Gray (VPAG) depending on the location and type of pain (Fig. 1). For Movement Disorders, Ventral Intermedius Nucleus (VIM) was lesioned. For Obsessive Compulsive Disorder (OCD) the Subcaudate portion of the Anterior Limb of the Interior Capsule (ALIC) was targeted (Fig. 2). Targets in Epilepsy were the Amygdala/Hippocampal complex and the Orbital Gyrus (OG). Coordinate determination was done in three planes and based upon atlas derived coordinates which were modified according to the anatomy visualized. The Anterior and Posterior Commissures, Internal Capsule, Third Ventricle and Aqueduct, and the Colliculi, could all be readily identified as well as adjacent blood vessels and pre-existing lesions or anatomic variations. Trajectories to the target were then adjusted to maximize safety and efficacy. All operations were done under local anesthesia, monitored and supplemented pharmacologically. Where feasible, e.g. thalamic and aqueductal region, targets were corroborated physiologically and clinically. Temporary recording electrodes were inserted in the case of Seizure Disorders for diagnostic purposes, permanent stimulating electrodes were placed in the case of Chronic Intractable Pain states and lesions were made for OCD and the Movement Disorders. Coordinate adjustments were made on the basis of clinical and physiological observations observed during the surgery. The somatopic patterns across the VPL and VPM and internal capsule were utilized to adjust coordinates.

Results

Clinical objectives were achieved in most cases. In chronic intractable pain, thirteen of sixteen patient experienced paresthesias in the painful areas and substantial pain relief; nine of the patients with movement disorders had good relief of tremor, although reoperation was necessary in one patient; obsessive compulsive behavior was substantially but not completely reduced in all six patients, reoperation was necessitated in two in order to enlarge the lesions. Satisfactory recording from the electrodes placed in seizure disorder cases was accomplished in all. There was no mortality. Transient minor neurologic deficits were

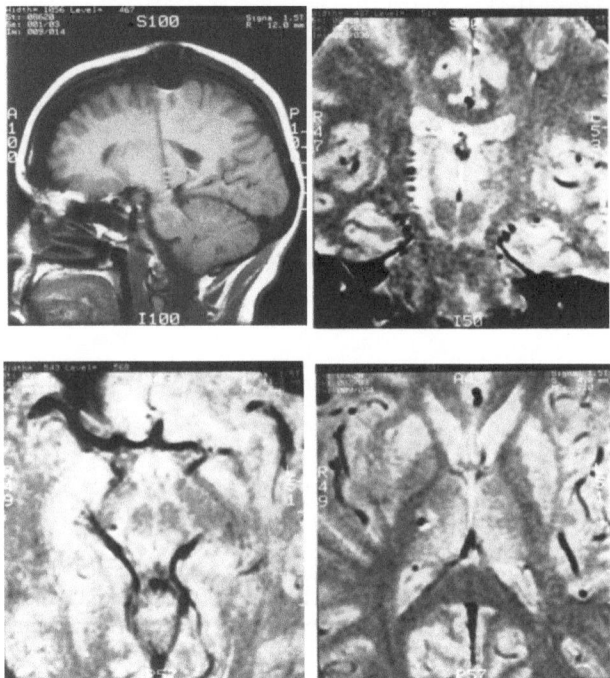

Fig. 1. FSE Images in sagittal coronal and two axial planes to demonstrate an electrode in the VPL next to internal capsule. Patients is treated for chronic intractable pain, deafferentation, of the left lower extremity

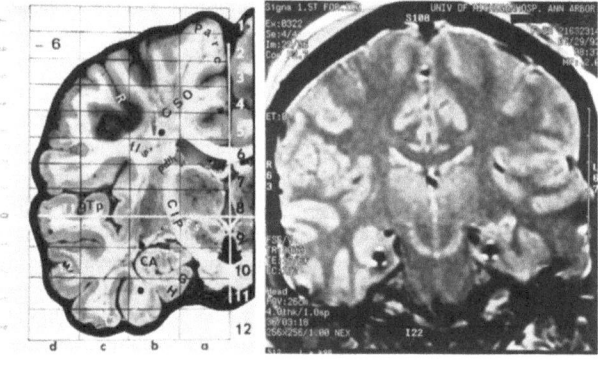

A   B

Fig. 2. Comparison of (A) a photograph from an atlas demonstrating Ammon's Horn (*CA*) and (B), an FSE image with bilateral recording electrodes in the amygdala/hippocampal complex. *CSO* Centrum semi-ovale; *Par c* Par. cortex; *R* Cenral sulcus; *fls* Fasc. long. sup.; *pdth* peduncle dorsal thalami; *T-1* sup. temp. gyrus; *T-2* med. temp. gyrus; *T-3* inf. temp. gyrus; *CIP* Post limb int. caps.; *gTP* Trans. temp. gyrus; *GH* para hipp. gyr.; *CS* subthal. corpus

experienced in four patients, one in each category. There was no morbidity related to the imaging. Postoperative images were obtained to confirm lesion or electrode placement and correlated well with the coordinates selected at the time of intervention.

Discussion

Anatomic and physiologic delineation of the target and its environs has been the "Holy Grail" of stereotactic neurosurgeons seeking to treat "functional" disorders. Fast SE MRI is better than conventional SE MR since it produces the superior tissue contrast of MRI in one quarter to one third the amount of time[1-4]. This permits the characterization of the target and it's environs by differentiation of gray matter, white matter, blood vessels, and ventricles. The stereotactically important landmarks; commissures, the internal capsule, the third ventricle and aqueduct as well nuclei and other pathways are readily distinguished and the rapid acquisition time minimizes movement artifact. Anatomic variation and the presence of co-existing lesions, all sources of error, are readily identified and can be compensated for. Trajectories are chosen to avoid dangerous structures or adjacent blood vessle. Rapid multiplanar crosschecking of the target and its environs can be accomplished.

References

1. Henning J, Naureth A, Friedburgh H, RARE imaging (1986) A fast imaging method for clinical MR. Magn Reson Med 3: 823–833
2. Henning J, Friedburgh H (1988) Clinical applications a methodological developments of the RARE technique. Magn Reson Imaging 6: 391–395
3. Mulkern RV, Wong STS, Winalski C, Jolesz FA (1990) Contrast manipulation and artifact assessment of 2D and 3D RARE sequences. Magn Reson Imaging 8: 557–566
4. Melki PS, Mulkern RV, Panych LP, Jolesz FA (1991) Comparing the FAISE Method with conventional dual echo sequences. J Magn Reson Imaging 1: 319–326

Correspondence: James Taren, M.D., Department of Neurosurgery, University of Michigan, Taufman Health Care Center, 2124B/0338-1500E Med Cntr Dr, Ann Arbor, MI48109, U.S.A.

Acta Neurochir (1993) [Suppl] 58: 61–64
© Springer-Verlag 1993

Accuracy of Ventrolateral Thalamic Nucleus Localization Using Unreformatted CT Scans and the B-R-W System
Experimental Studies and Clinical Findings During Functional Neurosurgery

I. R. Whittle, M. G. O'Sullivan, J. W. Ironside, and **R. Sellar**

Department of Clinical Neurosciences, Western General Hospital, Edinburgh, Scotland, U.K.

Summary

The accuracy with which the ventrolateral thalamic nucleus could be targetted using the BRW stereotactic system and unreformatted axial CT imaging was evaluated in cadaver brains mounted in a skull. Ball bearings (1.82 mm diameter) were placed in the nuclei ventro intermedius (Vim), ventro oralis anterior (Voa) and posterior (Vop) and the brains then sectioned, the locations of the ball bearings relative to the anterior (AC) and posterior commissures (PC) measured and evaluated with reference to a Schaltenbrandt and Bailey stereotactic atlas. Targetting of the ventrolateral thalamic region was frequently accurate (77%) but because of errors in estimation of the AC-PC plane (mean forward angulation of 9°) some ball bearings were placed too deeply in the Fields of and Forel, zona incerta and rostral subthalamic nucleus. Ventrolateral thalamotomy using the BRW stereotactic system and unreformatted axial CT imaging was undertaken in six patients with tremor due to a range of pathologies. The tremor was abolished or altered in four patients whilst no lesion was made in the other two patients since no site that reduced the tremor, could be located. The surgical and clinical implications of this study are discussed.

Keywords: Stereotactic surgery; CT scanning; thalamus; tremor.

Introduction

Whether ventriculography is essential for target coordinate estimation during functional neurosurgery is controversial[3]. Some surgeons use reconstructed CT images whilst others use axial MR images to obtain reference points[7]. The stimulus for the study arose from difficulties in defining the AC and PC from both sagittally reformatted axial CT scans and ventriculography[3]. Previous studies have suggested that for some diencephalic targets CT imaged directed target localization may be satisfactory and obviate the potential risks of ventriculography[2,3,6–8]. Since the ventrolateral (VL) thalamic nuclear complex is a common target during functional stereotactic surgery for tremor an experimental study was undertaken to evaluate the accuracy with which this diencephalic region could be located using unreformatted axial CT imaging and a BRW stereotactic system. Using similar techniques of target localization ventrolateral thalamotomy was undertaken in a small group of patients with tremor due to a diverse range of pathologies. The experimental and clinical results are compared.

Methods

Experimental Study

Cadaver brains were trimmed of their cortical mantle and placed in a human skull filled with warm gelatin. The calvarium was then replaced and secured, the BRW baseplate and CT localizer affixed and axial CT scans (GE8800, 3 mm slices) taken through the diencephalic region. After optimal visualization of the AC and PC target points were selected for the nucleus Ventro oralis anterior (Voa), 0 relative to midpoint AC-PC line, + 2 mm Vertical to AC-PC plane, and 9.5 mm Lateral from third ventricle wall, the nucleus ventro oralis posterior (Vop) − 3.5, + 2, 11, and the nucleus Ventro intermedius (Vim) − 5, + 2.5, 10.5. Ball bearings (1.82 mm) were then deposited at the target sites of ten hemispheres through a coronal burrhole. The diencephalon was then removed en bloc, sectioned in 5 mm slices parallel to the AC-PC plane and Xrays taken to localize the ball bearings. Final coordinates of the ball bearings with respect to the AC and PC were determined upon further sectioning and histology of the slices. The precise details of methodology are described elsewhere[10].

Clinical Study

Six patients with tremor of variable etiology (Table 1) were selected to undergo VL thalamotomy after failure of medical treatments. Procedures were performed under diazepam and droperidol neuroleptanesthesia. Target localization was performed as in the experimental study but additional targets were selected to encompass an area from + 2 to − 7 relative to the midpoint of the intercommissural line and from 8 to 13 mm lateral to the third ventricular wall. Impedance was measured as was clinical response to low (2 Hz) and high frequency (50–100 Hz) target stimulation. Temporary and final lesions were made with a Radionics radiofrequency system.

Table 1. *Clinical and Outcome Details of Patients Selected for VL Thalamotomy for Tremor Using the BRW System and Unreformatted Axial CT Scanning*

No.	Diagnosis	Clinical note	Outcome
1.	Multiple sclerosis	R. sided UL, LL "rubral" tremor. Spastic L. side	abolition of tremor (Fig. 2)
2.	Midbrain infarct	bilat. UL "rubral" tremor, severe LL myelopathy	no physiological locus found that altered tremor
3.	Post-traumatic tremor	R. UL, LL ballistic tremor,	"
4.	Multiple sclerosis	bilat. UL "rubral" tremor. bilat. LL spasticity	abolition R. UL tremor
5.	Multiple sclerosis	L. UL "rubral" tremor, spasticity R. UL, both LL	abolition pre-op tremor, unmasking of cerebellar intention tremor (Fig. 3)
6.	Parkinsonian tremor	bilat. UL Parkinsonian tremor with no other features of Parkinson's disease	abolition of R. UL tremor

R right; *L* left; *UL* upper limb; *LL* lower limb.

Results

Experimental Study

Thirty ball bearings were deposited into the VL complex with 23 (77%) lying within the boundaries of the Vim, Vop or Voa as defined by the Schaltenbrandt and Bailey Stereotaxic atlas. Five ball bearings were below the VL complex in the zona incerta, field of Foral or rostral subthalamic nucleus and two were positioned posteriorly (Fig. 1a and b). There was error

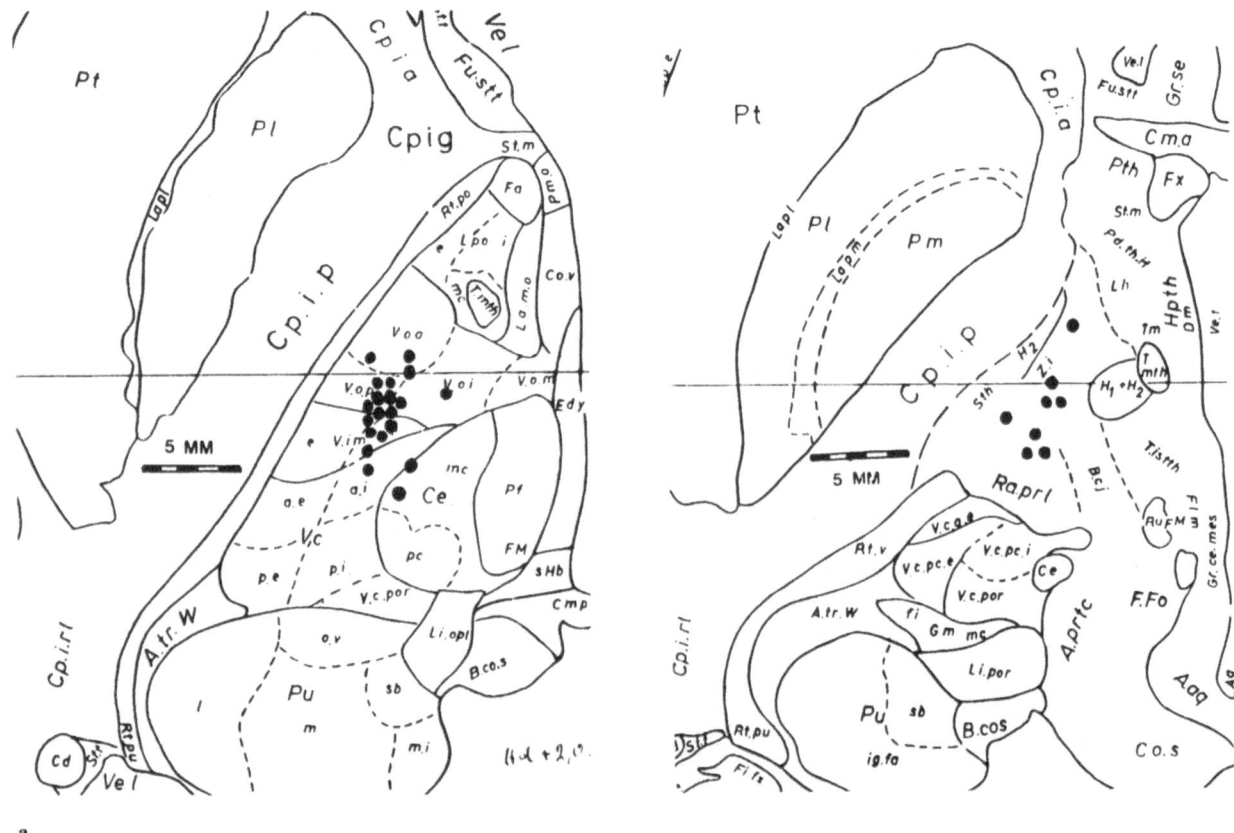

a   b

Fig. 1. The disposition of the ball bearings targetted at the VL subnuclei (Vim, Vop and Voa). Most were accurately placed in the VL complex. (a) Includes all those within ∓2 mm of the 2.5 axial slice from the Schaltenbrandt and Bailey human stereotactic atlas, whilst (b) shows the relative positions of those placed more than 2 mm below target point on the − 0.5 mm axial slice from the Schaltenbrandt and Bailey atlas

in the z coordinate estimation with 47% of the ball bearing locations being 2 mm or more (range + 2 to − 4 mm) from the correct vertical plane. Error in AP estimation of 2 or 3 mm occurred in 50% of cases, with 12 of the 15 errors being anterior overestimation. Lateral targetting was accurate to within 1 mm in 90% of placements, whilst 3 ball bearings were 2 mm from the targetted laterality.

Clinical Study

Tremor was abolished in three patients (Cases 1, 4 and 6). In each of these patients the amplitude of tremor was diminished with initial electrode insertion (− 3, + 2, 10) and transiently disappeared with temporary lesion making (Fig. 2). In one patient (Case 5) there was reduction in action tremor amplitude and frequency, however after surgery a cerebllar type intention tremor, which was not apparent preoperatively, was present

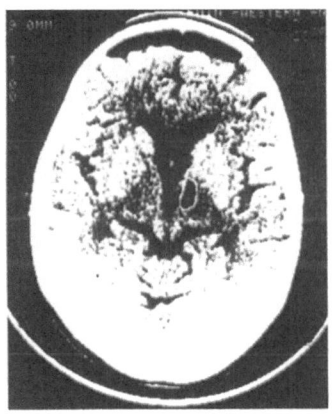

Fig. 2. Axial CT scan from the diencephalic region of case 1 after VL thalamotomy for a violent "rubral" type tremor caused by multiple sclerosis. The low density target lesion is clearly seen in the left thalamus and this lesion abolished the tremor

(Fig. 3). In two patients (Cases 2, 3) despite intraoperative stimulation at a number of target locations that covering a wide area of the VL thalamus and evoked a spectrum of neurological responses, no location that abolished tremor could be found and therefore no lesion was made. There was no morbidity associated with any procedure.

Discussion

This experimental study was part of a larger evaluation of the accuracy of diencephalic and pallidal target localization using the BRW system and axial CT imaging that suggested pulvinar, VL and pallidal targets could often be localized using unreformatted CT imaging but that errors in AP and vertical targetting occurred because of incorrect estimation (mean forward tilt of 9°) of the AC-PC plane[10]. The experimental results complement previous work evaluating the accuracy of CT imaged guided functional neurosugery for VL (errors of from 1 to 5 mm)[2,3] and pulvinar targets[6]. Given that (1) in this study CT imaging was performed on a relatively outmoded system and that greater accuracy could probably be obtained using either 9800 CT or MR imaging; (2) there is a wide (up to 7 mm difference) range of VL and subthalamic targets selected by eminent stereotactic neurosurgeons for lesioning to abolish Parkinsonian tremor and the disposition of the ball bearings encompassed 13 of these 16 preferred targets[5]; (3) if the disposition of the ball bearing placements is also compared with Hassler's clinicopathological studies 93% of them were placed within the region that caused 90–100% abolition of Parkinsonian tremor; and (4) intraoperative microelectrode recordings can be used to refine target nucleus location[8,9], then axial CT localization may be a satisfactory mode of anatomical tar-

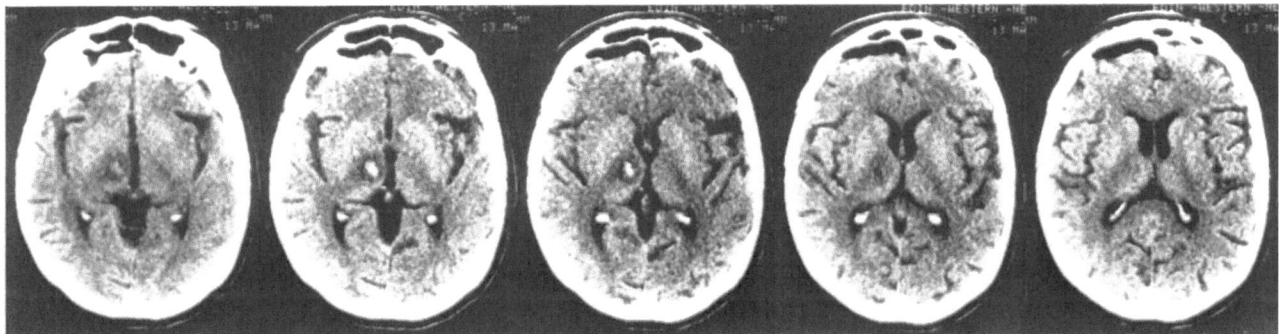

Fig. 3. Axial CT scans of the diencephalic region of case 5 after VL thalamotomy for "rubral" type tremor caused by multiple sclerosis. Although the frequency and intensity of the preoperative tremor was diminished post operatively there was unmasking of a cerebellar type of intention tremor. The lesion in the right VL thalamus is clearly seen

getting, and ventriculography superfluous for functional procedures targetting the VL thalamus[2,3,6,7].

The failure to find a satisfactory target that abolished tremor in two patients of the small clinical series may partly reflect the heterogenous causative pathology. It was assumed that a VL lesion would abolish non-Parkinsonian tremor in five of the six patients. Unsatisfactory surgical outcome could therefore have reflected more on patient selection rather than accuracy of target selection. Nonetheless in each case where a VL lesion was made tremor was either abolished or altered, and in the latter case post operative CT scan suggested the lesion encompassed the VL region (Fig. 3). In some patients with post-traumatic tremor intraoperative neurostimulation may not abolish tremor although a VL lesion at the site may later give a satisfactory result[1].

Conclusions

This study suggests that when the AC and PC are not well imaged on sagittal reformations of axial diencephalic CT scans estimation of the location of the VL subnuclei from reference points taken from the axial CT scans may be accurate enough to enable successful performance of functional stereotactic procedures for tremor.

Acknowledgement

This study was supported by a grant from the Scottish Home and Health Department. The BRW system was purchased by a generous grant from the Row-Fogo Trustees.

References

1. Bullard DE, Nashold BS (1988) Postraumatic movement disorders. In: Lunsford LD (ed) Modern stereotactic neurosurgery. Martinus Nijhoff, Boston, pp 341–352
2. Hadley MN, Shetter AG, Amos MR (1985) Use of the Brown-Roberts-Wells stereotactic frame for functional neurosurgery. Appl Neurophysiol 48: 61–68
3. Hariz MI, Bergenheim AT (1990) A comparative study on ventriculographic and computerized tomography-guided determinations of brain targets in functional stereotaxis. J Neurosurg 73: 565–571
4. Hassler R, Mundinger F, Riechert T (1979) Stereotaxis in Parkinson syndrome. Springer, Berlin Heidelberg New York, p 568
5. Laitinen LV (1985) Brain targets in surgery for Parkinson's disease. J Neurosurg 62: 349–351
6. Laitinen LV (1985) CT-guided ablative stereotaxis without ventriculography. Appl Neurophysiol 48: 18–21
7. Latchaw RE, Lunsford LD, Kennedy WH (1985) Reformatted imaging to define the intercommissurral line for CT guided stereotaxic functional neurosurgery. AJNR 6: 429–433
8. Nageseki Y, Shinazaki T, Hirai Bhauh, *et al* (1986) Long term follow up results of selective VIM thalamotomy. J Neurosurg 65: 296–302
9. Taskar RR, Yamashiro K, Lrnz F, *et al* (1988) Thalamotomy for Parkinson's disease: Microelectrode technique. In: Lunsford LD (ed) Modern stereotactic neurosurgery. Martinus Nijhoff, Boston, pp 297–314
10. Whittle IR, O'Sullivan MG, Ironside JW, *et al* (1993) An experimental study to evaluate the accuracy of diencephalic and pallidal target localization using the Brown-Roberts-Wells stereotactic system and unreformatted axial GE8800 CT scanning. Br J Neurosurg (in press)

Correspondence: I. R. Whittle, M.D., Departmental of Clinical Neurosciences, Western General Hospital, Edinburg, EH4 2XU, Scotland, U.K.

Acta Neurochir (1993) [Suppl] 58: 65–67

Functional Neurosurgery Using 3-D Computer Stereotactic Atlas

V. A. Shabalov, M. I. Kazarnovskaya, S. M. Borodkin, A. L. Kadin, V. Y. Krivosheina, and **A. V. Golanov**

Burdenko Neurosurgical Institute, Moscow, Russia

Summary

To facilitate a full-scale representation of thalamic nuclei and surrounding subcortical structures in the course of stereotactic procedures, a 3-D computer atlas of this region has been created, which permits visualization of the operative field including the involved structures and the position of the instrument. Necessary adaptation is performed according to the position of the intracranial reference points derived from CT-scan. Working with the atlas permitted more accurate and safer surgery. An IBM PC/AT computer was used.

Keywords: 3-D model; computer atlas; stereotactic operation.

Introduction

To facilitate computer assistance during functional stereotactic procedures, we produced a mathematical 3-D reconstruction of the anatomy of subcortical structures to obtain a 3-D model (computer atlas) of this region in the computer memory. The atlas makes it possible to synthesize arbitrarily oriented cross-sections during the operation, which reflect the position of the instrument and necessary digital parameters. Moreover, the standard anatomy can be adapted to individual brain reference points derived from CT-scan and the target point calculated more accurately.

Methods and Materials

The model created is based on the system of sagittal slices in the Shaltenbrandt-Wahren Atlas[5]. A complete 3-D reconstruction was performed to obtain 215 additional sagittal slices between the basic ones in 0.1 mm steps. As a result the 3-D model of the region was created in the longterm computer memory, which includes more than 15 million data points, each of them ascribed to a particular structure. One hundred and twelve structures and substructures were included[3,4]. Using the model it is possible to obtain arbitrarily oriented cross-sections of the region as coloured pictures, where each of the structures is coloured with a prescribed colour, so that the situation as a whole can be recognized.

A special marker is used to indicate the co-ordinates of the point and its relationship to the particular structure. In addition, the position of the electrode and the site of the possible lesion are superimposed on the image to allow an estimation of the situation and selection of the optimal procedure.

While elaborating the model we took into account the possible variability of structure size. To adapt the model to the individual brain we chose as reference points the anterior and posterior commissures obtained from CT-scan. We considered the distance between them the factor of proportional transformation along the x-axis. The width of the third ventricle represented the parallel displacement along the y-axis. The type of transformation along the z-axis was not significant, because we are dealing with levels near zero. However, we considered the possible adaptation along this axis as defined by the total height of the thalamus obtained from ventriculography or CT-scan. Our experience with the model for stereotactic operations confirmed this way of adaptating the model. This adaptation allows the position of the target point to be calculated in relation to real brain landmarks. It should be noted that the discrepancy between the traditional position of the target point and its position calculated according to the intracranial reference points can be significant and can amount to 1–4 mm, depending on the intercommissural distance and position of the target point.

The software also offers precise calculation of the stereotactic co-ordinates of the target point according to the orientation of the stereotactic frame with respect to intracranial reference points. In other words, we take into account possible deviations of the axes of the stereotactic frame from their corresponding intracranial positions to be able to obtain the exact co-ordinates of the target point. Our experience shows that deviations of the frame axes of just a few degrees may result in a significant discrepancy in the position of the target point.

The atlas was used during more than 30 stereotactic operations, which mostly included stereotactic thalamo- subthalamotomies and pallidotomies. Three of them were stereotactic pulvinotomies. The indications for these procedures were Parkinson's disease or other dyskinesias.

In the course of every functional stereotactic procedure we used functional diagnostic tests such as low and high frequency stimulation. In most cases we achieved good agreement between the 3-D computer atlas and functional signs. In two cases we used micro-electrode identification of deep brain structure and the data correlated well in both.

The following are two cases in which the 3-D atlas and the appropriate software were used:

Case 1. Male, 59-years old, PD with tremor and bradykinesia in left extremities. The first operation on the left VIM-VOP nuclei of the thalamus was performed 4 years before with good results— tremor and bradykinesia in contralateral extremities disappeared. The increase in symptoms on the left side was the reason for the next step of neurosurgical treatment. The second stereotactic procedure was performed on July 13, 1992. The length of the CA-CP line, obtained from CT-scan, was 26.3 mm and the target point was in the right subthalamic region and basal part of the VL-nuclei of the thalamus. The third ventricle was dilated about 11 mm. We adapted the 3-D computer model to these data and to the position of the stereotactic frame and obtained the corrected position of the target point comparing it with its position in the atlas. In particular, the distance from the midplane was 18 mm as opposed to 13.4 mm in the atlas (see Fig. 1). The intra-operative functional tests confirmed the accuracy of the target point. Anodic lesions in two planes abolished the tremor and bradykinesia in the left extremities.

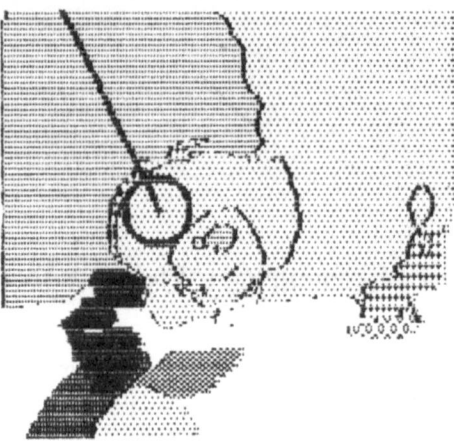

Fig. 3. Position of the second target point in case 2. This slice corresponds to an atlas distance of 23.9 mm and is the reconstruction between atlas slices 22 mm and 24.5 mm

Case 2. Male, 38-years old, dystonia musculum deformans with severe retrocollis, progressive disease with history longer than 10 years. Last year he had to hold his head still with both hands. Stereotactic surgery was performed on July 1, 1992, in the right pulvinar and in the right pallidum. The length of the CA-CP line was 28.5 mm and two target points in the pulvinar and pallidum were chosen. According to the standard stereotactic atlas the position of the target point in the pulvinar was 17 mm away from the middle of the CA-CP line, but taking into account the real CA-CP distance, we defined its x-co-ordinate as 21.1 mm (see Fig. 2). If the suggested method of measuring were incorrect, we would obtain too large a distance posteriorly with severe mesencephalic side-effects. The absence of such complications during our procedure proves the accuracy of our approach.

The computer simulation of the second target point in the pallidum shows that the location of the future lesion is near the tractus opticus. It was the reason for moving the electrode 2 mm upwards. After 2 lesions we have not observed any side-effects (see Fig. 3).

Summing up, it should be noted that this system not only helps determine the position of the electrode during the stereotactic procedure, but can help in planning the trajectory to provide the best approach to deep brain structures. It can also be used in functional neurosurgery training.

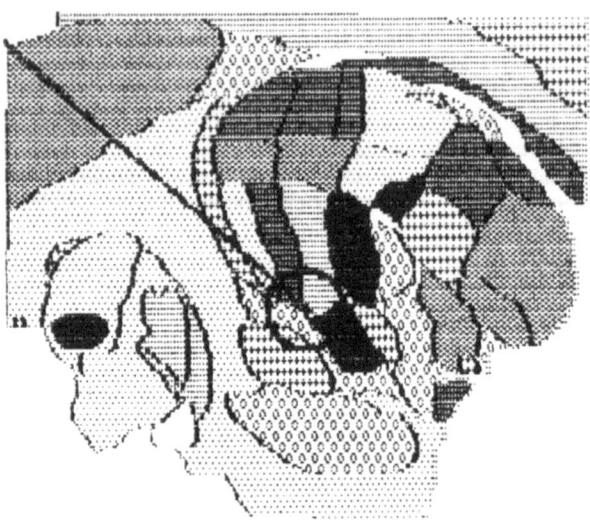

Fig. 1. Position of the lesion in case 1. This slice corresponds to an atlas distance of 13.4 mm from the midplane and it is the result of a mathematical reconstruction between slices 13.0 mm and 14.5 mm. The solid line is the projection of the electrode on the plane of the slice, the ring is the site of the possible lesion

Discussion

The approach described of computer assistance during stereotactic neurosurgery on subcortical brain structures supplements the existing ones[1,2]. We consider the full-scale reconstruction and creation of 3-D model in the long-term computer memory as a promising and convenient method of representing complex shaped objects. Further specification of the model will be accomplished in accordance with the experience with its use and accumulation of physiological data. But even now, the use of the system in more than 30 stereotactic operations has confirmed the correspondence of the

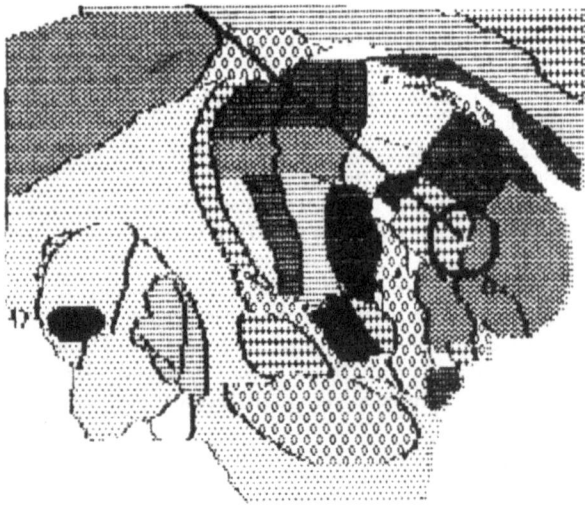

Fig. 2. Position of the first target point in case 2. This slice corresponds to an atlas distance of 13.6 mm, it is also the result of reconstruction between the same slices

model to the human brain and allowed more accurate and safer surgery.

References

1. Hardy T, *et al* (1982) Computer-assisted stereotactic surgery. Appl Neurophysiol (Basel) 45(4–5): 396–398
2. Hardy T, *et al* (1983) Computer graphics with computerised tomography for functional neurosurgery. Appl Neurophysiol (Basel) 46(1–4): 193–199
3. Kazarnovskaya M, Borodkin S, Shabalov V, *et al* (1991) 3-D computer model of subcortical brain structures for stereotactic neurosurgery. Materials of IX European congress of neurosurgery, Moscow, p 303
4. Kazarnovskaya M, Borodkin S, Shabalov V, *et al* (1991) 3-D computer model of subcortical brain structures. Comput Biol Med 21: 6
5. Schaltenbrandt G, Wahren W (1977) Atlas for stereotaxy of the human brain (Stuttgart) Plates 25–55

Correspondence: Vladimir Shabalov, M.D., Ph.D., Burdenko Neurosurgical Institute, Fadeev Str 5, 125047 Moscow, Russia.

Acta Neurochir (1993) [Suppl] 58: 68–70
© Springer-Verlag 1993

Stereotactic Biopsy in Cerebral Lesions of AIDS

P. L. Gildenberg[1], L. Langford[2], J. H. Kim[3], and R. Trujillo[4]

Houston Stereotactic Center, Houston, Texas, U.S.A.

[1] University of Texas Medical School, Houston, Texas, [2] University of Texas/M.D. Anderson Cancer Center, Houston, Texas, [3] Chonnam University, Kwangju, Korea, and [4] School of Public Health, Harvard University, Cambridge, Massachusettes, U.S.A.

Summary

The experiences with stereotactic biopsies in 121 patients with AIDS compared to 142 non-AIDS patients are presented.

In the AIDS group most of the tumors (38) were lymphomas (34). Other frequent diagnoses have been progressive multifocal leukencephalopathy (20) and toxoplasmosis (16)—although most AIDS patients already had been treated for toxoplasmosis, and those who responded to it did not undergo biopsy.

Initially among the AIDS patients there was a tendency of delayed intracranial bleeding (4 cases, 3 of them fatal). After initiation of a prophylactic coagulopathy protocol no other such complications have occurred in the following 70 biopsies.

Keywords: Stereotactic surgery; stereotactic biopsy; acquired immunodeficiency syndrome; AIDS; progressive multifocal leukoencephalopathy; primary cerebral lymphoma; toxoplasmosis; encephalitis; coagulopathy.

Introduction

Clinically significant cerebral lesions occur in 20%[7] to 40%[8] of AIDS patients. Although cerebral lesions associated with AIDS are commonly visualized on computerized tomography (CT) or especially magnetic resonance imaging (MRI), an accurate diagnosis often cannot be determined by diagnostic studies alone[1,2]. Since many cerebral lesions may have specific treatments dependent on an accurately documented diagnosis, and since establishment of prognosis, even in non-treatable lesions, may be an important factor in decisions concerning clinical management of non-cerebral problems, there is often a need for tissue diagnosis[6]. The safest and most effective way to obtain such diagnosis is imaging directed stereotactic biopsy[3,6].

Patients and Methods

Between November 28, 1984 and August 1, 1992, 121 stereotactic biopsies were performed in AIDS patients. This group was neither randomized nor selected, but was composed of those patients referred by internists and/or neurologists responsible for the overall management of each patient. All patients already had imaging studies to demonstrate the cerebral lesions, either CT scan, MR imaging, or both. The type of scan used to target the stereotactic biopsy was determined by which imaging procedure had already demonstrated the lesion(s). Consequently, 54 biopsies were directed by CT and 67 directed by MRI.

Except for the first 24 patients, most had been treated for toxoplasmosis for at least one and often two weeks prior to a decision for biopsy. Patients who responded to this empiric regimen by significant clinical and/or imaging improvement were considered to have toxoplasmosis and did not undergo biopsy, so were not included in this series.

All biopsies except 8 were performed under local anesthesia with sedation provided by an anesthesiologist in attendance, and the remainder under general anesthesia. Biopsies were performed at four different hospitals, depending on where the patient was first seen as an inpatient, the choice of the referring physician, and scheduling availability. Frozen sections or, more commonly, touch preparations were performed during the biopsies to confirm that potentially diagnostic material had been obtained while it was still possible to obtain additional tissue. In most cases, a small (2–3 mm) length of 3–0 stainless steel wire was left at the biopsy site to confirm its location on subsequent diagnostic studies.

There were four episodes (three fatal) of delayed bleeding between 12 and 24 hours after surgery among the first 32 patients, so all subsequent 89 patients underwent a protocol whereby, prior to and subsequent to the biopsy, fresh frozen plasma (usually four units each time), platelets (between 8 and 12 packs each time), and DDAVP (30 micrograms over 30 minutes), were administered intravenously, and there were no further episodes of bleeding when this protocol was employed.

Diagnoses as listed were all based on permanent sections and staining, many on special stains as well, and many by electron microscopy, depending on which studies the pathologists considered adequate to establish a diagnosis with reasonable clinical probability, ruling out all alternate diagnoses. In all except the first few cases, specimens were also sent for microbiological studies, including aerobic and anaerobic smears and cultures, acid fast smear and culture, as well as yeast, fungus, and viral cultures.

Results and Discussion

Because there was a great contrast between the distribution of diagnoses obtained in this group of AIDS patients when compared to a non-AIDS group of 142 non-selected biopsies, both patient populations are presented for comparison.

The lists of diagnoses in both the AIDS and non-AIDS groups are presented in Table 1. Several points are worthy of comment.

Although most AIDS patients had already been treated for toxoplasmosis, 16 patients who were either inadequately treated or did not tolerate treatment proved to have toxoplasmosis on biopsy.

The most common diagnosis among the AIDS patients was lymphoma, which represented 34 of the 38 tumors, and which is ordinarily treatable[4,5]. One lymphoma had superinfection with histoplasmosis, one with mycobacterium, and one also with cryptococcosis.

The incidence of one grade I astrocytoma, one anaplastic astrocytoma, one ependymoma (plus an additional ependymoma diagnosed at craniotomy and not included here), and one oligodendroglioma may be higher than one would suspect in the general population.

The diagnosis of progressive multifocal leukoencephalopathy was made in 33 patients. Three of these patients also had a secondary diagnosis, one each with also HIV encephalitis, toxoplasmosis, and diffuse thromboemboli.

There were 7 patients with abscess in both the AIDS and non-AIDS group. Among the AIDS patients, only two were identified with specific organisms, one with mycobacterium and one with multiple cytomegalic inclusion virus abscesses. In contrast, 2 non-AIDS patients grew Gram positive cocci from abscesses.

The most interesting contrast concerned tumors, which were the predominant diagnosis among the non-AIDS group with 100/142 among the non-AIDS patients versus 38/121 in the AIDS group. In the non-AIDS patients, 60 tumors were gliomas and 24 metastatic, whereas 33 of the 38 AIDS patients had lymphoma. Although there were 5 non-AIDS patients who had lymphomas, 2 were prior to 1986, following which the diagnosis of AIDS was made with greater accuracy. It is possible that these two might represent undiagnosed AIDS patients whose first manifestation was CNS lymphoma. Nevertheless, the incidence of 5% is greater than one would expect for non-AIDS primary CNS lymphoma.

There were 3 non-AIDS patients with inflammation consistent with encephalitis, and the organism was not identified in any. The diagnosis of HIV encephalitis was presumptive in 18 patients who were known to have AIDS and had localized encephalitis with no identifiable organism. Mycobacterium was cultured in one additional patient with diffuse encephalitis and CMV identified in another.

The patient with amaebic encephalitis was of particular interest. Motile Acanthamoeba rhysodes was present in the specimen. The encephalitis involved almost the entire brain, and the patient deteriorated rapidly and did not survive.

One patient demonstrated only gliosis, which did not progress to a definitive diagnosis. In addition, two specimens were frankly not diagnostic, suggesting only mild gliosis or inflammation.

It was necessary to abort the procedure in 2 patients, both of whom became so restless and unmanageable during MR scanning despite maximal sedation that it was considered wise to discontinue the procedure. Both patients were successfully biopsied under general anesthesia on subsequent days.

There were clinically apparent complications in 11 AIDS patients (with 6 deaths), in comparison with 14 complications (with 3 deaths) in the non-AIDS group. In 7 AIDS patients, progression of the clinical symptoms led to a diagnosis of local bleeding on a subsequent CT scan, although only 2 were evacuated surgically. In addition, 2 patients had progression of clinical symptoms caused by an increase in edema, although all were already being treated with corticosteroids.

There is a group of 4 patients with a complication not previously reported, although discussion with

Table 1. *Stereotactic Biopsy Diagnosis*

Lesion	Non-AIDS	AIDS	Total
Abscess	7	7	14
Tumor	100	38	138
Vascular	15	1	16
Inflammatory	—	1	1
Gliosis	4	1	5
Cysticercosis	1	—	1
Encephalitis	3	20	23
PML	—	33	33
Toxoplasmosis	—	16	16
Not diagnostic	3	2	5
Procedure aborted	2	2	4
Other	7	—	7
Total	142	121	263

colleagues performing biopsies elsewhere in the body suggests that it may occur. All patients were within the first 32 AIDS patients biopsied. All recovered from sedation following what appeared to be an uncomplicated biopsy. All had abrupt severe intracranial bleeding 12 to 24 hours later, which was fatal in 3. One patient illustrates this complication dramatically. Approximately 12 hours after the biopsy, the patient was alert and conversing with a nurse in the intensive care unit. She went to see another patient and returned within 5 minutes to find the patient comatose, unresponsive, with a dilated and fixed pupil. An emergency CT scan revealed a large acute subdural hematoma. The patient and family had previously agreed that nothing heroic would be done if such a devastating complication would arise, so the patient died without further management. All 4 bleeding complications occurred prior to initiation of the coagulopathy protocol outlined above, so that no such complications have occurred in the last 70 consecutive biopsies.

Two additional patients had diffuse mycobacterium infection, and 4 had abscesses in which no organism was identified. 19 patients had encephalitis, 18 from HIV (one also with PML) and one from CMV.

Conclusion

Stereotactic biopsy of cerebral lesions is both feasible and reasonably safe in AIDS patients, and yields a much different profile of diagnoses than in non-AIDS patients.

References

1. Anson JA, Glick RP, Reyes M (1992) Diagnostic accuracy of AIDS-related CNS lesions. Surg Neurol 37: 432–440
2. Chappell ET, Guthrie BL, Orenstein J (1992) The role of stereotactic biopsy in the management of HIV-related focal brain lesions. Neurosurgery 30: 825–829
3. Ciricillo SF, Rosenblum ML (1990) Use of CT and MR imaging to distinguish intracranial lesions and to define the need for biopsy in AIDS patients. J Neurosurg 73: 720–724
4. Goldstein JD, Dickson DW, Moser FG, Hirschfeld AD, Freeman K, Llena JF, Kaplan B, Davis L (1991) Primary central nervous system lymphoma in acquired immune deficiency syndrome. A clinical and pathologic study with results of treatment with radiation. Cancer 67: 2756–2765
5. Hochberg FH, Loeffler JS, Prados M (1991) The therapy of primary brain lymphoma. J Neurooncol 10: 191–201
6. Pell MF, Thomas DG, Whittle IR (1991) Stereotactic biopsy of cerebral lesions in patients with AIDS. Br J Neurosurg 5: 585–589

Correspondence: Philip L. Gildenberg, M.D., Ph.D., 6560 Fannin #1530, Houston, Texas, U.S.A.

Acta Neurochir (1993) [Suppl] 58: 71–74

Neuronavigator-Guided Cerebral Biopsy

J. Koivukangas, Y. Louhisalmi, J. Alakuijala, and **J. Oikarinen**

Department of Neurosurgery, University of Oulu, Finland

Summary

Neuronavigators are new dynamic interactive instruments that use on-line computers to orient imaging data to the surgical field and guide the neurosurgeon to his target. We have been working since 1987 on a neuronavigator that serves not only as a precise pointer, but also as a dynamic arm that can be used to hold instruments, such as biopsy guides. The neuronavigator arm consists of six joints with optical endoders and is attached to the Mayfield headholder. The arm is connected to a workstation running customized 3D image graphics software. Special instruments and surgical technique have been developed.

Here, we report on early clinical experience with ten biopsy procedures: 4 low-grade and 3 high-grade astrocytomas, one cranio-pharyngioma and one chronic intracerebral haematoma and intra-cerebral cyst, both of the latter with surrounding tumour suspect tissue. In all glioma cases serial biopsies were taken from optimal sites under ultrasound imaging control. Eight cases showed representative tumour tissue, while in two cases neoplasia was ruled out. The neuronavigator proved to be versatile, allowing comprehensive imaging data to be adapted to the surgical field.

Keywords: Neuronavigator; cerebral biopsy; three-dimensional digitizers.

Introduction

The term "neuronavigator" was originally coined in the neurosurgical literature by Watanabe *et al.*[7] to describe a "multijoint three-dimensional digitizer (sensor arm) and a microcomputer, which indicates the place of the sensor arm tip on preoperative CT images". Conventional stereotactic principles, including precise targeting, form the basis of neuronavigator surgery.

Stereotactic frames developed for clinical use since the 1940s were originally based on analog radiology, i.e. AP and lateral views. The frames represented state of the art precision mechanics. With the advent of computed tomography, and later magnetic resonance (MR) imaging, stereotactic frames were modified to accomodate digital radiology. Also, the concepts of volumetric stereotactic surgery and open stereotactic surgery, emphasizing the handling of target volumes instead of only points, were developed in an effort to accommodate the digital imaging methods. The general term, computer-assisted surgery, describes these and later systems incorporating on-line computers.

During the second half of the 1980s, several new systems no longer resembling the conventional stereo-tactic frames were developed rather independently, leading to different ways to bridge the imaging data to the surgical field[1-8]. Each system is unique, but all are based on the basic notion of stereotactic surgery: precise controlled 3D interaction with the brain. They incorporate recent advances in computers and special processors (speed, memory and cost), electromechanics, 3D graphics software and sen-sors, including potentiometers[7,8], optical encoders[1,3], magnetic field sensors[2], ultrasound receivers[6] and video cameras[4]. As opposed to the active robotic systems (such as 5), the so-called three-dimensional digitizers[1-4,7,8] are passive systems requiring movement of the instrument by hand. However, some of the neuronavigator systems can be made active with the use of actuators such as servo-motors.

While all three-dimensional digitizers incorporate on-line computers and sophisticated software, some use electromechanically jointed arms with position sensors, while others are armless, using magnetic fields, ultrasound, video cameras or infrared light, for examples, to sense the position and configuration of the end effector instruments in space.

The Present Oulu Neuronavigator

The Oulu Neuronavigator System (ONS) is a specially designed passive six-jointed electromechanical arm, with UNIX workstation (Hewlett-Packard 9000/720), customized software and interchangeable end effector in-

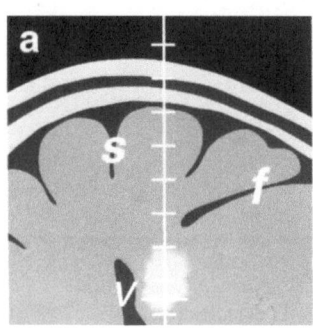
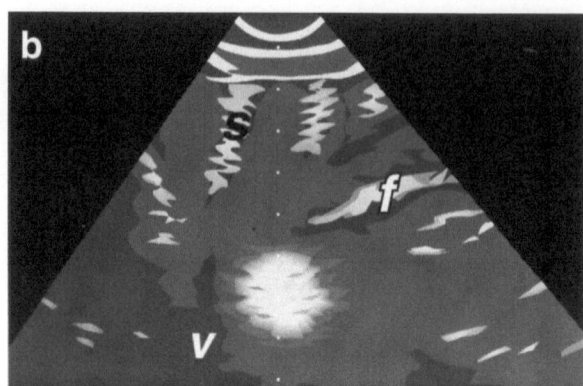
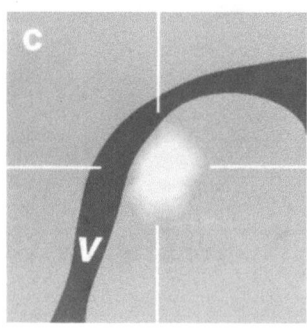

Fig. 1. The 2D images on the monitor during biopsy: The CT/MRI reconstruction (a) spatially corresponds exactly to the intra-operative ultrasound image (b). Note sulcus (*s*), Sylvian fissure (*f*), lateral ventricle (*v*) and tumour. As the ultrasound probe is moved in the burrhole, the CT/MRI reconstruction corresponding to the ultrasound image appears on the monitor. Also, a reconstruction (c) in the perpendicular to the axes shown in (a) and (b) is displayed at any desired depth, indicated in (a) as slightly longer crossline at lower end of axis

struments[3]. The present neuronavigator has no joint brakes. It is balanced using joint springs and light materials. Electromechanical brakes have been tried, but they were not found to be reliable. The neuronavigator can be stabilized during biopsy by hand, Leyla arm or other rigid means.

Previous experience during the last decade with intra-operative ultrasound affected the development of the graphics in the present system. In addition to the 3D image graphics, one of the two-dimensional reconstructions of CT/MRI data corresponds to the ultrasound image obtained with the ultrasound probe attached to the neuronavigator (Fig. 1). One of the other reconstructions is in the plane perpendicular to the plane of the ultrasound image at any desired depth.

In the present system, the principle of a common central axis[3] was employed to meaningfully align the neuronavigator pointer, ultrasound probe and biopsy guide. This same principle has also been used in other cases to align the suction tip, operating microscope and surgical laser[3]. As shown in Fig. 2, the central axis of the ultrasound probe coincided with the axis of the stereotactic biopsy forceps so that only the depth of the site of biopsy needed to be measured from the CT/MRI/US images. Also, simultaneous imaging and biopsy, possible only with real-time ultrasound imaging, was used for deep lesions (Fig. 3).

The present neuronavigator is a unique system, developed from "scratch": The neuronavigator arm was specified in our laboratory by the neurosurgeon and the engineers, the software was also written in our laboratory and the geometry and surgical principles were developed as the project progressed from labora-

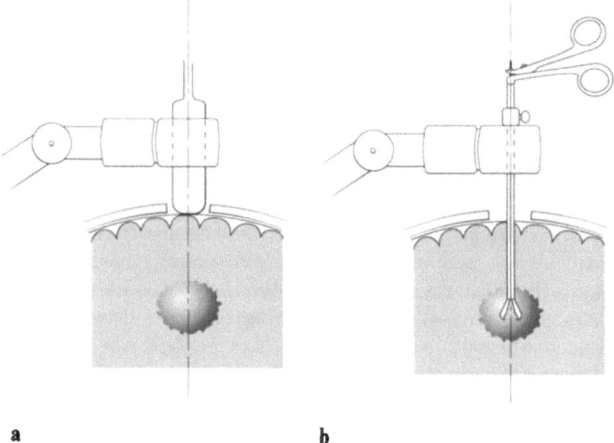

a   b

Fig. 2. The principle of the common central axis applied to sequential ultrasound imaging and biopsy procedure. The geometry of the interchangeable end piece of the arm aligns the ultrasound image to be coincident with the corresponding CT/MR image and the central axis of the ultrasound probe (a) to that of the biopsy forceps (b). The depth of the biopsy is then measured from the images

tory simulations to actual operations. For these reasons, intra-operative ultrasound was used to give state-of-the-art control and during all biopsy procedures the accuracy and functioning of the neuronavigator were first verified with on-line integrated ultrasound scanning.

Patients and Surgical Technique

Ten adult patients with intracerebral lesions were biopsied under ultrasound control with the neuronavigator (Table 1).

Most of the neoplasms were large diffuse subcortical masses, but the oligodendroglioma was in the thalamus, the craniopharyngioma extended up to the third ventricle and two of the low-grade astrocystomas were deep paraventricular lesions. The walls of the chronic

 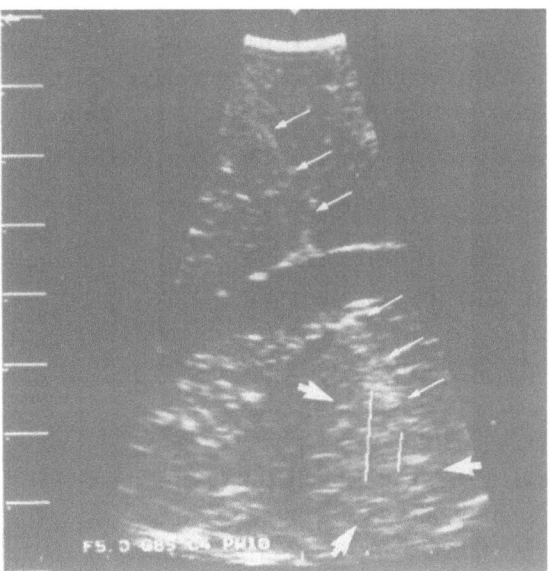

Fig. 3. Case NN. Thalamic oligodendroglioma biopsied (62 mm) under simultaneous real-time ultrasound control. Left: Nearly sagittal intra-operative MRI reconstruction showing tumour enhanced with colour (dark region, arrow) deep to lateral ventricle (v). Right: corresponding intra-operative ultrasound image showing echogenic tumour (arrows) within thalamus. Note that the tumour pixels enhanced in the MRI scans have been stereotactically transferred to the ultrasound image and appear as lines. Thus the information from the original MRI scans can be seen in not only MRI reconstructions but also in the ultrasound scans. Because this tumour was so deep and small, the biopsy forceps was passed through the lateral ventricle under simultaneous ultrasound imaging to the tumour: note track (small arrows) of forceps leading to tumour

Table 1

Histology	No.	Depth (mm)
Astrocytoma		
—low grade	4	16–65
—high grade	2	20–40
Oligodendroglioma	1	62
Craniopharyngioma	1	60
Intracerebral haematoma	1	30
Intracerebral cyst	1	35

intracerebral haematoma and the cyst were biopsied because of T2 signal indicating possible neoplasm.

The surgical technique developed for the neuronavigator included:

1. Simulation of the operation: Before the day of surgery the deep masses were visualized in the laboratory workstation using a "mouse navigator". This means that the image data cube could be turned and sliced to find the most suitable route to the lesion.
2. Site of burrhole: At operation the neuronavigator was used to interactively "image" the head to find the most suitable place for the burrhole.
3. Alignment of ultrasound and MRI/CT images: These image planes coincided with each other, helping to determine the optimal place to biopsy the tumour.
4. Biopsy: The biopsy forceps was guided by the neuronavigator by attaching the guide needle into the end piece of the neuronavigator. Biopsy could be taken either sequentially after ultrasound imaging or simultaneously under ultrasound control.
5. Endoscope alignment: In the case of the intracerebral cyst, the endoscope was passed through the guide needle placed into the end piece of the neuronavigator allowing visualization of the cyst.
6. Voice control: The calibration of the neuronavigator as well as some of the graphic software commands were done by voice, the neurosurgeon speaking into a microphone placed within the surgical mask.

Results

Representative samples were obtained in all cases of neoplasm. The walls of the haematoma and cyst were negative for neoplasm. There was one complication related to biopsy. Following endoscopy and biopsy of the suspect region of the cyst wall, filling of the cyst with blood could be noted in the ultrasound scan. With the neuronavingator still in place, a small piece of surgical applied to the bleeder through the biopsy guide resulted in cessation of bleeding.

The site of biopsy was optimally determined as the hyperechogenic region in the ultrasound scan corresponding to the contrast enhanced region in the CT/MRI images. In all of the present cases, visual comparison of the lesion and anatomical landmarks in the ultrasound scan and the corresponding CT/MR images showed striking similarity, confirming accuracy of the neuronavigator adequate for biopsy procedures.

Discussion

There are several reasons for developing electro-mechanical arms as opposed to armless systems. While the latter are handy as pointing instruments, they will be restricted somewhat by the tubing and cables needed with suction tips, endoscopes and ultrasound probes, for example. The arm, on the other hand, can be made rigid with joint brakes, so it can actually support instruments such as biopsy guides, endoscopes and ultrasound probes and their tubing and cables. Furthermore, the arm can be made active by adding servo-motors to the joints. In this way, rehearsed procedures can be rapidly and accurately repeated at the time of actual surgery, for example, by positioning the arm for biopsy. We have found that with proper design and materials and the use of counterbalances and springs, the arm can be flexible and versatile even when used as a pointer for purposes of spatial orientation.

In addition to allowing for real-time imaging control of the biopsy procedure, ultrasound imaging served to give supplementary information about the lesion (site of pulsating cerebral vessels, cystic and necrotic parts of tumours), to verify the accuracy of the neuronavigator, to check the site of biopsy and to rule out, or control, bleeding in the site.

The mechanical spatial accuracy of ± 1 mm achieved with the neuronavigator arm would be sufficient for biopsy of even deep lesions. The greatest source of inaccuracy in this and any system incorporating CT or MRI is along the axis perpendicular to the plane of the scan (in axial slices, the z-axis in many systems). This is due to the slice thickness, which can be 2–8 mm at our center, depending on the imaging sequence. The total spatial accuracy of the present system was ± 3–4 mm. This was one of the reasons we chose to use ultrasound control during biopsy procedures.

Finally, the terminology of the new systems needs to

be addressed: should we speak of stereotactic systems which are based on conventional frames, as distinct from three-dimensional digitizers and robots. This would be a traditional approach. Or, rather, should we not speak of stereotactic frames, stereotactic arms and other digitizer systems, and stereotactic robots? Should not the term stereotactic be applied only to all systems accurate enough to allow biopsy and/or functional procedures. Following the original terminology of Watanabe *et al.*[7], we have called our three-dimensional digitizer arm a neuronavigator. The issue is surely more than only semantics because the purpose of many of the new systems is to bring the principles of stereotactic neurosurgery to procedures traditionally considered nonstereotactic, such as removal of intracerebral tumours.

Acknowledgements

This work has been supported by the Technological Development Center of Finland, the Academy of Finland and Inari and Reijo Holopainen Fund. It is a EUREKA project. Artistic work in Figs. 1 and 2 by Mr. Brian Beardsley, University of Minnesota.

References

1. Guthrie BL, Kaplan R, Kelly PJ (1989) Freehand stereotaxy: Neurosurgical stereotactic operating arm. Proceedings of the Neurosurgical Society of America, pp 59–61 (abstract)
2. Kato A, Yoshimine T, Hayakawa T, Tomita Y, Ikeda T, Mitomo M, Harada K, Mogami H (1991) A frameless, armless navigational system for computer-assisted neurosurgery. J Neurosurg 74: 845–849
3. Koivukangas J, Louhisalmi Y, Alakuijala J, Oikarinen J (1993) Ultrasound-controlled neuronavigator-guided brain surgery. J Neurosurg (accepted)
4. Krybus W, Knepper A, Adams L, Ruger R, Meyer-Ebrecht D (1991) Navigation support for surgery by means of optical position detection. In: Proceedings of the International Symposium CAR'91 Computer Assisted Radiology. Springer, Berlin Heidelberg New York Tokyo, pp 362–366
5. Lavallee S (1990) Computer assisted medical interventions. In: Hohne KH, *et al* (eds) 3D Imaging in medicine, NATO ASI Series, Vol F60. Springer, Berlin Heidelberg New York Tokyo, pp 301–312
6. Roberts DW, Strohbehn JW, Hatch JF, Murray W, Kettenberger H (1986) A frameless stereotaxic integration of computerized tomographic imaging and the operating microscope. J Neurosurg 65: 545–549
7. Watanabe E, Watanabe T, Manaka S, Mayanagi Y, Takakura K (1987) Three-dimensional digitizer (neuronavigator): New equipment for computed tomography-guided stereotaxic surgery. Surg Neurol 27: 543–547
8. Watanabe E, Mayanagi Y, Kosugi Y, Manaka S, Takakura K (1991) Open surgery assisted by the neuronavigator, a stereotactic, articulated, sensitive arm. Neurosurgery 28: 792–800

Correspondence: John Koivukangas, M.D., Visiting Professor, Department of Neurosurgery, University of Minnesota, Box 96, 420 Delaware St., Minneapolis, MN 55455, U.S.A.

Acta Neurochir (1993) [Suppl] 58: 75–76

A Computer Controlled Stereotactic Arm: Virtual Reality in Neurosurgical Procedures

C. Giorgi, M. Luzzara, D. S. Casolino, and **E. Ongania**

Neurosurgery, Istituto Neurologico "C. Besta", Milano, Italy

Summary

A computer controlled mechanical arm for stereotaxy is presented. It has 5 free joints and can be attached to a stereotactic frame. High precision digital encoders register the angular position of each joint and the computer determines the position of the tip of the instrument in the stereotactic space. Accuracy and usefulness are discussed.

Keywords: Stereotactic instrumentation; computer guided stereotaxy.

Highly detailed, multimodal morphological data in stereotactic coordinates are available to the surgeon, using "localizers" in MRI, CT DSA examinations. These data, possibly integrated with functional information provided by stereotactic PET or atlases, contribute to the identification of most significant target sites and safest trajectory of approach in tumour biopsies and interstitial radiotherapy. In general, when stereotaxy is applied to reach target points within the brain, arc-centered and interlocked-angles aiming instruments continue to adequately serve the purpose for which they were conceived.

When stereotaxy is used to guide microsurgical removal of cerebral lesions, detailed information about the spatial morphology of the lesion and surrounding healthy tissue, in stereotactic co-ordinates, is particularly important[1]. In this situation the surgical intervention is not limited to the identification of a trajectory to reach a target point, but it comprises a series of manoeuvres aiming at the progressive reduction of the tumour volume, which implies removal of pathological tissue alternating with dissection of its margins from surrounding tissue. In this situation special stereotactic instrumentation is needed[2-4].

The aim of this work has been to develop a guiding instrument that relieves the surgeon from the necessity of setting different target points during the operation, allowing fast and reliable identification of new trajectories without losing the precision otherwise supplied by the usage of a stereotactic arc.

Method

We have developed a mechanical arm with 5 free joints, attached to a stereotactic frame in a known position. The angular position of each joint is detected by a high precision digital encoder, kept at a desired angle by fail-safe electromagnetic brakes (Fig. 1).

The encoders send a stream of readings to the graphic processor through a multiplexer. These values determine the position of the tip of the instrument in the stereotactic space. The accuracy of the instrument has been tested by comparing the error distribution over a three-dimensional grid of 2.5 mm mesh, obtained with a high precision digital x, y, z positioner, with the position calculated with the direct kinematic model. The results disclose an error ranging from 0.5 to 0.8 mm, the highest values being found at the maximum distance from the origin of the calibration mesh.

Using the "phantom" localizer of a stereotactic system, the arm has been calibrated in a series of preset positions. At this point, a common reference system between the frame and the arm has been identified, and the angular values from the encoders can be utilized to

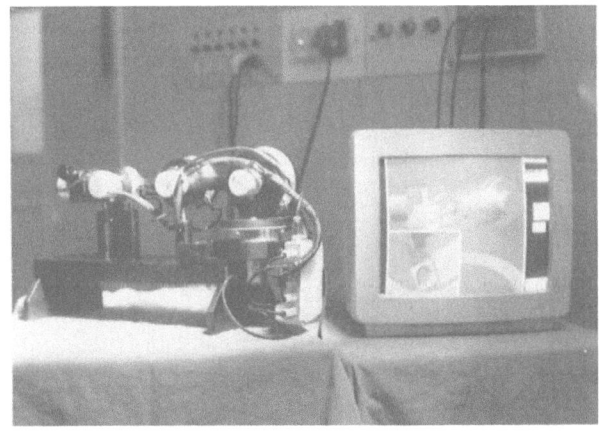

Fig. 1. The stereotactic arm, resting on a fast calibration bench; its position is documented on a graphic monitor

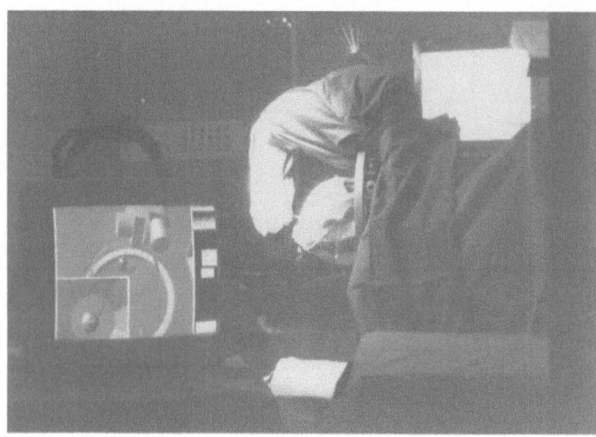

Fig. 2. The surgical arm, wrapped in sterile drapes, mounted on a stereotactic frame. Note the graphic monitor, showing the arm orientation and tip position. The insert in the lower left quadrant of the monitor shows a "probe eye view" to the lesion

graphically represent the configuration of movement when the instrument is operated by the surgeon. CAD modelling techniques display a pictorial description of the arm, which is moved in the space defined by the stereotactic frame.

Shaded surface rendering, transparencies and stereoscopic representation offer an effective description of the arm position with respect to the reconstructed anatomy. The surgeon can plan the trajectories of approach, the extent of the bone flap, and evaluate the possible functional effects related to different surgical strategies.

During surgery, the stereotactic arm serves as a tissue retractor holder, or a laser pointer. Ongoing development will utilize it as a support for an intra-operative echographic probe, to obtain stereotactic images giving information about the possible dislocation of tissue as a result of the ongoing surgery and monitoring the phases of lesion removal (Fig. 2).

Discussion

The development of this instrument introduces neurosurgery into the broad field of virtual reality applications, because it allows the surgeon an immediate visual feedback of the movements of his hand related to a realistic representation of the anatomy. Several devices can be added to enhance the efficacy of visual and proprioceptive feed-back to the surgeon but, in our opinion, this instrument comprises what is necessary to "navigate in dangerous waters with no visibility". This is, in fact, what the surgeon is asked to do in removing deep seated cerebral lesions.

References

1. Giorgi C, Ongania E, Casolino SD, Riva D, Cella G, Franzini A, Broggi G (1991) Deep seated cerebral lesion removal, guided by volumetric rendering of morphological data, stereotactically acquired. Clinical results and technical considerations. Acta Neurochir (Wien) [Suppl] 52: 19–21
2. Kwoh YS, Hou J, Jonckeere EA, Hayati S (1988) A robot with improved absolute positioning accuracy for CT guided stereotactic brain surgery. IEEE Trans Biomed Eng 35: 153–160
3. Reinhardt HF, Landolt H (1989) CT-guided real time stereotaxy. Acta Neurochir (Wien) [Suppl] 46: 107–108
4. Watanabe E, Watanabe T, Manaka S, Mayanagi Y, Takakura K (1987) Three-dimensional digitizer (neuronavigator): New equipment for computer tomography guided neurosurgery. Surg Neurol 27: 543–547

Correspondence: C. Giorgi, M.D., Istituto Nationale, Neurologico "C. Besta", Via Celoria 11, I-20133 Milano, Italy.

Radiosurgery

Acta Neurochir (1993) [Suppl] 58: 79–84
© Springer-Verlag 1993

Brachytherapy—Interstitial Implant Radiosurgery

Ch. B. Ostertag

Abteilung Stereotaktische Neurochirurgie, Neurochirurgische Universitätsklinik, Freiburg, Federal Republic of Germany

Summary

With both surgery and radiosurgery a selective destruction and removal of a lesion is undertaken. Using interstitial irradiation a calculated volume of necrosis is induced, which is subsequently removed by macrophage activity. Interstitial irradiation is therefore considered a radiosurgical method. To better understand the side effects of radiosurgical lesions in the brain, we studied the effects of three physically distinct gamma emitters iridium-192, gold-198 and iodine-125 with respect to the morphological development of radionecroses and the consecutive vasogenic oedema using different dose rates and application times. Experimental dose effects on normal brain (beagle dogs) revealed that depending on the energy range, application time and dose rate, calcified or liquified necroses develop from the center around the seed into the periphery. These well-defined radionecroses are surrounded by reactive glia and, temporarily, by fibrinoid necrosis of endothelial cells. In contrast to permanent implants, effects of temporary implants on tissue are characterised in the early stages by oedema, in later stages by progressive resorption of necrotic tissue and less permanent damage. Dose rate and energy range determine the volume of the necroses and the magnitude of vasogenic oedema. It is the vasogenic oedema which primarily determines the extent of demyelination and reactive glioses in the vicinity of the radionecrosis. Toxic side effects of interstitial implant radiosurgery can be minimised by using temporary implants with moderate dose rates, in which case repair processes are free to take place after removal of the radioactive source from the brain. Based on experimental and clinical findings a biological dosimetry has evolved which favours temporary implants of iodine-125 with dose rates of 10 cGy/h.

Keywords: Radiosurgery; radiation tolerance; brachytherapy; radiobiology; iodine-125.

Introduction

The discovery of radium by Marie and Pierre Curie in 1898 was promptly followed by the clinical application of this radio-isotope in the treatment of various cancers. Alexander Graham Bell suggested in 1903 that "there is no reason why a tiny fragment of radium sealed-up in a fine glass tube should not be inserted into the very heart of the cancer, thus acting directly on the diseased material"[3]. The principal advantage of an interstitial radioactive implant in neoplastic tissue over external beam irradiation is that the dose of radiation delivered to the normal surrounding brain is considerably lower than that delivered to the neoplasm. The amount of radiation reaching normal brain tissue varies inversely with the square of the distance from the interstitial source. The technique is based on the principle of selectivly irradiating a tumour while protecting the healthy surrounding brain from radiation energy. Talairach in 1953 was the first to implant radioactive sources (gold-198 grains) stereotactically into brain tumours. But even before Talairach, the method had its ups and downs[3]. The biological basis was not fully understood, as can be seen in the confusing terminology (brachytherapy, curietherapy, brachycurietherapy, interstitial radiotherapy). Furthermore, a number of questions remains unanswered. Are we talking about radiation therapy or radiosurgery? Is it based on a differential effect between normal and neoplastic cells? Is the toxicity radiation-induced or attributable to secondary effects? Which radio-isotope, total dose and dose rate is most effective and, at the same time, best tolerated by the brain. I would like to describe the experimental effects of the various radio-isotopes and dose rates, and then correlate them with clinical observations.

Experimental Observations

In order to elucidate the tissue response to interstitial irradiation we studied consecutively the morphological and physiological effects caused by gamma emitters with different energies and different dose rates in

Table 1. *Principle Physical Properties of Radio-isotopes Used in the Experiment*

Isotope	Half-life	Energy		
		Alpha	Beta	Gamma (MeV)
Gold-198	2.7 d	—	(+)[a]	0.41
Iridium-192	74.2 d	—	(+)[a]	0.30–0.61
Iodine-125	60.2 d	—	—	0.028–0.035[b]

[a] Beta energy shielded by platinum coating.
[b] Photon.

an animal model. Experimental dose effects on the brain (beagle) caused by the radioactive iodine-125 (I-125, half-life 60.2 days, 0.028 to 0.035 MeV, iridium-192 (Ir-192, half-life 74.2 days, 0.30 to 0.61 MeV), gold-198 (Au-198, half-life 2.7 days, 0.41 MeV) were examined consecutively over 5 to 365 days (Table 1). Permanent low activity, low-dose rate implants were compared over time with temporary high-activity, high-dose rate implants, both delivering the same accumulated total radiation dose. Particular attention was paid to the morphological development, size, and composition of the radiation necrosis, as well as the changes in the blood-brain barrier. Single seeds of I-125 and Au-198 or wire pieces, respectively, of Ir-192 were stereotactically implanted into the left frontal centrum semiovale of six-month old Beagle dogs. The implants were either permanent (lost implants) or temporary removable implants, i.e. the radioactive source was held in the tip of a teflon catheter. After observation periods of between 5 and 365 days, the animals were sacrificed. The procedures for implantation, tissue perfusion, histological and immunohistological staining are described elsewhere[14,23–25,27]. The dosimetry for permanent low-dose rate and for temporary high-dose rate implants was calculated such that a 26,000 (10,000) c Gy isodose was achieved at a 5 (10) mm distance from the source. Isodose distribution and accumulated tissue dosage were calculated on the assumption of a point source of irradiation. For a detailed description of quantitative measurement of capillary transfer rates, i.e. quantitative autoradiographic and computer tomographic methods employed in the different experimental series see[10,11,33].

Four animals served as controls, i.e. an inactive seed of I-125 or Au-198 was implanted into the left centrum semiovale using the same experimental procedures of physiological measurements and histopathology.

There was a characteristic morphological response to all three radiation sources in the brain (Fig. 1). Radionecroses were manifest as early as 10 days after implantation. Within 25 days, a coagulation necrosis developed in the immediate vicinity of the radioactive source, which later increased in size and became progressively mineralised (Fig. 2). Staining for serum

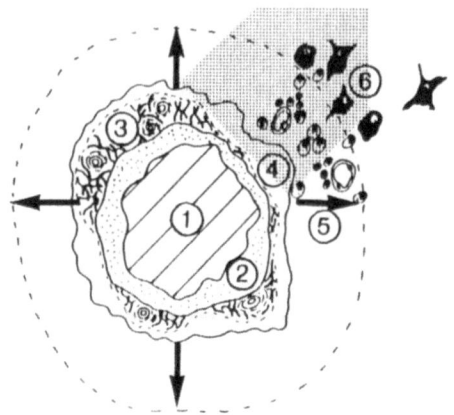

Fig. 1. Schematic illustration of morphological changes to gamma sources in the brain. The radionecrosis evolves from the center into the periphery (arrows) in the brain. The dotted line indicates the final size of radionecrosis. *1* Central coagulation necrosis. *2* Calcified rim of necrosis. *3* Perinecrotic zone of gliosis with fibrinoid vessel wall necrosis. *4* Spongiform "lucent" zone. *5* Zone of organization with macrophages, capillary proliferation, perivascular inflammatory infiltrates. *6* Reactive gliosis and demyelination. Directional perifocal vasogenic oedema (dotted area)

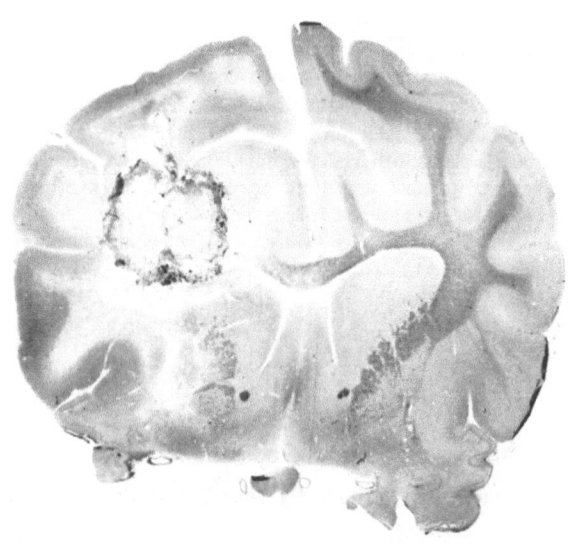

Fig. 2. Coronal brain section (Klüver-Barrera stain) through an experimentally induced radionecrosis with temporary (10-days) gold-198 implant (31.9 mCi) after 46 days' survival. A sharply demarcated, marginally calcified necrotic zone can be recognized. Note the considerable mass effect with midline shift due to vasogenic oedema and the pallor of the ipsilateral white matter

proteins revealed the presence of perifocal oedema, which extended into the white matter of the entire ipsilateral hemisphere. Vessels within the developing radionecrosis showed the typical hyaloid vessel wall degeneration. Vessels outside the radionecrosis, however, even within a distance of 1 mm, appeared normal. A reactive gliosis with a dense fibrillary network of GFAP-positive astrocytes was found around the radionecrosis in the surrounding white matter. GFAP-positive astrocytes in later stages were also found in the contralateral hemisphere as a reaction to vasogenic oedema which spread via the corpus callosum. Demyelination was always restricted to the ipsilateral centrum semiovale without affecting the internal capsule or the contralateral hemisphere. The neurons next to but not incorporated in the radionecrosis, however, appeared to be morphologically intact even at a distance of 0.2 mm. A remarkable factor was the time course of the lesion growth. Whereas the size of the radiolesion resulting from low-dose rate permanent implants had not reached its maximum until after 70 days, the radionecrosis after high-dose rate temporary implants showed its maximum size after 25 days, after which the lesion decreased relatively quickly in comparison with the radionecrosis induced by permanent implants.

A high dose delivered over a short period caused a larger volume, thus resulting in an increased capillary surface of impaired endothelial cells with the inevitably induced vasogenic oedema. Under low doses and long-term exposure small necroses were produced along with moderate, but long-lasting vasogenic oedema. When temporary implants were used, lively macrophage acitivity set in after the emitter had been removed and. the zone of necrotic tissue then decreased. The extent of demyelination in the white substance was directly related to the intensity of the vasogenic oedema. The glial reaction and demyelination clearly exceeded the physical penetration depth of the gamma radiation, at least that of I-125 photon radiation, and were therefore considered secondary effects. The effects of scattering and attenuation in tissue vary considerably for the three isotopes used in the experiment. The specific dose rate factor, i.e. the dose rate per unit activity, divided by the geometrical attenuation is not a constant with distance, particularly not for I-125[2,17,30]. We, like other investigators of the subject, found that the volume of the radionecrosis and the extent of the vasogenic oedema can be highly influenced by the total dose and the application time[8,9,32]. Lower dose rates caused small radionecroses,

which did not increase in size after 100 days. When the dose is delivered continuously, typical repair mechanisms of the cells such as reproduction, redistribution, and repair of sublethal damage are obstructed. The continuous low-dose rate, however, was effective enough to inhibit macrophage activity, which is why the radionecrosis had not decreased after an observation period of at least one year. Radionecroses were accompanied by an extensive vasogenic oedema. However, after removal of the radiation source, repair processes such as macrophage activity were able to take action and the lesion almost completely disappeared.

The volume of radionecroses in normal brain tissue therefore depends on the total dose, the dose rate and energy of the implanted radioactive source.

In neoplastic tissue other factors such as neovascularisation and susceptibility of tumour cells to radiation are other contributing factors. In a previous experiment we studied the necrotising effect of I-125 seeds in neoplastic tissue[28]. Avian-sarcoma-virus (ASV)-induced dog brain tumours were chosen to serve as a primary model for human brain tumours. ASV-tumours are autochthonous, i.e. their blood supply and growth are characteristic of a primary brain tumour. The model is extensively studied with respect to its biological and physiological characteristics[11]. The tumour type most often induced is an anaplastic glioma. The I-125 implants produced sharply delineated calcifying necroses with vital unaffected tumour outside the necroses. The necrotising and calcifying effect was already apparent after 18 days and complete after 90 days. The response of neoplastic tissue to interstitial irradiation in terms of the volume of necrosis was 3 to 5 times greater when compared with I-125 radionecroses in normal brain tissue. The necrobiotic characteristics, however, were similar.

The temporary, geographically confined opening of the blood brain barrier is considered a specific side-effect of interstitial radiosurgery in the brain. The amount of blood-born fluids that cross the blood-brain barrier depends on the total surface area of the irradiated defective capillaries. Experimental and clinical results from blood to tissue transport studies are usually expressed in terms of a transfer constant K1, which characterises the unidirectional transfer across capillaries[10,11].

Using C-14-alpha-amino-isobutyric acid (AIB) autoradiography it could be shown that in comparison with normal brain the layer of irradiated brain tissue on the surface of a necrotic volume develops higher capillary permeability (up to 7.7 times higher) and maintains this over a period of several months after low-dose rate permanent interstitial irradiation[10]. More recently, quantitative computer tomographic methods as described by Groothuis et al.[11] were used, which allow consecutive quantitative in vivo measurements of individual animals. The values for K_1 after temporary interstitial high-dose rate irradiation exceeded those found after low-dose rate permanent irradiation for all three gamma emitters. With Au-198 implants, whether permanent or temporary, the highest K_1 values were found. K_1 values for Ir-192 temporary implant were slightly lower. All of the experimental animals showed nodular lesions (in the early stages) or ring-shaped lesions of the capillary function in standard computed tomography by the time the first examination was carried out. The capillary permeability in a localised zone surrounding the implant was temporarily 10 to 40 times higher than that in normal brain tissue. The vascular volume did not differ from normal brain tissue. Whereas after permanent interstitial irradiation the impaired capillary permeability (K_1) remained rather constant over a period of up to one year, the capillary permeability after temporary irradiation had a peak after the maximum volume of radionecrosis had developed and after that gradually returned to normal. For a detailed description of the physiological effects see article by Warnke et al. in this volume[33].

Clinical Findings

Conventional radiotherapy (teletherapy) is based on the difference in radiosensitivity of tumour and brain. The treatment volume exceeds the target volume, which exceeds the tumour volume. The dose rate is high and intermittent. The dose limit is set by the tolerance of the surrounding healthy brain[13]. The cells most sensitive to radiation exposure are those still capable of replication. This includes glial elements, particularly in the younger age groups, and endothelial cells in all age groups[5]. The most serious late effects of radiotherapy are caused by diffuse radiation damage to the capillary endothelium resulting in brain swelling due to disturbance of blood brain barrier function[5,20]. Fraction size, irradiated volume, and age play a significant role and there are variations in individual susceptibility[4,5]. In interstitial implant radiosurgery (brachytherapy), however, treatment volume, target volume and tumour volume are essentially identical. Interstitial implant radiosurgery

is not based on a differential sensitivity. Using interstitial implants a high, i.e. a necrotising radiation dose is delivered to a precisely predefined tissue volume without affecting surrounding normal brain. Generally, there should be no toxicity, however, clinical follow-up studies have revealed considerable toxicity which largely depends on the irradiated volume, the energy range, the total dose and the applied dose rate[12]. Clinical experience with interstitial implant dosimetry supports the concept that tissue tolerance is a function of both dose rate and total dose.

Clinical follow-up examinations using computed tomography and more recently magnetic resonance have demonstrated that there can be an extensive although temporary oedema surrounding the radionecrosis[6,7,22,26,31]. Temporary vasogenic oedema was seen after gamma-knife radiosurgery, after linear convergent beam radiosurgery and after interstitial implant radiosurgery[19,21]. It usually is expressed more in white matter lesions than in the grey matter. Treatment with high dose glucocorticoids usually results in improvement of vasogenic oedema. CT or MR findings alone, however, can be misleading since the discrimination of white matter oedema and tumour invasion might be difficult, as is demonstrated by stereotactic biopsies[22].

Autopsy data on the response of the brain to interstitial implants are rare and inconsistent, since most patients were treated with external beam radiation in addition to interstitial implants thus demonstrating cumulative radiation toxicity[15]. There are indications that peritumoural oedema may enhance radiosensitivity of the surrounding white matter and that mid-brain structures exhibit an increased tendency for the development of delayed radionecrosis. The previously used permanent iridium implants and also permanent iodine-125 implants caused marked tumour regression and in some cases neoplastic cells were no longer detectable. In some cases, however, delayed radionecrosis developed in the adjacent normal brain and these radionecroses were particularly extensive in basal mid-line structures. The radionecrosis, however, did not conform to isodose lines, but to anatomical structures i.e. necroses were not caused by a direct radiation effect but by toxic vasogenic oedema.

Discussion

The purpose of interstitial implant radiosurgery is to deliver a focal necrotising dose of radiation to a predefined volume of neoplastic tissue without affecting surrounding normal brain. Prerequisite is a precise knowledge of the relationship between the radiation dose, the volume treated and the subsequent response of normal tissue surrounding the lesion. According to the definition given by Larsson, the implantation of radioactive sources is but one method of radiosurgery[18]. Radiosurgical toxicity, however, seems to be the same whether cobalt-60-gamma-knife, a linear accelerator, heavily charged particles or similar techniques are used[1]. Radioactive interstitial implants have been systematically used since the 1950s and most data on brain tolerance have come from treated patients. Follow-up studies using CT and MR have shown extensive temporary oedema formation in the surrounding white matter of the ipsilateral hemisphere. Experimental studies have confirmed that this oedema formation is largely vasogenic oedema due to impairment of blood brain barrier function, i.e. in other words, that it does indeed result from impairment of capillary function. It is the volume-dependent vasogenic oedema i.e. the total surface area of damaged capillaries which accounts for the volume of serum proteins released into the interstitium and the resulting toxicity.

The finding of increased capillary permeability, although geographically and temporarily restricted in radiosurgery, has far reaching implications for clinical application. Under clinical conditions the time factor cannot be changed when permanent (lost) implants or single shots are used. With lost implants no corrections can be made. This also applies to overdosing, which in the past has been observed in clinical situations with serious consequences. Underdosing, on the other hand, often requires re-implantation or results in incomplete control of tumour growth. Clinical experience has shown that the oedema phase can be shortened by using temporary implants with moderate dose rates. After removal of the radioactive source repair processes are free to take place. Based on experimental findings and clinical observation a biological dosimetry has evolved which uses iodine-125 temporary seeds with the preferred dose rate of 10 cGy/h and cumulative reference dose on the tumour margin of 60 Gy. The clinical results so far obtained demonstrate that toxic vasogenic oedema no longer poses a serious problem. In a series of more than 500 gliomas adverse effects of irradiation was as low as 3.1% morbidity, with no case of radionecrosis which required removal due to a space-occupying effect[29]. The effectiveness and low toxicity have also been demonstrated in a series of solitary brain metastases[16].

Conclusion

Interstitial implant radiosurgery is a careful balance between necrosis and repair. It primarily causes an expanding lesion which histologically is characterised by the energy range of the radio-isotope, the activity and the dose rate used. Vasogenic oedema is a secondary effect which is volume-dependent and expressed more in white matter lesions. Using temporary implants a rapidly shrinking lesion can be observed due to macrophage activity in the absence of radiation. Stereotactic interstitial implant radiosurgery in our hands has replaced open surgery for a variety of brain tumors. However, it is still a treatment modality with many unanswered questions as to the optimum radioactive sources, dose rates, iso-effects and the tissue tolerance doses.

Acknowledgements

The author gratefully acknowledges the support of P. Kleihues, Abt. Neuropathology, Universität Zürich and of M. Kiessling, Abt. Neuropathology, Universität Heidelberg. Thanks go to H. K. Leetz, Institut für Radiologische Physik der Universität des Saarlandes, Homburg/Saar, W-Germany for providing the dosimetry calculations. The author is indebted to his collaborators P. C. Warnke, F. J. Hans and Mrs. A. Korst for their technical assistance.

The studies were supported by the Deutsche Forschungsgemeinschaft (OS 58/2-1 and OS 58/2-2).

References

1. Andersson B, Larsson B, Leksell L, Mair W, Rexed B, Sourander P, Wennerstrand J (1970) Histopathology of late local radiolesions in the goat brain. Acta Radiol 9: 305–394
2. Anderson LL, Hsin MK, Ing-Yuan D (1981) Clinical dosimetry with I-125. In: George FW (ed) Modern interstitial and intracavitary radiation cancer management. Masson, New York
3. Bernstein M, Gutin PH (1981) Interstitial irradiation of brain tumors: A review. Neurosurgery 6: 741–750
4. Burger PC, Mahaley MS, Dudka L, Vogel FS (1979) The morphologic effects of radiation administered therapeutically for intracranial gliomas. Cancer 44: 1256–1272
5. Caveness WF (1980) Experimental observations: Delayed necrosis in normal monkey brain. In: Gilbert HA, Kagan AR (eds) Radiation damage to the nervous system. Raven, New York, pp 1–38
6. Daumas-Duport C, Blond S, Vedrenne Cl, Szikla G (1984) Radiolosion versus Recurrence: Bioptic data in 39 gliomas after interstitial, or combined interstitial and external radiation treatment. Acta Neurochir (Wien) [Suppl] 33: 291–299
7. Davis RL, Barger GR, Gutin PH, Philips TL (1984) Response of human malignant gliomas and CNS tissue to brachytherapy: A study of seven autopsy cases. Acta Neurochir (Wien) [Suppl] 33: 301–305
8. Fike JR, Cann ChE (1984) Contrast medium accumulation and

washout in canine brain tumors and irradiated normal brain: A CT study of kinetics. Radiology 151: 115–120

9. Fike JR, Cann CE, Philips TL, Bernstein M, Gutin PH, Turowsky K, Weaver KA, Davis RL, Higgins RJ, Da Silva V (1985) Radiation brain damage induced by interstitial I-125 sources: A canine model evaluated by quantitative computed tomography. Neurosurgery 16: 530–537

10. Groothuis DR, Wright DC, Ostertag CB (1987) The effect of I-125 interstitial radiotherapy on blood brain barrier function in normal canine brain. J Neurosurg 67: 985–902

11. Groothuis DR, Lapin GD, Vriesendorp FJ, Mikhael MA, Patlak CS (1991) A method to quantitatively measure transcapillary transport of iodinated compounds in canine brain tumors with computed tomography. J Cereb Blood Flow Metab 11: 939–948

12. Gutin PH, Leibel SA (1985) Stereotactic interstitial irradiation of malignant brain tumors. Neurol Clin 3: 883–893

13. Hall EJ (1978) Radiobiology for the radiologist. Harper and Row, Hagerstown, Maryland

14. Janzer RC, Kleihues P, Ostertag CB (1986) Early and late effects on the normal dog brain of permanent interstitial Iridium-192 irradiation. Acta Neuropath 70: 91–102

15. Kiessling M, Kleihues P, Gessaga E, Mundinger F, Ostertag CB, Weigel K (1984) Morphology of intracranial tumours and adjacent brain structures following interstitial iodine-125 radiotherapy. Acta Neurochir (Wien) [Suppl] 33: 281–189

16. Kreth FW, Warnke PC, Ostertag CB (1993) Interstitial implant radiosurgery of cerebral metastases. Acta Neurochir (Wien) [Suppl] 58: 112–114

17. Krishnaswamy V (1978) Dose distribution around an 125-I seed source in tissue. Radiology 126: 489–491

18. Larsson B (1992) Radiobiological Fundamentals in Radiosurgery. In: Steiner L, et al (eds) Radiosurgery: Baseline and trends. Raven, New York

19. Lindquist C, Hindmarsh T, Kihlström L, Mindus P, Steiner L (1992) MRI and CT Studies of Radionecrosis Development in the Normal Human Brain. In: Steiner L, et al (eds) Radiosurgery: Baseline and trends. Raven, New York

20. Leibel SA, Sheline GE (1987) Radiation therapy for neoplasms of the brain. J Neurosurg 66: 1–22

21. Marks LB, Spencer DP (1991) The influence of volume on the tolerance of the brain to radiosurgery. J Neurosurg 75: 177–180

22. Oppenheimer JH, Levy ML, Sinha U, El-Kadi Hikmat, Apuzzo MLJ, Luxton G, Petrovich Z, Zee CS, Miller CA (1992) Radionecrosis secondary to interstitial brachytherapy: Correlation of magnetic resonance imaging and histopathology. J Neurosurgery 31(2): 336–343

23. Ostertag CB, Groothuis D, Kleihues P (1984) Effects on tumour and brain: Experimental data on early and late morphologic effects of permanently implanted gamma and beta sources (Iridium-192, Iodine-125 and Yttrium-90) in the brain. Acta Neurochir (Wien) [Suppl] 33: 271–280

24. Ostertag CB, Hossmann KA, vd Kerckhoff W (1982) Radiation effects of Iridium-192 implants in the cat brain. Nucl Med 21: 99–104

25. Ostertag CB, Weigel K (1982) Three-dimensional CT scanning of the dog brain. J Comput Assist Tomogr 6(5): 1036–1037

26. Ostertag CB, Weigel K, Birg W (1979) CT-changes after long-term interstitial iridium-192-irradiation. In: Stikla G (ed) Stereotactic cerebral irradiation. Elsevier, Amsterdam

27. Ostertag CB, Weigel K, Warnke P, Lombeck G, Kleihues P (1983) Sequential morphological changes in the dog brain after interstitial Iodine-125 irradiation. Neurosurgery 13: 523–528

28. Ostertag CB, Warnke P, Kleihues P, Bigner D (1984) Iodine-125 interstitial irradiation of virally induced dog brain tumors. Neurol Res 6: 176–180

29. Ostertag CB, Kreth FW (1992) Stereotactic interstitial radiotherapy in the treatment of gliomas. Acta Neurochir (Wien) 119: 53–61

30. Sondhaus CA (1981) I-125: Physical properties, photon dosimetry and effectiveness. In: George FW (ed) Modern interstitial and intracavitary radiation cancer management. Masson, New York

31. Szikla G, Betti O, Blond S (1979) Data on late reactions following stereotactic irradiation of gliomas. In: Szikla G (ed) Stereotactic cerebral irradiation. Elsevier, Amsterdam

32. Turowsky K, Fike JR, Cann CE, Higgins RJ, Davis RL, Gutin PH, Phillips TL, Weaver KA (1986) Normal brain Iodine-125 radiation damage: effect of dose and irradiated volume in a canine model. Radiology 158: 833–838

33. Warnke PC, Hans FJ, Ostertag CB (1993) Impact of stereotactic interstitial irradiation on brain capillary physiology. Acta Neurochir (Wien) [Suppl] 58: 85–88

Correspondence: Prof. Dr. Christoph B. Ostertag, Abteilung Stereotaktische Neurochirurgie, Neurochirurgische Universitätsklinik, D-79106 Freiburg, Federal Republic of Germany.

Acta Neurochir (1993) [Suppl] 58: 85–88

Impact of Stereotactic Interstitial Radiation on Brain Capillary Physiology

P. C. Warnke, F. J. Hans, and **Ch. B. Ostertag**

Abteilung Stereotaktische Neurochirurgie, Neurochirurgische Universitätsklinik Freiburg, Federal Republic of Germany

Summary

rCBF, capillary permeability and vascular volume have been measured during the time course of interstitial stereotactic radiosurgery in normal and tumour-bearing dog brain. For rCBF measurement the stable Xenon-CT-technique with a modified Kety-Schmidt equation has been used, and for measurement of blood-to-brain transport of meglumine iothalamate the two-compartment CT-method as developed by Groothuis, which also reflects vascular volume. Anaplastic gliomas had been induced by intracerebral injection of avian sarcoma virus. Radiation sources have been 192-iridium and 198-gold.

Both of the isotopes caused spherical blood-brain-barrier lesions with a more than 10-fold blood-to-brain transport increase. These effects occurred earlier and more pronounced with 192-iridium, but much longer lasting with 198-gold.

Interstitial radiosurgery of the tumours led to a further increase of capillary permeability.

Blood flow was significantly lowered not only within radionecrosis but also in the adjacent brain areas.

Keywords: Interstitial radiation; radiosurgery; animal experiments; rCBF; capillary permeability; vascular volume; 192-iridium; 198-gold.

Introduction

Stereotactic interstitial irradiation as one way of radiosurgery is used for a variety of primary brain tumours and metastases[1,8,9,14,15]. Although a dosimetry based on clinical efficacy is fairly established, a dosimetric data base relying on a quantitative biological dose-effect relationship in vivo in the brain is still lacking. With the advent of modern techniques like CT, PET and MRI sophisticated tools are at hand to measure in vivo a number of physiological parameters of brain and/or brain tumours and the effect of radiosurgery upon those[2]. Parameters of interest are regional cerebral blood flow, blood-brain barrier permeability, vascular space and extracellular space. All of these parameters are involved in or interfered by oedema formation often concomitant with interstitital irradiation[4].

In order to gather the aforementioned biological data base for refinement of dosimetry and to study the effect of interstitial radiosurgery on capillary physiology we measured rCBF, capillary permeability and vascular volume in normal and tumour-bearing dog brain during the time course of interstitial stereotactic radiosurgery.

Material and Methods

Six-month-old beagle dogs of both sexes were used for the experiments. 192-Iridium and 198-gold were implanted stereotactically into the left frontal centrum semiovale of six dogs each. Regional cerebral blood-flow was measured using the stable Xenon-CT technique and a modified Kety-Schmidt equation[10]. Blood-to-brain transport of meglumine iothalamate reflecting capillary permeability was measured applying a two-compartment CT-method as developed by Groothuis[7]. This method also rendered simultaneously the brain-to-blood-transport and the vascular volume of the brain compartment examined. Both isotopes were implanted temporarily within Silicon catheters. Dosimetry was performed as described by Ostertag in this volume[13]. All radionuclides were removed 10 days after implantation and the aforementioned measurements of blood-flow and capillary permeability were performed at various time points. The measurements were repeated four days after continuous dexamethasone application (2.5 mg/kg/day).

Interstitial radiosurgery was also performed in six dogs bearing anaplastic gliomas induced by intracerebral injection of avian sarcoma virus (10[5] focus forming units) into the right subependymal plate[3]. In these dogs Iodine-125 was used as temporary implant for 10 days. The implanted seeds had an activity of 25 mCi resulting in a 6000 cGy isodose line at the tumour margin as defined by CT and MRI scanning. The effective dose rate was 10 cGy/h. Measurements of blood-flow and capillary permeability were done before and 10 days after completion of radiosurgery.

Results

All animals developed spherical blood-brain-barrier lesions as judged from the CT-scan after contrast enhancement. Quantitation of blood-brain-barrier transport of meglumine iothalamate revealed a more than 10-fold increase in blood-to-brain transport immediately after completion of the radiosurgery. Although the blood-brain-barrier breakdown was seen with both isotopes, there was a conspicuous difference concerning the extent of increased capillary permeability as well as its time course. The results concerning these two variables are presented in Table 1 and 2. Whereas 192-iridium did increase blood-brain-barrier permeability much earlier and more pronouncedly than 198-gold we could observe a reclosure of the blood-brain-barrier with a slow decrease of capillary permeability up to 70 days after radiosurgery. This was not the case with the higher energy radiation source of gold-198. Even up to 200 days after completion of the radiosurgery we could still measure an increase of capillary permeability measured as K_1. The changes in permeability were due to transport of the water soluble compound into the extracellular space of the irradiated brain as can also be seen from the vascular volume measured concomitantly. With both isotopes we did not get a significant or directional of V_p over time.

Blood flow in the radionecrosis itself was as low as 8.4 ± 2.7 ml/100 g/min reflecting tissue necrosis. In the surrounding area of vasogenic oedema caused by the spherical blood-brain-barrier breakdown adjacent to the radionecrosis we did find conspicuously lower blood flow values (medium 16.88 ± 9.4 ml/100 g/min) than in contralateral and ipsilateral non-affected white matter. The spatial configuration of the blood flow changes is shown in Fig. 1. The results of the rCBF measurements are summarized in Table 3.

Treatment with dexamethasone for four days led to a 40% decrease of capillary permeability as measured in the animals treated with gold-198 20 days after implantation ($K_1 = 6.06 \pm 0.86$ before, $K_1 = 3.71 \pm 0.49$ after).

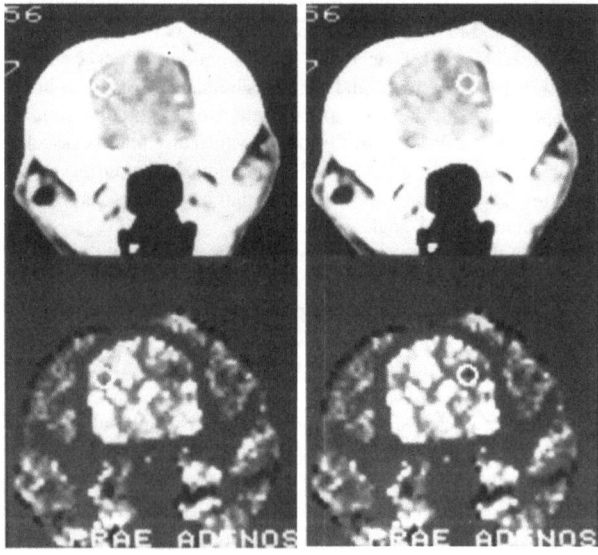

Fig. 1. Stable Xenon blood flow images of normal dog brain 10 days after interstitial 192-iridium radiosurgery. Top: CT-image, Bottom: black and white coded flow image. Quantitative flow values in encircled regions (radionecrosis and contralateral identical region of interest) are given as F in ml/100 g/min. Note the widespread rCBF depression in the oedematous white matter adjacent to the radionecrosis

Table 1. *Time Course of Blood-to-Brain Transport* (K_1), *Brain-to-Blood Transport* (K_2) *and Vascular Volume* (V_p) *After 192-Iridium Radiosurgery in Normal Dog Brain*. Dimensions are as follows: $K_1 = \mu l/g/min$, $K_2 = m^{-1}$ and $V_p = \%$ of 100 ml tissue volume

Days	K_1	K_2	V_p
Iridium-192			
10	36.8 ± 17.2	0.08	
12	12.8 ± 4.8	0.05	1.37
20	10.5 ± 3.9	0.04	0.9
46	9.6 ± 3.7	0.04	1.5
60	3.8 ± 1.1	0.03	1.2
70	3.0 ± 1.2	0.03	1.42
Normal brain	1.28	—	1.1

Table 2. *Time Course of* K_1, K_2 *and* V_p *After 198-gold Radiosurgery in Normal Dog Brain*. Dimensions are identical with those in Table 1

Days	K_1	K_2	V_p
Gold-192			
10	4.8 ± 0.02	0.03	1.2
18	5.2 ± 0.2	0.04	0.5
46	5.8 ± 0.2	0.03	0.4
70	11.5 ± 1.1	0.05	2.2
100	12.8 ± 0.7	0.06	1.7
200	14.8 ± 1.2	0.06	1.9
Normal brain	1.28	—	1.1

Table 3. *Effect of Intestitial Radiosurgery on Regional Cerebral Blood Flow After 192-Iridium Implantation in Normal Dog Brain*. Oedema is defined as the region adjacent to the area of blood-brain-barrier breakdown with significantly decreased Hounsfield units as compared to contralateral brain. Values are given as means $\pm$ standard deviation

rCBF (ml/100g/min)	
Oedema	16.88 ± 9.41
White matter	33.5 ± 10.7
Basal ganglia	63.5 ± 10.7
Cortex	68.7 ± 17.3

Also blood-flow did increase significantly in the oedematous brain area from 16.88 ± 9.4 to 26.15 ± 12.29 ml/100 g/min after dexamethasone treatment ($p < 0.05$, t-test).

Interstitial radiosurgery of avian sarcoma virus induced tumours led to a further increase of capillary permeability by a factor of 9.4. As in normal brain the area of increased permeability was spherical surrounding the central radionecrosis. rCBF except for the radionecrosis induced was not lowered significantly within the tumour but in the surrounding vasogenic oedema (15.78 ± 5.5 versus 10.12 ± 4.8 ml/100 g/min).

Discussion

Stereotactic interstitial radiosurgery does produce circumscribed alterations of capillary physiology in the tissue surrounding the definite radionecrosis. These alterations are reflected as changes in capillary permeability and regional cerebral blood flow. Although a qualitative description of blood-brain-barrier breakdown with temporary interstitial radiosurgery has been done[5,16], a quantitative measurement of blood-brain-barrier permeability has yet not been undertaken. Still quantitation is necessary to generate a dose effect relationship in radiosurgery. Furthermore as attempted in this paper quantitation is absolutely necessary to study the differential effects of isotopes of different energy on the cerebral vasculature.

The only quantitative study employing quantitative autoradiography has been undertaken in permanent iodine-125 implants so far[6]. Due to the radiobiological characteristics of permanent implants these data can not be extrapolated onto temporary implants. Indeed our data show that temporary implants with various energies do behave conspicuously differently.

Whereas iridium-192 does cause an immediate blood-brain-barrier breakdown with an enormous increase of capillary permeability which then within a time frame of about 100 days allows the endothelial cells to repair and to become functionally intact again, the opposite holds true for gold-198. Here we see a small but steady increase of impaired capillary function expressed as an increase of blood-to-brain transport of the water soluble molecule meglumine iothalamate. These data concerning capillary function correlate quite well with the velocity of the formation of the radionecrosis in temporary versus permanent implants as described by Ostertag[12] as well as they take into consideration the different efficacy of iridium-192 and gold-198 in terms of creation of a temporary radionecrosis and its resolution. Interestingly during the increase and following decrease and/or the long lasting increase in capillary permeability demonstrating an enormous impairment of capillary function we did not see a significant or directional change of the vascular volume. This correlates favourably with the histological description of the brain tissue surrounding the radionecrosis[11,12]. There was no significant angiogenesis detectable nor an increased number of pathological vessels with increased capillary surface. The brain-to-blood transport rate described here by the dimension of reciprocal time changed concomitantly with the K_1 value in our animals as expected for a compound which is not actively transported but passes the blood-brain-barrier merely by passive diffusion as has been shown for meglumine iothalamate[7].

Having both, the K_1 and the K_2 value at hand, it will be possible for future experiments to calculate the ratio of K_1 divided by K_2 to calculate the size of the compartment within which the marker distributes itself i.e. the extracellular space. The aforementioned alterations of capillary permeability result in widespread vasogenic oedema which leads to depression of regional cerebral blood flow as has been shown consistently in all our experimental animals. It is the combination of both the change in capillary permeability and the decreased regional cerebral blood-flow which besides the size of the induced radionecrosis will limit the application of interstitial radiosurgery in the brain. But the methodology used in this experimental setting, applying in vivo methods, can also be used to establish the time course of changes in capillary physiology during interstitial radiosurgery of patients harbouring brain tumours. As the methods renders quantitative data, this can be correlated with the total dose, the effective dose rate and the energy range of the radionuclide employed. Thus the limitations of a given dosimetry in an individual patient can be defined much more accurately and a "biological dosimetry" which takes acute and delayed tissue reactions into consideration can be designed to refine radiosurgery even more into the direction of an accurate and reproducible neurosurgical tool.

References

1. Bernstein M, Gutin PH (1981) Interstitial irradiation of brain tumors: A review. J Neurosurg 6: 741–750
2. Bernstein M, Marotta T, Stewart P, Glen J, Resch L, Henkelman M (1990) Brain damage from 125-I brachytherapy evaluated by

MR imaging, a blood-brain barrier tracer, and light and electron microscopy in a rat model. J Neurosurg 73: 585–593

3. Bigner DD, Vick NA, Kvedar JP, Mahaley MS ED (1972) Virus-cell relationships in dog brain tumors induced with Schmidt-Ruppin rous sarvoma virus. Progr Exp Tumor Res 17: 40–58

4. Caveness WF, Kemper TL, Brightman MW, Blasberg RG, Fenstermacher JD (1978) Directional character of vasogenic edema. Adv Neurol 20: 271–291

5. Fike JR, Cann CE, Phillips TL, Bernstein M, Gutin PH, Turowski K, Weaver KA, Davis RL, Higgins RJ, DaSilva V (1985) Radiation brain damage induced by interstitial 125-I sources: A canine model evaluated by quantitative computed tomography. Neurosurg 16: 530–537

6. Groothuis DR, Wright DC, Ostertag CB (1987) The effect of 125-I interstitial radiotherapy on blood-brain barrier function in normal canine brain. J Neurosurg 67: 895–902

7. Groothuis DR, Lapin GD, Vriesendorp FJ, Mikhael MA, Patlak CS (1991) A method to quantitatively measure transcapillary transport of iodinated compounds in canine brain tumors with computed tomography. J Cereb Blood Flow Metab 11: 939–948

8. Gutin PH, Phillips TL, Hosobuchi Y, Wara WM, Mackay AR, Weaver KA, Lamp S, Hurst S (1981) Permanent and removable implants for the brachytherapy of brain tumors. J Radiat Oncol Biol Phys 7: 1371–1381

9. Hosobuchi Y, Phillips TL, Stupar TA, Gutin PH (1980) Intersti-tial brachytherapy of primary brain tumors. J Neurosurg 53: 613–617

10. Meyer JS, Hayman LA, Yamamoto M, Sakai F, Nakajima S (1980) Local cerebral blood flow measured by CT after stable xenon inhalation. AJNR 1: 213–225

11. Ostertag CB (1989) Stereotactic interstitial radiotherapy for brain tumors. J Neurosurg Sci 33(1): 83–89

12. Ostertag CB (1991) Experimental central nervous system injury from implanted isotopes. In: Gutin PH, Leibel SA, Sheline GE (eds) Radiation injury to the nervous system. Raven, New York, pp 183–190

13. Ostertag CB (1992) Brachytherapy-interstitial implant radio-surgery. Acta Neurochir (Wien) [Suppl] 58: 79–84

14. Prados M, Leibel S, Barnett CM, Gutin P (1989) Interstitial brachytherapy for metastatic brain tumors. Cancer 63: 657–660

15. Sewchand W, Amin PP, Drzymala RE, Salazar OM, Salcman M, Samaras GM, Botero E (1984) Removable high intensity iridium-192 brain implants. J Neuro Oncol 2: 177–185

16. Turowski K, Fike JR, Cann CE, Higgins RJ, Davis RL, Gutin PH, Phillips TL, Weaver KA (1986) Normal brain iodine-125 radiation damage: Effect of dose and irradiated volume in a canine model. Radiology 158: 833–838

Correspondence: Prof. Dr. Christoph B. Ostertag, Neurochirur-gische Universitätsklinik, Hugstetter Strasse 55, D-79106 Freiburg, Federal Republic of Germany.

Acta Neurochir (1993) [Suppl] 58: 89–91

Linac-Accelerator-Radiosurgery

V. Sturm[1], W. Schlegel[2], O. Pastyr[2], H. Treuer[1], J. Voges[1], R. P. Müller[3], and W. J. Lorenz[2]

[1] Department of Stereotactic and Functional Neurosurgery, University of Cologne, [2] Department of Radiology, German Cancer Research Center, Heidelberg, and [3] Department of Radiotherapy, University of Cologne, Federal Republic of Germany

Summary

A survey is given of the actual possibilities and limitations of the use of linear accelerators (Linac radiosurgery systems) for intra = cranial radiosurgery.

Depending on the collimator size, spherical fields from 5–54 mm in diameter can be irradiated with dose gradients from 10% (large fields) to 20% (small fields) per millimeter distance between surface and treatment volume. This is comparable to the possibilities of Gamma-Knife and Proton-irradiation.

Optimal mechanical adjustment of gantry and linac table are necessary for the required stability of the isocenter. Mechanical inaccuracy should be smaller than 0.8 mm. Advanced computerized 3D-treatment planning systems are indispensable prerequisites for accurate treatment and use of the flexibility of the linac system. Future developements are outlined.

Keywords: Radiosurgery; linear accelerator; Linac system.

Introduction

In the early fifties Leksell[9] coined the term "radio-surgery" for stereotactically guided, extremely focussed convergent beam irradiation with high single doses. From the Swedish group an irradiation unit with multiple isocentrically oriented Cobalt-60-sources has been developed. The "Gamma-Knife" enables single dose irradiation of small, more or less spherical fields with high precision and steep dose gradients. Although a few other centers developed facilities for stereotactic proton- and heavy ion irradiation[4,6], "Gamma-Knife-Radiosurgery" has remained the "golden standard" for Radiosurgery for many years. Despite excellent clinical results, gained in the treatment of small arterio-venous malformations and various benign intracranial lesions[10,12,14] its use has been confined to a few centers until recently, mainly because of the high costs of the equipment.

In order to overcome the main physico-technical drawbacks of Gamma-Knife-Radiosurgery (limitation to small irradiation fields, with maximum diameters of 25 mm) and to reduce the costs of the equipment, some groups developed systems enabling the use of Linear accelerators for radiosurgery in the early eighties[1–3,5]. In recent years a steadily growing number of radiological and neurosurgical hospitals has been equipped with Linac-Radiosurgery systems. The present status of Linac-Radiosurgery (LRS), advantages and drawbacks as well as future prospects, will be discussed and described.

Present Status

Comparable to the Gamma-Knife, the basic principle of LRS is beam convergence. In our system[5] beam-convergence is achieved by use of 10 arcs of Linac-table rotation in steps of 18° and rotation of the gantry from 20–160°. Depending on the size of the collimator, spherical fields from 5–54 mm in diameter can be irradiated with dose gradients from 10% (large fields) to 20% (small fields) per millimeter distance from the surface of the treatment volume. In comparison, the dose gradients achievable with Gamma-Knife, LRS and Proton-irradiation do not differ significantly.

Mechanical adjustment of both gantry and linac-table is absolutely necessary to achieve the required stability of the isocenter, i.e. the point, in which all converging collimated beams cross. It must be stressed, that no commercially available Linear-accelerator has an isocenter, which is sufficiently stable for radiosurgical procedures and that time consuming and tedious mechanical work is necessary to achieve the required precision and stability. The same is true for the alignment of the positioning lasers[11].

A mechanical overall accuracy has to be reached, which does not permit deviations of the calculated target point from the actual crossing point of all converging beams by more than 0.8 mm. Considering the pixel- and voxel-size of advanced CT and MR devices as well as the inaccuracy of X-ray systems, which limit the precision of target localisation to plus/minus 1–2 mm under optimal conditions, a mechanical inaccuracy equal to or smaller than 0.8 mm is tolerable.

Advanced computerised 3D-treatment planning systems, based on CT, MRI, PET and X-ray data have been developed[13], which are an indispensible prerequisite for accurate treatments. The planning-systems permit free use of the flexibility of Linac-systems to choose the proper collimators for irradiation of spherical fields from 5–54 mm in diameter. A limited modification of the shape of the irradiation fields is feasible by changing the angles of both gantry and table rotation. The shape of the isodose surface, covering the target volume, can be modified to a lesser degree than the steepness of the dose gradients in the directions to organs at risk.

Proper 3D-treatment planning, the use of the flexibility of Linear accelerators and a variety of easily changeable collimators make precise matching of the treatment volume to spherical target volumes with diameters between 5 and 54 mm possible. Multiple field-irradiation can be avoided in the majority of cases. Thus, treatment of larger target volumes with homogeneous dose distribution is easily feasible. Possibilities to "tailor-shape" the therapeutic isodose-surfaces are given to a minor degree. Dose-gradient conformation, enabling to better spare organs at risk is more easily achievable by changing the treatment parameters.

Indications for LRS

LRS has been shown to be adequately precise and to yield the same steepness of dose gradients as other modalities of radiosurgery. Thus, there is no principle difference with regard to indications. The possibility to treat larger target volumes enables the use of LRS for larger intracranial lesions than previously treated and additionally to treat selected cases of naso-pharyngeal tumours, which are localisable in the stereotactic coordinate system.

Future Prospects

Variation of the treatment parameters (choice of collimators of various sizes, modification of table and gantry rotation) enables one to change the geometry of the treatment volume according to the individual requirements to a certain degree, but real conformation-therapy is still impossible. The development of techniques which would allow "tailor-shaped" irradiation of non spherical, irregularly shaped target-volumes homogeneously with one irradiation field is urgently required. Exact matching of target and treatment volume would reduce overdosage to organs at risk and underdosage in the target volume and eliminate the need to use multiple treatment-volumes with dose-overlapping and radiobiologically disadvantageous field-inhomogeneities.

A promising solution is the development of multi-leaf collimators. The shape of their opening can be precisely matched to the very surface of the treatment-volume in the "beam's eye view" of the collimator, which changes with each angle of table-rotation.

A small, extremely precise multi-leaf collimator with tungsten-lamellae of 1 mm thickness has been developed by our group (Pastyr *et al.*, unpublished). It will be in use after completing the treatment-software and dosimetry (Schlegel *et al.*, unpublished). Linear accelerators excellently fulfill the requirements for conformation-therapy with multi-leaf collimators.

The second main drawback of Radiosurgery is the radiobiological disadvantage of single dose-irradiation. The application of high single doses limits the use of all presently available RS-systems to the treatment of small fields. Except for arteriovenous malformations[2-4,8,14] and brain metastases[15] single dose irradiation is unfavourable because of its inherent risk of radionecrosis and damage to healthy tissue[7] and the inability to profit by the clearly established radiobiological advantages of fractionation.

To enable the application of small numbers of fraction relocatable stereotactic frames, and different mask-system have been developed. A more promising, non-invasive method is the development of integrated computerised hard- and software systems enabling one to monitor movements of the patient during irradiation and to correct for them. Such a system is presently being developed by our group.

Discussion

LRS meets all requirements for radiosurgery, i.e. steepness of dose gradients and precision of dose-application. It optimally meets the requirements for the most important future treatment techniques, i.e. conformation therapy and fractionated radiosurgery. Compared to Gamma-Knife Radiosurgery, the main drawback of LRS is the necessity to mechanically stabilize the isocenter of the accelerator to a precision of less than 0.8 mm and to perform very accurate tests for quality assurance routinely throughout the period of use. Accurate performance of these tests and of the treatment procedures requires a highly specialised staff of dedicated physisists, neurosurgeons and radiotherapists. LRS is dangerous and should only be performed in centers which do fulfill these requirements.

References

1. Barcia-Solario JL, Hernandez G, Broseta J, Gonzalez-Darder J, Ciudad J (1982) Radiosurgical treatment of carotid-cavernous fistula. Appl Neurophysiol 45: 520–522
2. Betti O (1987) Treatment of arteriovenous malformations with the linear accelerator. Appl Neurophysiol 50: 262
3. Colombo F, Benedetti A, Pozza F, Marchetti C, Chierego G (1989) Linear accelerator radiosurgery of cerebral arteriovenous malformations. Neurosurgery 24: 833–840
4. Fabrikant JI, Lyman JT, Hosobuchi Y (1984) Stereotactic heavy ion Bragg peak radiosurgery for intracranial vascular disorders: Method for treatment of deep arteriovenous malformations. Br J Radiol 57: 479–490
5. Hartmann GH, Schlegel W, Sturm V, Bernd K, Pastyr O, Lorenz WJ (1985) Cerebral radiation surgery using moving field irradiation at a linear accelerator facility. Int J Radiation Oncol Biol Phys 11: 1185–1192.
6. Kjellberg RN, Preston WM (1961) The use of the Bragg peak of a proton beam for intracerebral lesions. In: Second International Congress of Neurological Surgery. Washington DC, Excerpta Medica, Amsterdam, E103-4
7. Kjellberg RN (1979) Isoeffective dose parameters for brain necrosis in relation to proton radiosurgical dosimetry. In: Szikla G (ed) Stereotactic cerebral irradiations. Elsevier, Amsterdam, pp 157–166
8. Kjellberg RN, Hanamura T, Davis KR, Lyons SL, Adams RD (1983) Bragg peak photon beam therapy for arteriovenous malformations of the brain. N Engl J Med 309: 269–274
9. Lecksell L (1951) The stereotaxic method and radiosurgery of the brain. Acta Chir Scand 102: 316–319
10. Noren G, Arndt J, Hindmarsch T, Hirsch A (1988) Stereotactic radiosurgical treatment of acoustic neurinomas. In: Lunsfford LD (ed) Modern stereotactic neurosurgery. Martinus Nijhoff, Boston, pp 481–489
11. Pastyr O, Hartmann GH, Schlegel W, Schabbert S, Treuer H, Lorenz WJ, Sturm V (1989) Stereotactically guided convergent beam irradiation with a linear accelerator: Localization-technique. Acta Neurochir (Wien) 99: 61–64
12. Rähn T, Arndt T, Thoren M, Backlund EO (1977) Stereotactic radiosurgery in pituitary dependent Cushing's disease and Nelson's syndrome. Proc 6th Int Congr Neurol Surg, Sao Paolo
13. Schlegel W, Scharfenberg H, Doll J, Hartmann G, Sturm V, Lorenz WJ (1984) Three dimensional dose planning using tomographic data. In: IEEE Comp Society (eds) Proc 8th Int Conf on the Use of Computers in Radiation Therapy, IEEE Comp Soc Press, Silver Spring, pp 191–196
14. Steiner L (1988) Stereotactic radiosurgery with the cobalt 60 Gamma Unit in the surgical treatment of intracranial tumours and arteriovenous malformations. In: Schmidek HH, Sweet WH (eds) Operative neurosurgical techniques, Vol 1. Grune and Stratton, Orlando, pp 515–529
15. Sturm V, Kober B, Höver KH, Schlegel W, Boesecke R, Pastyr O, Hartmann G, Schabbert S, Zum Winkel K, Kunze S, Lorenz WJ (1987) Stereotactic percutaneous single dose irradiation of brain metastases with a linear accelerator. Int J Radiation Oncol Biol Phys 13: 279–281

Correspondence: J. Voges, M.D., Department of Stereotactic and Functional Neurosurgery, University of Cologne, Josef-Stelzmann-Strasse 9, D-50931 Köln, Federal Republic of Germany.

Acta Neurochir (1993) [Suppl] 58: 92–97

Radiosurgery for Vascular Malformations of the Brain Stem

Ch. M. Duma, L. D. Lunsford, D. Kondziolka, D. J. Bissonette, S. Somaza, and **J. C. Flickinger**

Departments of Neurosurgery, Radiation Oncology and the Specialized Neurosurgical Center, University of Pittsburgh Medical Center, Pittsburgh, PA, U.S.A.

Summary

The challenges associated with microsurgery of vascular malformations located in the midbrain, pons and medulla have promoted the development of alternative therapeutic techniques. To assess the efficacy and safety of radiosurgery in the management of brain stem vascular malformations we reviewed our 5-year experience in 50 patients evaluated between 4 and 51 months (mean, 25 months) after radiosurgery.

Twenty-eight patients (56%) underwent gamma unit radiosurgery for symptomatic arteriovenous malformations (AVMs), and 22 patients (44%) for angiographically occult vascular malformations (AOVMs). Patients varied in age from 7 to 76 years (mean, 39 years). Forty-one patients (82%) had from 1 to 5 hemorrhages prior to gamma knife radiosurgery. Ten (20%) had one or two prior unsuccessful operations, and 37 (74%) presented with a neurological deficit. Of the patients with AVMs, 6 were considered Spetzler Grade III, and 22 (79%) Grade VI (inoperable: major component within the brain stem parenchyma). Forty-four malformations (88%) were adjacent to or within the midbrain and pons; the remainder involved the medulla. Average malformation diameters varied from 6 to 30.4 m (mean, 20.6; mean volume 4614 mm³). The minimal radiation dose to the margin of the malformations ranged from 12 to 25.6 Gy (mean, 18.9 Gy).

Of the 28 patients with AVMs, 8 had follow-up angiograms at a minimum of 2 years after radiosurgery (or sooner if their MRIs suggested obliteration). Of these patients, 7 (88%) showed complete obliteration of their malformations. No patients with AOVMs re-hemorrhaged if more than 15 months elapsed after radiosurgery. Four patients (8%; two with AVMs and two with AOVMs) sustained a symptomatic re-hemorrhage during the follow-up period. New neurologic deficits related to radiosurgery occurred in 10 patients. Three were temporary with return to baseline function at an average of 15 months post-radiosurgery. Of the remaining 7 patients with new deficits (14%), 3 were improving at last follow-up, and one patient died 35 months after radiosurgery of presumed radiation injury.

Precise gamma unit radiosurgery provides a safe and effective management strategy for brain stem vascular malformations. Its use is particularly valuable for malformations considered unsuitable for microsurgical resection.

Keywords: Radiosurgery; vascular malformations; brain stem; AVM; results.

Introduction

Surgical removal of vascular malformations of the brain stem can be hazardous. Their arterial supply is often contiguous with that of the parenchyma, and they lie within the most vital region of the brain[16]. With improvements in brain stem microsurgical technique, successful obliteration of some arteriovenous malformations (AVMs) in this location has been reported rarely. In such reports there is often a subgroup of AVMs which are incompletely resectable or "inoperable". Resection of angiographically occult vascular malformations (AOVMs or cavernous type malformations) within the parenchyma of the brain stem is also treacherous. AOVMs with normal brain intervening between the pial surface and the malformation may also be considered "inoperable" if the risks of removal are excessive[17].

Stereotactic radiosurgery offers a unique ability to obliterate AVMs within the brain stem without the risks inherent to microsurgery. We have also used the technique to manage AOVMs in this location – particularly those considered "inoperable". In this report we present out 5-year results using gamma knife radiosurgery in 50 patients harboring brain stem vascular malformations, and we provide a new dose-volume guideline for minimizing radiation injury risk to this critical area of the brain.

Clinical Material and Methods

Patient Population

Between August 1987 and August 1992, 50 patients with vascular malformations of the brain stem (28 with AVMs, and 22 with AOVMs) were managed using the 201-source cobalt-60 gamma knife. Most patients had vascular malformations deemed unsuitable for microsurgical excision at our multidisciplinary conference attended by neurosurgeons, neuroradiologists, radiation oncologists and physicists.

All patients with AVMs (10 females and 18 males) exhibited well-circumscribed lesions on high-resolution multiplane angiography. The patients' mean age was 37 years (range: 7 to 76 years). Treatment criteria for patients with AOVMs were a minimum of two clinically defined hemorrhages, angiographically invisible pathology and characteristic lesions on magnetic resonance imaging (MRI). The mean age was 42 years (range: 14 to 70 years). The clinical characteristics of the 50 patients in this series are presented in Table 1.

Vascular Malformation Location and Grade

Of the 28 AVMs, 26 were located within the midbrain and pons (Fig. 1), and 2 involved the medulla. The Spetzler and Martin[17] grading system was used for AVM classification. Twenty-two AVMs (79%) were considered Grade 6 (inoperable), and 6 were Grade 3. Three AVMs were in the medulla, 13 in the pons, and 6 in the midbrain.

Radiosurgical Technique

All adult patients had the application of the Leksell Model "G" stereotactic coordinate frame (Elekta Instruments, Georgia) and underwent cerebral angiography and/or MR imaging and radiosurgery under local anesthesia supplemented by intravenous sedation. General endotracheal anesthesia was used for children less than 14 years of age. Biplane high-resolution magnification subtraction stereotactic angiograms were obtained to define the nidus of the AVM and determine target coordinates.

For patients with AOVMs MRI scans were performed and used for stereotactic coordinate planning in all patients using T1, T2 and T1 with gadolinium-enhanced images; in some, computed tomographic (CT) scans were used adjunctively.

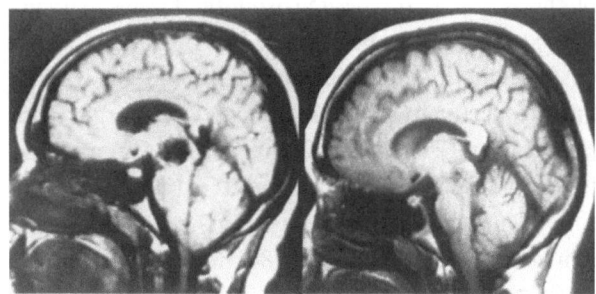
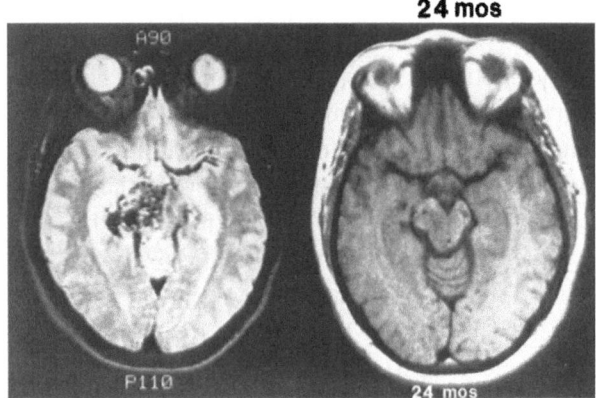

Fig. 1. Sagittal (top) and axial (bottom) magnetic resonance images of a midbrain/pontine AVM pre- (left images) and 24 months post- (right images) radiosurgery. This patient suffered no complications from radiosurgery

Radiosurgical dosimetry was performed on a high-speed Micro Vax (Digital Equipment Corporation, Westminster, Massachusetts) computer system. The treatment isodose, central dose and dose to the margin were determined by the neurosurgeon, the radiation oncologist, and the radiation physicist. Selection of the dose was based on knowledge gained by the treatment of conventional AVMs and AOVMs, by the location of the lesions, and their size. Patients received a single intravenous dose of 40 mg. of methylprednisolone immediately after the radiosurgical dose was delivered. All patients were discharged from the hospital on the day after radiosurgery.

Follow-up Evaluation

Follow-up imaging studies (MRI) were scheduled at 6-month intervals for all 50 patients for the first 2 years after radiosurgery. A high-resolution angiogram was requested of all patients with AVMs 2 years after radiosurgery (or sooner if their MRIs suggested obliteration).

Results

Radiation Dosimetry for AVMs (Table 2)

The treatment volumes ranged from 134 to 14,800 cu mm (mean: 5400 cu cm). Average malformation diameters ranged from 6 to 30.4 mm (mean: 18 mm). Seventy-one percent of the AVMs treated were between 1,000 and 10,000 cu mm. Only one patient had subtotal coverage

Table 1. *Patient Characteristics (N = 50)*

	AVM	AOVM
Females	10	13
Males	18	9
Mean age in years (range)	37 (7–76)	42 (14–70)
Mean time in years from diagnosis (range)	2 (1–12)	3 (1–15)
Number of prior bleeds	20	21
Mean no. bleeds per pt (range)	1 (1–4)	2 (2–5)
Signs and symptoms		
Headache	12 (43%)	9 (41%)
Neurological deficit	18 (64%)	19 (86%)
Prior embolization	2 (7%)	0
Prior operation	6 (21%)	4 (18%)
Karnofsky performance score		
100	10	3
90	7	6
80	2	4
70	5	7
60	4	2

of the AVM at the desired isodose. Twelve patients (43%) were treated at the 50%, 5 at the 55%, 4 at the 60%, 1 at the 70%, 4 at the 80% and 2 at the 90% isodose lines respectively. The minimum AVM margin dose ranged from 12 Gy to 25.6 Gy (mean: 20.6 Gy), and that to the center ranged from 22.2 to 50 Gy (mean: 35.7 Gy). The 18-mm collimator was used in 7 patients, the 14-mm collimator in 16, and the 8-mm collimator in 10.

Radiation Dosimetry for AOVMs (Table 2)

The treatment volumes ranged from 373 to 11,000 cu mm (mean: 3615 cu mm). Average AOVM diameter ranged from 9 to 22 mm (mean: 15 mm). All 22 patients had complete coverage of the AOVM at the desired isodose. Nine patients (41%) were treated at the 50%, 7 at the 60%, 4 at the 70% and 2 at the 90% isodose lines respectively. The minimum AOVM margin dose ranged from 13 Gy to 20 Gy (mean: 16.8 Gy), and that to the center ranged from 20 Gy to 40 Gy (mean: 28.6 Gy). The 18-mm collimator was used in 2 patients, the 14-mm collimator in 12, and the 8-mm collimator in 10.

Clinical and Radiologic Evaluation

No immediate postoperative complications occurred. Mild headache associated with the application of the stereotactic frame was managed with oral analgesics. All patients left the hospital within 24 hours of radiosurgery.

The mean follow-up period for all 50 patients was 25 months (range: 10 to 51 months). Headache was

improved in 6 patients, and neurologic deficits were improved in 7 patients. Of the 28 patients with AVMs, 8 had follow-up angiograms at a minimum of 2 years after radiosurgery. Of these, 7 (88%) showed complete obliteration of their malformations. Nine of 10 patients with transient or permanent complications related to radiosurgery exhibited increased signal on T2 MR images consistent with edema.

Rehemorrhage

Two (7%) of the 28 patients with AVMs rehemorrhaged in the follow-up period. One occurred 18 months after radiosurgery contributing to a mild increase in an oculomotor paresis. This patient's AVM was not obliterated at a 37-month angiogram and he has been rescheduled for repeat radiosurgery. A second patient sustained a rehemorrhage 3 months after radiosurgery resulting in a left facial paresis and poor upward gaze.

Two (9%) of the 22 patients with AOVMs rehemorrhaged in the follow-up period. One occurred 12 months after radiosurgery resulting in an increased left hemiparesis. This patient was evaluated at a 2-year office visit and was found to have his pre-radiosurgical baseline examination. A second patient sustained a rehemorrhage 6 months after radiosurgery resulting in a worsening of a hemiparesis and proprioceptive deficit. At a 40-month follow-up this patient had markedly improved from baseline. No patients with AOVMs rehemorrhaged if more than 15 months elapsed after radiosurgery.

Transient Deficits Related to Radiosurgery

One patient with a midbrain/pontine AVM developed a new hypesthesia and ataxia 24 months after radiosurgery. An angiogram showed complete obliteration of the malformation, and at a follow-up examination at 42 months the patient had no neurological deficits. A second patient with a pontine AOVM developed an increased ataxia and a Vth and VIth nerve paresis 6 months after radiosurgery. He had a normal neurological examination at 24 months. A third patient with a midbrain/pontine AOVM developed ataxia 4 months after radiosurgery. At her 17 month follow-up examination she had no neurologic deficits.

Persistent Deficits Related to Radiosurgery

Seven patients (14%) developed persistent deficits related to radiosurgery. Three patients had AVMs and

Table 2. *Radiation Dosimetry (N = 50)*

	AVMs (N = 28)	AOVMs (N = 22)
Mean avg. diam in cm (range)	18 (6–30.4)	15 (9–22)
Mean vol. in cu mm (range)	5400	3615
	(134–14,800)	(373–11,000)
Mean min margin dose (range)	20.6 (12–25.6)	16.8 (13–20)
Mean dose to center (range)	35.7 (22.2–50)	28.6 (20–40)
Pts treated at various isodose lines		
50%	12	9
55%	5	0
60%	4	7
70%	1	4
80%	4	0
90%	2	2
Number of various collimators used		
18 mm	7	2
14 mm	16	12
8 mm	10	10
4 mm	0	0

Table 3. *Permanent or Improving Complications Related to Radio-surgery (N = 7, 14%)*

Type	Avg. diam	Margin dose	Onset	Complication
AVM	14.3 mm	25 Gy	18 mos	ataxia, dysarthria
AVM	19.3 mm	20 Gy	12 mos	INO
AVM	11.6 mm	25 Gy	18 mos	bilat VI paresis downbeat nystagmus
AOVM	14.3 mm	16 Gy	7 mos	bilat VI, LV, RIII, ataxia, nystagmus
AOVM	15.3 mm	16 Gy	8 mos	dysarthria, ataxia
AOVM	15.7 mm	16 Gy	19 mos	worsened diplopia increased ataxia
AOVM	12.3 mm	16 Gy	9 mos	inc. ataxia, proprio-ceptive deficit

4 had AOVMs. Their symptoms and time of onset are listed in Table 3. Within 6 to 12 months of onset 4 patients had gradual improvement in their deficits, but were left with deficits not present before radiosurgery. One of the seven patients with a midbrain AVM died 35 months after radiosurgery (without new hemorrhage); histologic evaluation showed malformation obliteration but significant radiation effect at the target volume. The dose-volume prescription used was standard in that patient.

Discussion

Management of Brain Stem AVMs

We have shown that "inoperable" AVMs of the brain stem can undergo successful gamma knife stereotactic radiosurgery. Spetzler and Martin's grading system defines inoperable AVMs as those that are "large, diffuse and are dispersed through critical or neurologically eloquent areas, or [which have] a diffuse nidus that encompass critical structures such as the hypothalamus or brain stem". They state that surgical resection of these lesions "would almost unavoidably be associated with a totally disabling deficit or death"[17]. Seventy-nine percent of AVMs in the present series satisfied these criteria.

The natural history of unoperated AVMs has been scrutinized. Hemorrhage rates in *unruptured* malformations vary from 2.1 to 4% per year; the risk of death from rupture varies from 13–29% and long-term morbidity in survivors varies from 23% to 27%[1,2,21]. Hemorrhage from posterior fossa AVMs may occur more frequently and more often prove fatal than hemorrhage from other intracranial AVMs[5]. In those patients presenting with hemorrhage, the risk of rebleed in the first year ranges from 6 to 17% with a 1 to 4% per year risk of rehemorrhage thereafter. The mortality rate from hemorrhage is approximately 1% per year[6,7,15]. In our series of 28 patients who underwent stereotactic radiosurgery, the rebleed rate was 2 of 28 (7%) over an average follow-up of 25 months (or approximately 3.5% per year). This appears to be consistent with the expected natural history of non-obliterated arteriovenous malformations. No hemorrhage has occurred after documented obliteration.

Surgical resection of brain stem AVMs has been reported infrequently. Drake et al.[3], in their series of 66 posterior fossa AVMs, described 15 AVMs within the brain stem as "treacherous to treat". This subset of brain stem malformations had a 27% mortality. Nine of the 15 AVMs were unsuccessfully explored. Only 2 were excised safely. Other successes with brain stem AVMs have been reported as case reports only. Sugiura[18] reported successful obliteration of an AVM within the middle cerebellar peduncle, embedded in the floor of the fourth ventricle. Yonekawa et al.[22] described staged obliteration of an AVM partially embedded in the midbrain.

We have one of the largest reported series of brain stem AVMs. Our total obliteration rate was 88% in the eight patients who have reached the two-year post radiosurgery mark. This is similar to our experience in managing AVMs in all brain locations of similar size[12]. We expect a similar trend in the remaining 20 patients in the series. Permanent morbidity was seen in 3 of the 28 patients (11%). One patient died 35 months after radiosurgery of presumed radiation injury. Since 79% of the AVMs in our series were considered Spetzler and Martin grade VI (inoperable) our results compare favorably with the reports of successful surgical obliteration of those brain stem AVMs deemed "operable".

Management of Brain Stem AOVMs

Reports of successful excision of brain stem AOVMs are rare[10,11,14,19,23,24]. It has been suggested that the only safely "operable" AOVMS of the brain stem are those which come in contact with the pial surface. This directed the approach for 16 successfully treated cavernous malformations by Zimmerman et al.[24]. Kashiwagi et al.[8] reported 11 patients with brain stem AOVMs, 7 of whom underwent successful surgery. Yoshimoto and Suzuki[23] reported successful excision of cavernous angiomas within the brain stem in 3 of 4 patients. The fourth patient died.

In a comparison of radiosurgery versus microsurgery for cavernous malformations of the brain stem Weil *et al.*[20] reported a 50% morbidity in 6 patients managed with stereotactic radiosurgery, and a 33% rehemorrhage rate within a 2 year interval after radiosurgery. In contrast, all 7 patients managed microsurgically had total excision of their lesions with a 14% major morbidity rate. We have not had such an experience. Our results show a 9% rehemorrhage rate during a mean follow-up interval of 25 months (approximately 4.5% per year), and a persistent, but gradually improving complication incidence from delayed radiation effects of 18% (4 of 22). None of our patients suffered rehemorrhage if more than 15 months elapsed after radiosurgery.

The advantages of radiosurgery for brain stem AOVMs include: 1) a near zero risk of operative morbidity, 2) immediate return to full preoperative function after treatment, and 3) the ability to manage malformations deep within the brain stem parenchyma[9,13]. Current concerns related to AOVM radiosurgery include: 1) patient selection is difficult due to a poorly-defined natural history and variable clinical presentation, 2) the limits of the actual nidus are difficult to determine even with high-resolution MR imaging, 3) no current standardized imaging modality can confirm obliteration of an AOVM, and 4) the minimum effective and safe stereotactically administered radiation dose is unclear[9].

Dose-Volume Considerations for Brain Stem Malformations

The integrated logistic formula prediction for a 3% risk of permanent symptomatic radiation necrosis of the brain has been used by us in the past as a guideline for dose selection for radiosurgery of AVMs[4,12]. Figure 2 shows the relatively linear portion of the 3% risk line (upper solid line) superimposed on the scattergram plot of the dose-diameter relationships of our 50 patients in this present series. The squares represent patients with AVMs or AOVMs in this series who suffered radiation injuries which appear persistent. We are unable to demonstrate a clear dose-volume relationship to complications with this data.

Conclusions

1) "Inoperable" brain stem vascular malformations may be successfully managed with gamma knife radiosurgery with an acceptably low morbidity. 2) The rehemorrhage rate of unobliterated AVMs and AOVMs managed with radiosurgery is similar to their natural history (at least in the first 18 months). 3) Brain stem AVMs have an obliteration rate similar to AVMs elsewhere in the brain. 4) No hemorrhages have occurred in patients with AOVMs 15 months or more after radiosurgery. 5) The complication rate may be decreased by application of new dose prescription guidelines.

References

1. Brown Jr. RD, Wiebers DO, Forbes G, *et al* (1988) The natural history of unruptured intracranial arteriovenous malformations. J Neurosurg 68: 352–357
2. Crawford PM, West CR, Chadwick DW, *et al* (1986) Arteriovenous malformations of the brain: Natural history in unoperated patients. J Neurol Neurosurg Psychiatry 49: 1–10
3. Drake CG, Friedman AH, Peerless SJ (1986) Posterior fossa arteriovenous malformations. J Neurosurg 64: 1–10
4. Flickinger JC (1989) An integrated logistic formula for prediction of complications from radiosurgery. Int J Radiat Oncol Biol Phys 17: 879–885
5. Fults D, Kelly DL (1984) Natural history of arteriovenous malformations of the brain. Neurosurgery 15: 658–662
6. Graf CJ, Perret GE, Torner JC (1983) Bleeding from cerebral arteriovenous malformations as part of their natural history. J Neurosurg 58: 331–337
7. Itoyama Y, Uemura S, Ushio Y, *et al* (1980) Natural course of unoperated intracranial arteriovenous malformations: Study of 50 cases. J Neurosurg 71: 805–809
8. Kashiwagi S, van Loveren HR, Tew JM, *et al* (1990) Diagnosis and treatment of vascular brain-stem malformations. J Neurosurg 72: 27–34
9. Kondziolka D, Lunsford LD, Bissonnette DJ, *et al* (1990) Stereotactic radiosurgery of angiographically occult vascular malformations: Indications and preliminary experience. Neurosurgery 27: 892–900

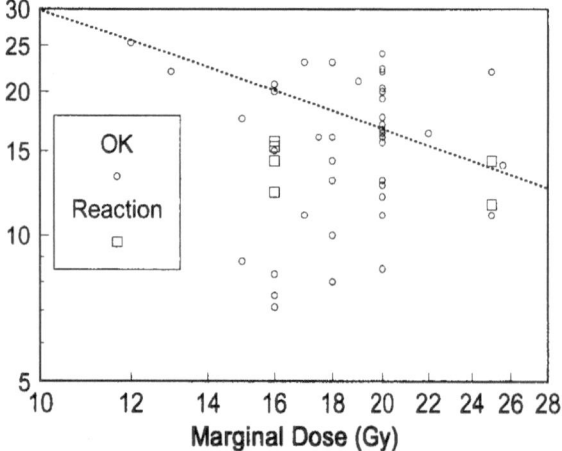

Fig. 2. Scatterplot of dose versus average diameter for 50 patients with AVMs and AOVMs. Squares represent permanent or improving neurologic deficits related to radiation injury. The upper solid line represents a relatively linear segment of the isoeffect curve for a 3% risk of permanent symptomatic radiation necrosis of the brain predicted by the integrated logistic equation for radiosurgery of AVMs. No cleare dose-volume relationship to complications was observed

10. LeDoux MS, Aronin PA, Odrezind GT (1991) Surgically treated cavernous angiomas of the brain stem: Report of two cases and review of the literature. Surg Neurol 35: 395–399
11. Lobato RD, Perez C, Rivas JJ, *et al* (1988) Clinical, radiological, and pathological spectrum of angiographically occult intracranial vascular malformations. Analysis of 21 cases and review of the literature. J Neurosurg 68: 518–531
12. Lunsford LD, Kondziolka D, Flickinger JC, *et al* (1991) Stereotactic radiosurgery for arteriovenous malformations of the brain. J Neurosurg 75: 512–524
13. Ogilvy CS (1990) Radiation therapy for arteriovenous malformations: A review. Neurosurgery 26: 725–735
14. Ondra SL, Doty JR, Mahla ME, *et al* (1988) Surgical excision of a cavernous hemangioma of the rostral brain stem: Case report. Neurosurgery 23: 490–493, 1988
15. Ondra SL, Troupp H, George ED, *et al* (1990) The natural history of symptomatic arteriovenous malformations of the brain: A 24-year follow-up assessment. J Neurosurg 73: 387–391
16. Solomon Ra, Stein B (1986) Management of arteriovenous malformations of the brain stem. J Neurosurg 64: 857–864
17. Spetzler RF, Martin N (1986) A proposed grading system for arteriovenous malformations. J Neurosurg 65: 476–483
18. Sugiura K, Baba M (1990) Total removal of an arteriovenous malformation embedded in the brain stem. Surg Neurol 34: 327–330
19. Wakai S, Ueda Y, Inoh, S, *et al* (1985) Angiographically occult angiomas: A report of thirteen cases with analysis of the cases documented in the literature. Neurosurgery 17: 549–556
20. Weil S, Tew JM, Steiner L (1990) Comparison of radiosurgery and microsurgery for treatment of cavernous malformations of the brainstem. J Neurosurg 72: 336A (Abstract)
21. Wilkins RH: Natural history of intracranial vascular malformations: A review. Neurosurgery 16: 421–430, 1985
22. Yonekawa Y, Handa H, Taki W (1983) Total removal of a brain stem arteriovenous malformation: Case report. Neurosurgery 13: 443–446
23. Yoshimoto T, Suzuki J (1986) Radical surgery on cavernous angiomas of the brain stem. Surg Neurol 26: 72–78m
24. Zimmerman RS, Spetzler RF, Lee KS, *et al* (1991) Cavernous malformations of the brain stem. J Neurosurg 75: 32–39

Correspondence: Christopher M. Duma, M.D., Apt. 501, 1550 Clarendon Blvd., Arlington, VA 22209, U.S.A.

Acta Neurochir (1993) [Suppl] 58: 98–100

Stereotactic Radiotherapy plus Radiosurgical Boost in the Treatment of Large Cerebral Arteriovenous Malformations

J. L. Barcia-Salorio[1,2], J. A. Barcia[1,2], F. Soler[3], G. Hernández[4], and J. M. Genovés[2]

[1] Servicio de Neurocirugía, Hospital Clínico Universitario, Valencia, [2] Departamento de Cirurgía, Universidad de Valencia, [3] Servicio de Radiodiagnóstico, Hospital Clínico Universitario, Valencia, and [4] Servicio de Terapéutica Física, Hospital Clínico Universitario, Valencia, Spain

Summary

Small sized AVMs respond well to stereotactic radiosurgery, while larger AVMs do poorly with stereotactic radiosurgery or stereotactic fractionated radiotherapy. A combination of both methods is proposed for the treatment of these larger lesions.

Keywords: Cerebral arteriovenous malformations; stereotactic radiosurgery; radiotherapy.

Introduction

Stereotactic radiosurgery is generally effective for the treatment of small (less than 30 mm in diameter) cerebral arteriovenous malformations, attaining 80% closure in most series. On the contrary, larger AVMs are considered nontreatable by stereotactic radiosurgery[3].

One of the strategies for the treatment of these AVMs is to reduce both their size and blood flow before the definitive treatment, regardless of whether it is open surgery or radiosurgery. This has been attempted with several techniques, including intravascular embolization[5].

Large AVMs have been treated by stereotactic fractionated radiotherapy alone, with discouraging results[1]. Nevertheless, in our experience with 42 patients, we have observed that some reduction in the size and the flow of these large AVMs can be achieved with this method, although not complete closure.

Based on their results with fractionated radiotherapy on these large AVMs, the authors have applied a combination of both methods (radiotherapy followed by a radiosurgical boost) to treat 13 AVMs.

Clinical Material and Methods

A total of 42 patients harbouring cerebral AVMs greater than 30 mm in diameter underwent fractionated radiotherapy. Total doses varied between 50 to 60 Gy in sessions of 2 Gy each. Of these, 17 AVMs showed a marked reduction to less than 30 mm. In 15 of these cases, a second single session of stereotactic radiosurgery was performed with doses ranging between 25 to 40 Gy. Two patients were lost to follow-up.

Results

Of the 13 cases irradiated and studied, six (46.15%) closed angiographically during the first year. Nine (69.2%) AVMs had closed after the second year. Two (15.3%) AVMs had almost closed at the end of the second year, with only an early filling vein remaining. Three patients (23.07%) rebled during the first year after irradiation, and another three showed transient seizures during the early postoperative period, now being symptom free. Nine patients (69.2%) showed preoperative neurological deficits (mainly contralateral spastic hemiparesia or hemianaesthesia); four of them improved postoperatively. No permanent worsening of the neurological status was recorded.

Figure 1 shows the pre-operative angiogram in one case, Fig. 2 the angiograms 1 year after fractionated radiotheraphy with 60 Gy, and Fig. 3 the angiograms 2 years after stereotactic radiosurgery with 40 Gy.

Discussion

Most of the reports on radiosurgery of AVMs state that its efficacy depends on the previous size of the

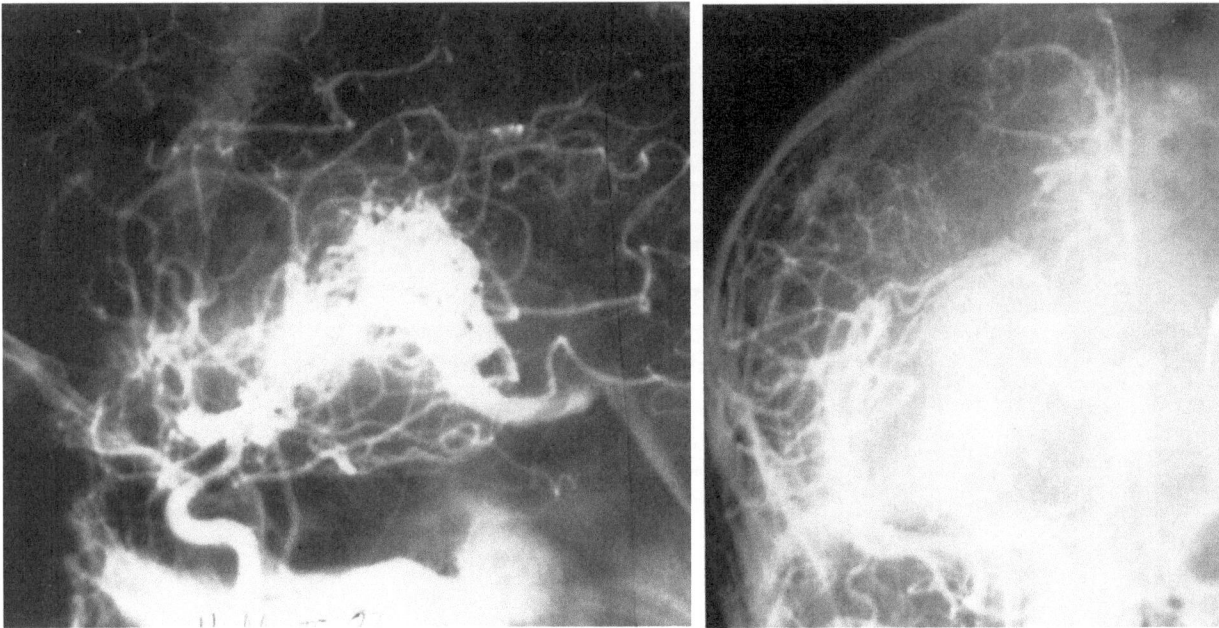

Fig. 1. Pre-operative angiograms in one of the cases

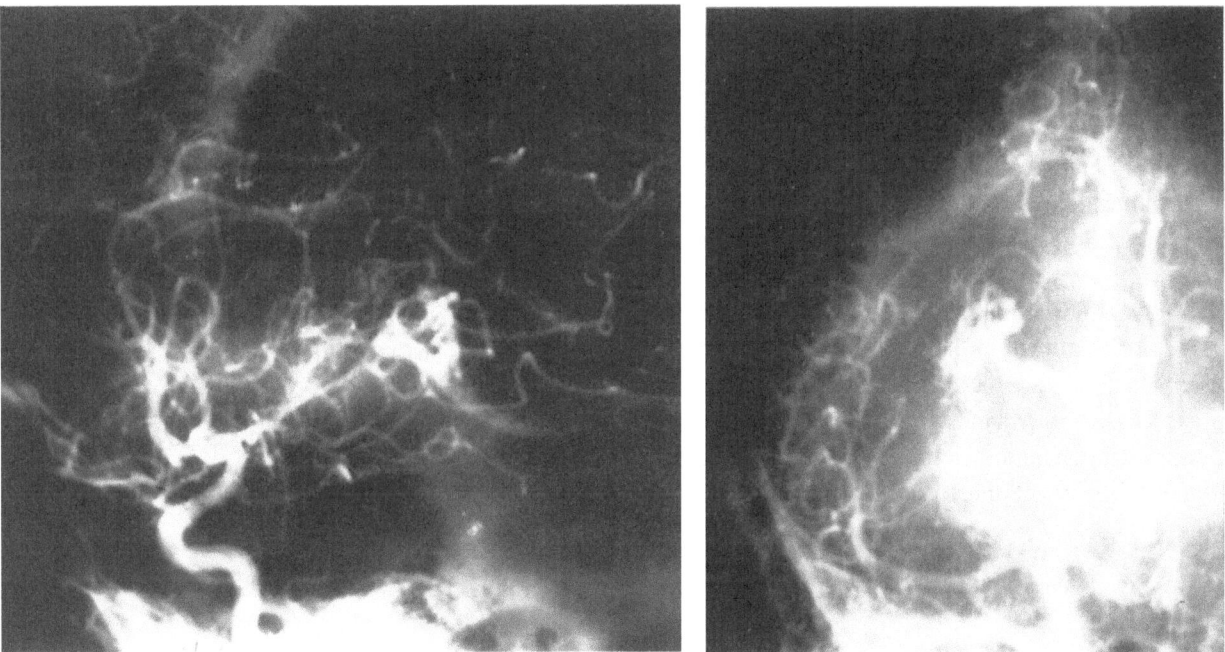

Fig. 2. Angiograms 1 year after fractionated radiotherapy with 60 Gy in the same case as Fig. 1

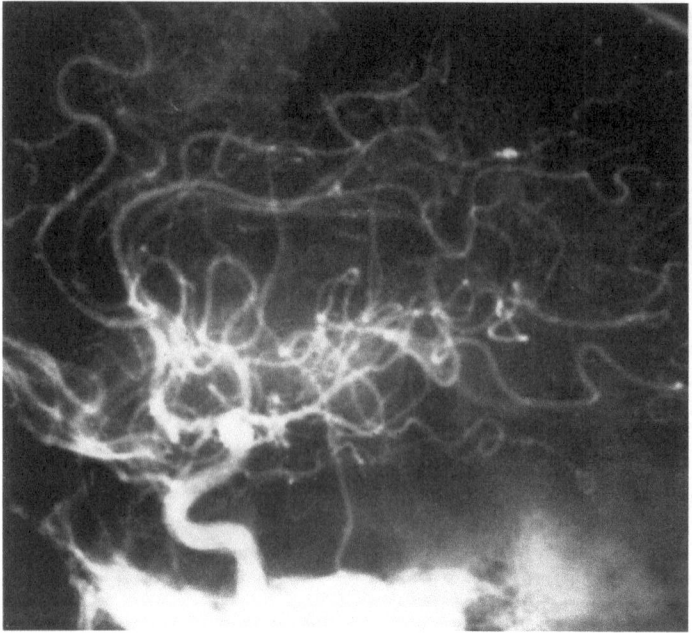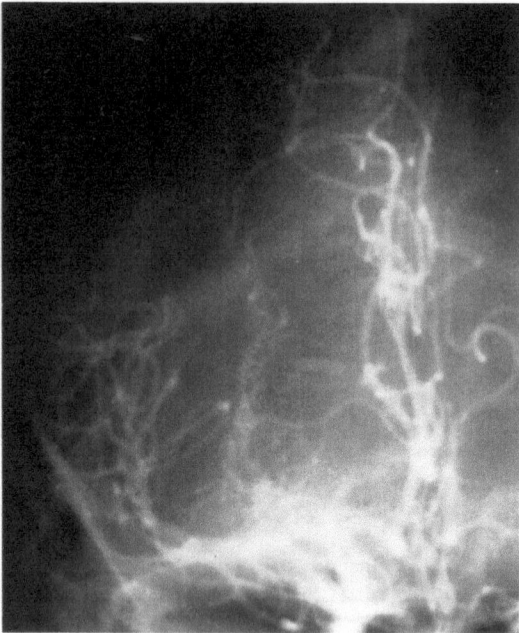

Fig. 3. Angiograms 2 years after stereotactic radiosurgery with 40 Gy in the same case as Fig. 1

AVM[2]. We have obtained the same results in our series.

The explanation for this fact may be the small size of irradiation volumes obtained by the collimators normally used in radiosurgery. Nevertheless, when larger collimators are used, there seems to be no relationship between the size of the AVMs and the results, within certain limits[4].

Kjellberg[2] stated that fractionated therapy is of no value for the treatment of large AVMs. This result was confirmed by Greitz *et al*.[1]. Nevertheless, we have found a significant reduction of the size and the flow of large AVMs in 17 of 42 cases (40.47%) who received fractionated radiotherapy. Of these, 84.5% of the cases were closed or almost closed by a further radiosurgical "boost".

Intravascular embolization is probably more effective as a prior treatment for large AVMs, but it cannot be used when they are located at functionally important areas. On the other hand, closure by radiation therapy develops slowly and therefore does not cause haemodynamic alterations due to sudden obliteration. These considerations make prior fractionated radiotherapy a rational alternative for the treatment of large AVMs in combination with radiosurgery.

References

1. Lindquist Ch, Steiner L (1988) Stereotactic radiosurgical treatment of arteriovenous malformations. In: Lunsford LD (ed) Modern stereotactic neurosurgery. Martinus Nijhoff, Boston, p 491
2. Greitz T, Lax, I, Bergstrom L, *et al* (1986) Stereotactic radiation therapy of intracranial lesions: Methodologic aspects. Acta Radiol (Stockh) 25: 81
3. Kjellberg RN, Davis KR, Lyons S, *et al* (1983) Bragg-peak proton beam therapy for arteriovenous malformation of the brain. Clin Neurosurg 31: 248
4. Sturm V, Kober B, Höver KH, *et al* (1987) Stereotactic percutaneous single dose irradiation of brain metastases with a linear accelerator. Int J Radiat Oncol Biol Phys 13: 279
5. Vinuela F, Dion JE, Fox AJ, *et al* (1990) Interventional neuroradiology for intracranial arteriovenous malformation. In: Barrow DL (ed) Intracranial vascular malformations. AANS Park Ridge, Illinois, p 169

Correspondence: J. A. Barcia, M.D., Servicio de Neurocirugía, Hospital Clínico Universitario, Av. Blasco Ibañez, 8, E-46010 Valencia, Spain.

Acta Neurochir (1993) [Suppl] 58: 101–103
© Springer-Verlag 1993

Radiosurgery of Cavernous Sinus Tumors

A. A. F. De Salles[1], C. L. Bajada[2], S. Goetsch[2], M. Selch[2], F. E. Holly[2], T. Solberg[2], and D. P. Becker[1]

[1] Division of Neurosurgery and [2] Department of Radiation Oncology, School of Medicine, University of California, Los Angeles, U.S.A.

Summary

Stereotactic radiosurgery was applied to concentrate high dose of photon irradiation to tumors enclosed in the cavernous sinus. We have treated 10 meningiomas, five pituitary tumors, and three metastatic lesions in the past two years. Follow-up time ranges from 3 to 18 months with an average of 10 months. Tumor volume ranged from 0.92 to 53.31 cc with an average 10.79 cc. CT scans and MRIs were used to demonstrate the tumor relationship to cranial nerves and structures of the brain. One to 6 isocenters were used. Collimator size varied from 7 to 28 mm, and the dose to the margin of the tumor ranged from 1400 to 2500 cGy with an average of 1650 cGy. Maximal dose range of 1575 to 5000 cGy. The margin of the tumor was encompassed within the 50 to 95% isodose volume with an average treatment prescribed to the 57% isodose volume. Symptomatic response was seen in 12 patients, and stabilization of symptoms in three patients with progression of symptoms observed in three patients. Radiographic imaging revealed response in eight patients, unchanged in three patients and progression in three patients. This report shows the feasibility of applying linear accelerator radiosurgery to the treatment of cavernous sinus tumors. This form of therapy promises to play an important role in complementing tumor resection.

Keywords: Stereotactic radiosurgery; cavernous sinus; tumors.

Introduction

Since the studies of the cavernous sinus surgical anatomy and the description of the surgical approach to lesions enclosed in this tight compartment, the results of resections of meningiomas, pituitary adenomas, and malignancies infiltrating the cavernous sinus have improved considerably[2,6,21]. Recent reported series of cavernous sinus tumor resections showed that aggressive removal of these tumors is possible, however, the increased neurological dysfunction after the surgery may be unacceptable[1,2,6,19]. Therefore, partial removal of the cavernous sinus tumor requires completion with adjuvant therapy[18].

Historically, conventional fractioned schemes of radiation have been the adjuvant form of therapy[3,4,23]. However, undue effects of radiation to the normal brain, and failure to control tumor recurrence have been the limitations of this therapy[10,11]. Radiosurgery is a natural candidate to complement microsurgical resections. Residual irregular tumors can be included in a high radiation volume with avoidance of important structures in close proximity to the tumor[9,12,13]. This report describes manipulation of linear accelerator based radiosurgery parameters, for example, arc weight, collimator size, and number of isocenters, to conform the radiation volume to the irregular residual tumors within the cavernous sinus.

Materials and Methods

Eighteen lesions and seventeen patients were included in this study. There were 11 females and six males. The ages range from 25 to 75 years old with a mean of 56. One patient had 2 tumors. All patients had previous operation for tumor resection or histologic diagnosis. There were ten meningiomas, five pituitary adenomas, and three metastatic squamous cell carcinomas. Patients were not admitted to the hospital. The mean tumor volume was 10.79 cc with a range of 0.92 cc to 53.31 cc. The dose to the periphery of the tumors varied from 1400 cGy to 2500 cGy, with a mean prescribed dose of 1650 cGy. The maximal dose to the tumor varied from 1575 to 5000 cGy with an average of 3590 cGy. The number of isocenters varied from 1 to 6. The average isodose volume of dose prescription was 57%, ranging from 50% to 95%. The irradiation dose to the optic nerve was kept below 800 cGy. The cranial nerves inside of the cavernous sinus, and fifth nerve usually received less than 2000 cGy.

Patients were treated with a Varian Clinac-18* adapted to radiosurgery by a Philips SRS200 radiosurgery system**[8]. Radiation arcs were restricted to avoid the eyes and air cavities. Relative dose from each arc was changed to avoid maximal dose to critical structures within or in close relationship to the tumor. Arcs were also designed

---

* Varian Associates, Palo Alto, CA, U.S.A.

** Philips Medical Systems, Shelton, U.K.

to avoid crossing important structures such as the brainstem or optic pathways. Computed tomography (CT) and magnetic resonance images (MRI) were obtained the same day of the treatment with a BRW stereotactic frame*** affixed to the patient's head. These scans underwent volumetric and multiplanar reconstruction for detailed analysis of the tumor-structures relationship to permit optimal dosimetry and target calculation. All patients returned to their usual activities the day after radiosurgery.

Results

Follow up ranged from 3 to 18 months with an average of 10 months. Improvement of the symptoms was observed in 12 patients, and the symptoms were unchanged in 2 patients, progression was observed in 3 patients. Radiological regression of the tumors was observed in eight patients, the lesions were unchanged in 2 patients, and progressed in 4 patients. Three patients did not have imaging follow-up. Tumor progression was noted through the cranial nerves passing by the cavernous sinus in two patients who had metastatic squamous cell carcinoma. A Nelson's tumor progressed in the direction of the optic pathway one year after radiosurgery, the radiosurgical dose was less than optimal close to the optic tract due to a history of previous radiation therapy. A patient with a left Meckel's cave meningioma invading the cavernous sinus presented persistent left trigeminal territory pain. An MRI performed 6 months after radiosurgery showed slight enlargement of the tumor. She underwent tumor resection, the tumor could be easily removed with the ultrasound aspirator. The histopathology disclosed areas of coagulative necrosis and coloid material accumulation in the vascular wall of the tumor vessels. The two patients with metastatic squamous cell carcinoma underwent conventional radiation therapy after progression of the disease.

Discussion

This study demonstrates that radiosurgery can be performed to irregular intracavernous sinus tumors. This technique serves as an adjuvant therapy for partial resections which are common, and probably should be more frequent to avoid severe neurological dysfunction after aggressive surgery[17,18,20]. Radiosurgery has been previously applied in the treatment of benign tumors[5,16]. However, only recently, after the development of radiosurgery techniques that rely upon multiplanar and volumetric reconstructions of CT and MRI, as well

as multiple isocenters, could irregular tumors or partially resected tumors be approached by radiosurgery[5,7,10,12,14]. Additional subtle improvement of dose distribution based on manipulation of arc weight and collimator size allows for avoidance of cranial nerves closely related to the tumor or even involved within the tumor[9].

Radiosurgery can be applied safely. Patients followed closely after radiosurgery can be promptly treated with surgical resection or additional conventional radiation therapy if necessary. Radiosurgery may in fact facilitate surgical resection of difficult tumors because of its ability to cause central tumor necrosis without changes in the tumor surrounding tissues. The necrotic central core of the tumor can be easily aspirated, facilitating the microsurgical detachment of the tumor from the important structures in close proximity to or encompassed by the tumor. This was demonstrated in one of our cases operated on after radiosurgery in which the tumor was within the cavernous sinus.

In conclusion, CT and MRI volumetric and multiplanar reconstructions allow for development of radiosurgical plan with selected number of isocenters, collimator size, arc restriction and selective arc weight that offer tailored dose distribution to irregularly shaped or partially resected tumors. Patients can have an effective and non aggressive therapy returning promptly to their activities.

References

1. Al-Mefty O, Fox JL, Smith RR (1988) Petrosal approach for petroclival meningiomas. Neurosurgery 22: 510–517
2. Al-Mefty O, Smith RR (1988) Surgery of tumors invading the cavernous sinus. Surg Neurol 30: 370–381
3. Barbaro NM, Gutin PH, Wilson CB, Sheline GE (1987) Radiation therapy in the treatment of partially resected meningiomas. Neurosurgery 20: 525–528
4. Carella RJ, Ransohoff J, Newall J (1982) Role of radiation therapy in the management of meningioma. Neurosurgery 10: 332–339
5. De Salles AAF, Asfora WT, Abe M, Kjellberg RN (1987) Transposition of target information from the magnetic resonance and computed tomography scan images to conventional x-ray stereotactic space. Appl Neurophysiol 50: 23–32
6. Dolenc V (1983) Direct microsurgical repair of intracavernous vascular lesions. J Neurosurg 58: 824–831
7. Flickinger JC, Lunsford LD, Coffey RJ (1991) Radiosurgery of acoustic neuromas. Cancer 67: 345–353
8. Friedman WA, Bova FJ (1989) The University of Florida radiosurgery system. Surg Neurol 32: 334–342
9. Goetsch SJ, De Salles AAF, Holly FE (1992) Treatment planning for stereotactic radiosurgery with linear accelerators. J Neurosurg (submitted)
10. Guy J. Mancuso A, Beck R (1991) Radiation-induced optic neuropathy: A magnetic resonance imaging study. J Neurosurg 74: 426–432

*** Radionics, Inc., Burlington, MA, U.S.A.

11. Harrison MJ, Wolfe DE, Lau T-S (1991) Radiation-induced meningiomas: Experience at the Mount Sinai Hospital and review of the literature. J Neurosurg 75: 564–574
12. Kondziolka D, Lunsford LD (1991) Stereotactic radiosurgery for squamous cell carcinoma of the nasopharynx. Laryngoscope 101: 519–522
13. Kondziolka D, Lunsford LD, Coffey RJ (1991) Stereotactic radiosurgery of meningiomas. J Neurosurg 74: 552–559
14. Kondziolka D, Lunsford LD, Flickinger JC (1991) The role of radiosurgery in the management of chordoma and chondrosarcoma of the cranial base. Neurosurgery 29: 38–46
15. Kooy HM, Nedzi LA, Loeffler JS (1991) Treatment planning for stereotactic radiosurgery of intracranial lesions. Intl Radiation Oncol Biol Phys 21:683–693
16. Leksell L (1971) A note on the treatment of acoustic neuromas. Acta Chir Scand 137: 763–765
17. Mayberg MR, Symon L (1986) Meningiomas of the clivus and apical petrous bone. Report of 35 cases. J Neurosurg 65: 160–167
18. Ojemann RG (1992) Skull-base surgery: A perspective. J Neurosurg 76: 569–570
19. Samii M, Ammirati M, Mahran A (1989) Surgery of petroclival meningiomas: Report of 24 cases. Neurosurgery 24: 12–17
20. Sekhar LN, Pomeranz S, Janecka IP (1992) Temporal bone neoplasms: A report on 20 surgically treated cases. J Neurosurg 76: 578–587
21. Sekhar LN, Sen CN, Jho HD, Janecka IP (1989) Surgical treatment of intracavernous neoplasms: A four-year experience. Neurosurgery 24: 18–30
22. Spetzler RF, Daspit CP, Pappas CTE (1992) The combined supra- and intratentorial approach for lesions of the petrous and clival regions: Experience with 46 cases. J Neurosurg 76: 588–599
23. Taylor BW, Marcus RB, Friedman WA (1988) The meningioma controversy: Postoperative radiation therapy. J Rad Oncol Biol Phys 15: 299–304

Correspondence: Antonio A. F. De Salles, M.D., Ph.D., Division of Neurosurgery, 300 U.C.L.A. Medical Plaza, Los Angeles, CA 90024, U.S.A.

Acta Neurochir (1993) [Suppl] 58: 104–107
© Springer-Verlag 1993

Gamma Knife Surgery in Acoustic Tumours

G. Norén[1], D. Greitz[2], A. Hirsch[4], and I. Lax[3]

Departments of [1]Neurosurgery, [2]Neuroradiology, and [3]Hospital Physics, Karolinska Hospital, Stockholm, and [4]Department of Audiology, Centrallasarettet, Västerås, Sweden

Summary

Presentation of the experiences with 254 acoustic neurinomas, treated at the Karolinska Gamma Knife Center from 1969 to 1991, with a minimum follow-up of 12 months.

Early loss of contrast enhancement on CT or MRI was seen in 70%. Unilateral tumours showed size decrease in 55%, no change in 33%, and increase in 12%. NF 2 tumours had decrease in 33%, no change in 43%, and increase in 24%. Some degree of facial weakness was seen after 17% of treatments, but always with later improvement of function. The incidence of trigeminal neuropathy was 19%. Preservation of hearing was 77%.

Gamma knife treatment is as efficient as microsurgery, but without risk of infection, bleeding or CSF leak. It requires no hospitalisation. The patient can go back to work after a few days. It therefore should be offered as an alternative to every acoustic neurinoma patient.

Keywords: Acoustic neurinoma; radiosurgery; gamma knife; results.

Introduction

If somebody had stated some twenty years ago that a technique involving irradiation would be used for the routine treatment of acoustic neurinomas that person would probably have been regarded as insane. Still Lars Leksell was successful in introducing Gamma Knife radiosurgery for the treatment of acoustic neurinomas in 1969[8]. In recent years a number of reports on this treatment has been published[2,3,9]. This presentation is based on the experiences gained at the Karolinska Gamma Knife Center in Stockholm, in the treatment of acoustic neurinoma patients during the period 1969 to 1991.

Clinical Material and Methods

Since June 1969, 254 acoustic neurinomas were treated and subsequently followed for a minimum period of 12 months (range 12–206,

mean 54). There were 193 procedures for unilateral tumours and 61 for neurofibromatosis 2 (NF2) neurinomas.

The target was located and defined by means of CT or MRI performed with a stereotactic frame attached to the patient's head. The planning of the treatment was supported by a computer programme. By this, the radiation dose to all parts of the tumour and the surrounding structures was defined with a high degree of accuracy.

In the Gamma Knife treatment, the converging gamma beams from 201 ^{60}Co sources were collimated to a focus giving rise to a volume of sharply defined radiation. This volume was made to coincide exactly with the target by positioning the patient's head by means of the stereotactic frame in the Gamma Knife according to the precalculated co-ordinates. The whole radiation was delivered in one session usually taking 15 to 25 minutes. The dose delivered to the periphery of the tumour was previously usually 18 to 20 Gy but was reduced during the last three years of this study to 10 to 15 Gy with a maximum of 15 to 25 Gy at the centre. The patient was usually discharged from the hospital a few hours after treatment.

Aim of Treatment

It is important to define the goal of the treatment with Gamma Knife radiosurgery as compared to microsurgery, since the tumour is clearly not removed.

DiTullio and co-workers have given the following definition of the "ideal" result of microsurgery: total tumor removal, normal facial nerve activity, and no complications[1]. Depending on the approach, "preservation of hearing" could be added to this list.

Our own proposed corresponding definition of the "ideal" result of radiosurgery is: no further growth, normal facial nerve activity, preservation of hearing, and no complications[10].

Result

Tumour Size

An early finding on CT after Gamma Knife surgery, seen in 70% of the cases, was loss of contrast enhancement starting 6 to 12 months after the treatment (Fig. 1). A corresponding decrease of signal was seen on MRI with approximately the same frequency.

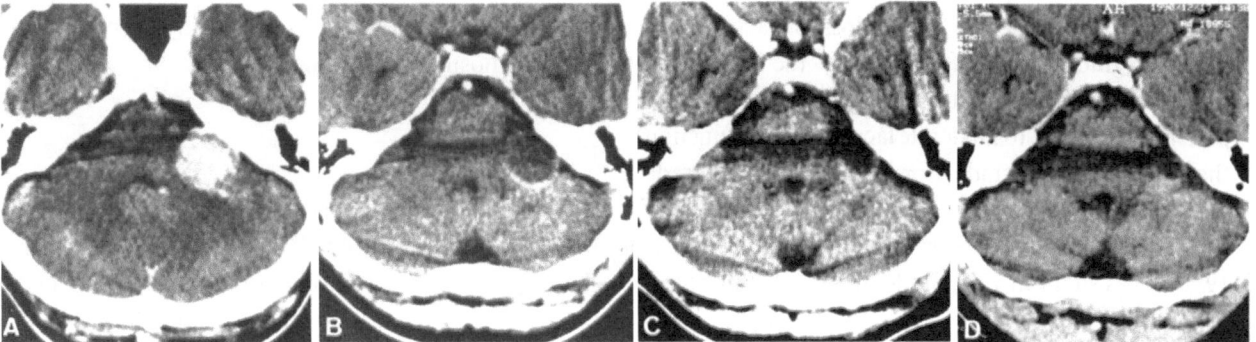

Fig. 1. Acoustic neurinoma with 2.3 cm maximum intracranial diameter at the time of radiosurgery (A), 6 months (B), 11 months (C), and 30 months (D) after the treatment. Note the loss of contrast enhancement and the ensuing very evident shrinkage

The unilateral tumours showed decrease of size in 55%, no change in 33%, and increase in 12%. Half of the tumours in this latter group (6%) had a temporary swelling with increase followed by decrease at different stages of follow-up but usually during the period 12 to 18 months after the radiosurgery. If adding the tumours with decrease to those with no change and temporary swelling the response rate could be defined as 94%. The situation concerning change of size of the NF2 tumours was somewhat different with decrease in 33%, no change in 43%, and increase in 24%. Of these latter, one third (8%) showed temporary swelling. The percentage of NF2 tumours showing response, defined similarly, was 84%.

The percentage of tumours showing shrinkage at different time intervals in the first 10 years is shown in Fig. 2.

The six percent non-responding unilateral tumours, in the majority of cases, showed the signs of continued growth two to three and always within five years after the treatment. There were no recurrences beyond that point in this group.

The situation for the non-responding 16% NF2 tumours was more complex with recurrences sometimes as late as ten years after the treatment or even later.

Facial and Trigeminal Nerve Function

Some degree of facial weakness was seen after 17% of the treatments. It was slight in 6% (House-Brackmann

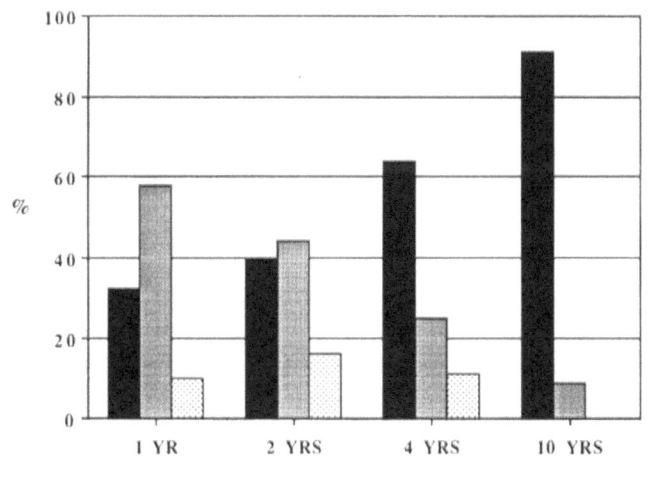

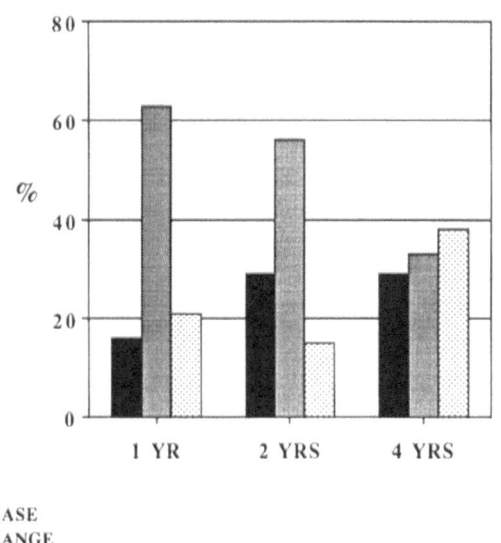

■ DECREASE
▨ NO CHANGE
▨ INCREASE

Fig. 2. Change of tumour size at different time intervals after radiosurgery. The percentage of tumours with decrease, no change, and increase within each group is shown. The percentage of unilateral acoustic neurinomas (left) showing shrinkage increases gradually over the years. The response of the NF2 tumours (right) is less evident and also less permanent. The 10 years follow-up observations in this group were too few to be included

grade 2–3) and moderate in 7% (grade 4)[5]. It was always temporary but in 4% of severe loss of facial nerve function (grade 5 to 6), after recovery, the function in this group of patients returned to grade 2 or 3. In the patients with slight to moderate facial nerve dysfunction the final result was grade 1 or in some patients grade 2 after recovery.

The total incidence of trigeminal neuropathy was 19%. It was slight in 12%, moderate in 3%, and severe in 4%.

Most of the cases with trigeminal and/or facial neuropathy occurred in two periods, namely in 1975 and 1988 in connection with new Gamma Knife installations at the Karolinska Hospital. The new units had different radiophysical properties as compared to the old ones. The number of facial as well as trigeminal nerve involvements decreased when this was realised, further experience was gained, and measures were taken accordingly.

In 1990, one patient out of 39 had a slight facial weakness lasting for a few weeks giving a total incidence of facial nerve involvement of 2.5% in that year. In 1991, again one case of slight temporary facial weakness occurred in 55 treated tumours, giving an incidence of 1.8%. No trigeminal neuropathy induced by the treatment occurred in these years.

Hearing

In patients (132 tumours) with a hearing loss of less than 90 dB pure tone average before the radiosurgery, the hearing was unchanged within narrow limits in 22%, slightly to moderately impaired in 55%, and severely impaired or lost in 23% one year after the treatment[4]. This corresponded to a percentage of ears with preserved hearing of 77%.

Oedema and CSF Circulation

A reaction around the tumor usually as a thin zone of cerebellar peritumoral oedema occurred in 8% of the cases. Clinically these patients usually complained of a slight to moderate temporary balance disturbance reaching a peak 6 to 12 months after the treatment. The oedema was in 4 patients or 1.5% so pronounced that it led to a CSF circulation disturbance necessitating a shunting procedure[12].

A more important factor in producing CSF circulation problems was the sometimes very pronounced increase of the protein concentration caused by the neurinoma itself. Because of this a shunt operation had to be performed in 6% before and 3% after the Gamma Knife procedure. The need of a CSF diversion procedure in these patients was consequently not correlated to the radiosurgical treatment.

Discussion

This study shows that acoustic neurinomas respond to single dose gamma irradiation and to a lower dose than that necessary to obliterate arterio-venous malformations, minimum 20–25 Gy[11], and to eradicate cerebral metastases, at least 25–30 Gy[6]. Neurinomas may consequently be more radiosensitive than generally assumed.

No growth over 1–2 years or even more may sometimes reflect the spontaneous course of an acoustic neurinoma. A study by Laasonen and Troupp[7] shows that growth occurs in about 90% of untreated tumours if followed for a period of 12–24 months. No growth of a radiosurgically treated tumour 12 and even more so 24 months after the procedure would consequently with a high degree of probability reflect a response to the Gamma Knife treatment.

An acoustic neurinoma that does not grow or is not extremely large is never a threat to the patient's health or life except on those rare occasions when the tumour has caused hydrocephalus, usually by increasing the CSF protein concentration. A permanent arrest of growth is consequently a fully sufficient goal of the Gamma Knife treatment. As shown in this study, a majority of the tumours, after some time, shows signs of shrinkage.

Conclusions

Gamma Knife surgery has, since the mid-seventies, gradually replaced microsurgery as the main modality for the treatment of acoustic neurinomas at our centre. It is offered to all patients with a tumour size of less than approximately 4 cm.

The method is as efficient as microsurgery without the risks of infection, bleeding, CSF leak, or mental depression. The patient can leave the hospital a few hours after the treatment and go back to work within one or two days. Gamma Knife surgery should be offered every acoustic neurinoma patient as an alternative to microsurgery.

References

1. DiTullio MV, Malkasian D, Rand RW (1978) A critical comparison of neurosurgical and otolaryngological approaches to acoustic neuromas. J Neurosurg 48: 1–12

2. Flickinger JC, Lunsford LD, Coffey RJ, Linskey ME, Bissonette DJ, Maitz AH, Kondziolka D (1991) Radiosurgery of acoustic neurinomas. Cancer 67: 345–353

3. Ganz JC, Myrseth E, Thorsen F, Backlund EO (1992) Acoustic schwannoma: early results of radiosurgical treatment. In: Tos M, Thomsen J (eds) Acoustic neuroma. Kugler, Amsterdam, pp 301–304

4. Hirsch A, Norén G (1992) Audiological evaluation after stereotactic radiosurgery in acoustic neurinomas. In: Tos M, Thomsen J (eds) Acoustic neuroma. Kugler, Amsterdam, pp 293–295

5. House WF, Brackmann DE (1985) Facial nerve grading system. Otolaryngol Head Neck Surg 93: 184–193

6. Kihlström L, Karlsson B, Lindquist C, Norén G, Rähn T (1991) Gamma knife surgery for cerebral metastasis. Acta Neurochir (Wien) [Suppl] 52: 87–89

7. Laasonen EM, Troupp H (1986) Volume growth rate of acoustic neurinomas. Neuroradiology 28: 203–207

8. Leksell L (1971) A note on the treatment of acoustic tumours. Acta Chir Scand 137: 763–765

9. Norén G, Greitz D, Hirsch A, Lax I (1992) Gamma knife radiosurgery in acoustic neurinoma. In: Steiner L (ed) Radiosurgery: Baseline and trends. Raven, New York, pp 141–148

10. Norén G, Greitz D, Hirsch A, Lax I (1992) Gamma knife radiosurgery in acoustic neurinomas. In: Tos M, Thomsen J (eds) Acoustic neuroma. Kugler, Amsterdam, pp 289–292

11. Steiner L (1988) Stereotactic radiosurgery with the cobalt 60 gamma unit in the surgical treatment of intracranial tumors and arteriovenous malformations. In: Schmidek HH, Sweet WH (eds) Operative neurosurgical techniques, Vol 1. Grune and Stratton, Orlando, pp 515–529

12. Thomsen J, Tos M, Børgesen SE (1990) Gamma knife: Hydrocephalus as a complication of stereotactic radiosurgical treatment of an acoustic neuroma. Am J Otol 11: 330–333

Correspondence: Georg Norén, M.D., Ph.D., Department of Neurosurgery, Karolinska Hospital, S-10401 Stockholm, Sweden.

Acta Neurochir (1993) [Suppl] 58: 108–111

Interstitial Irradiation of Cerebral Gliomas with Stereotactically Implanted Iodine-125 Seeds

J. Voges[1], **H. Treuer**[1], **W. Schlegel**[2], **O. Pastyr**[2], and **V. Sturm**[1]

[1] Department of Stereotactic and Functional Neurosurgery, University of Köln, Köln and [2] German Cancer Research Center, Heidelberg, Federal Republic of Germany

Summary

Ninetyseven consecutive patients with primarily inoperable or only partially resectable cerebral gliomas have been analysed retrospectively. Mean tumour surface doses of 70 Gy (low grade gliomas) and 56 Gy (high grade gliomas) have been applied with stereotactically implanted Iodine-125 seeds at low dose rates. Patients with a glioma grade III or grade IV and permanent seed implantation additionally received a fractionated external beam irradiation. With mean follow-up times of 55.8 months (glioma grade I), 51 months (glioma grade II) and 59.6 months (glioma grade III) the estimated mean survival probabilities are 105 months, 102 months and 65.7 months respectively. In the glioma grade IV group the estimated mean survival time has been 15.6 months after continuous interstitial irradiation (response rate: 36%). Temporary interstitial irradiation in cases with a glioma grade IV (dose rate: 2.1 Gy/day) caused initial tumour shrinkage in 77%. Neurological deficits following radiation induced vasogenic oedema were reversible in 2 patients and irreversible in another 2 patients. 6 years after the Iodine-125 implantation and continuous interstitial irradiation 1 patient developed a severe localised radiation necrosis.

Keywords: Glioma; interstitial irradiation; radiotherapy; stereotactic technique.

Introduction

The local cure of cerebral gliomas by surgery alone often is precluded by the invasion of tumour cells into surrounding, macroscopic healthy brain tissue. Considering, that most gliomas do not metastasize within the central nervous system[2] and grow in a circumscribed area of the brain[3], alternative treatment modalities are logical. To achieve local control with optimal sparing of the peritumoural healthy tissue, interstitial irradiation with implanted isotopes was established during the past 30 years[7,14].

On the base of a stereotactic device, originally described by Riechert and Mundinger[10] our group has developed from 1979 onwards an integrated CT-guided and CT-based computerised stereotactic treatment planning system for interstitial irradiation with stereotactically implanted Iodine-125 seeds[1,11–13]. 110 out of 220 patients with inoperable or partially extirpated cerebral gliomas, treated with this technique from 1981 through 1992 were selected for a retrospective analysis.

Methods and Material

After fixation of the patient's head in a modified, CT-compatible Riechert-Mundinger stereotactic frame[13] and administration of contrast medium (Solutrast 300[R]) a CT-examination was performed. CT-data were transfered by magnetic tape into a computer (VAX 11/700 Computer, VAX station VS 3500, Digital Equip. Corp., USA) and the tumour borders were outlined manually at the computer screen. Determination of target points (tip of the Iodine-125 seeds) as well as the treatment planning for interstitial irradiation were carried out with a special software described elsewhere[1,11,12].

110 consecutive patients with cerebral glioma, treated since October 1981 with stereotactic Iodine-125 seed implantation and continuous or temporary interstitial irradiation, have been selected for a retrospective analysis. At the end of the evaluation (March 1992) the follow-up times should be at least 2 years (glioma grade I–III) or 18 months (glioma grade IV) respectively. The lesions of these patients were sufficiently demarcated in contrast enhanced CT scans, the maximum diameter not exceeding 5 cm. The histopathological diagnoses were established before seed implantation either by stereotactic biopsy or partial microneurosurgical resection. From 1981 through 1987 the material biopsied stereotactically in our department was graded according to the Kernohan classification[6]. Since 1988 the WHO grading system has been used[15]. 4 patients were lost to follow-up. 9 patients with a modified treatment schedule (glioma grade I–III: temporary seed implantation, glioma grade I/II: combined interstitial and external irradiation) were excluded.

The treatment parameters are listed in Table 1. Patients with high grade gliomas and permanent seed implantation additionally received external fractionated beam irradiation (boost dose 10–40 Gy) during the first three weeks postoperatively in order to increase the dose rate to values, equivalent to those achieved with conventional radio-

Table 1. *Characteristics of 97 Patients Treated with Interstitial Irradiation*

	Grade I	Grade II	Grade III	Grade IV
Localization (no. pat.)				
—Hemispheres	4	23	12	15
—Dienceph./basal ganglia	10	5	11	11
—Brain stem	1	—	3	1
—Cerebellum	1	—	—	—
Histology (no. pat.)				
—Astrocytoma	16	22	23	5
—Oligoastrocytoma	—	4	3	—
—Oligodendroglioma	—	2	—	—
—Glioblastoma	—	—	—	22
Treatment				
Permanent seed implant.:				
—Tumour dose	60–110 Gy (mean: 70 Gy)		40–70 Gy (mean: 56 Gy)	
—Mean dose rate	0.8 Gy/day		0.6 Gy/day	
Temporary seed implant.:				
—Mean tumour dose		—	—	60 Gy
—Mean dose rate		—	—	2.1 Gy/day

therapy. The tumours of 13 patients (grade IV gliomas located in the cerebral hemispheres) were treated with temporary interstitial irradiation (42–120 days)

3, 6 and 9 months postoperatively and later on at yearly intervals, CT-scans without and with contrast-medium as well as neurological examinations have been carried out. Survival was defined as the time interval between seed implantation and the end of the retrospective analysis (March 1992) or the patient's death.

Results

97 evaluated patients (40 females and 57 males) ranged in age at the time of seed implantation from 1 to 76 years with a mean age of 31 years. The tumour locations are listed in Table 1.

The tumour volumes, measured at the day of seed implantation, ranged from 0.9 to 150 cc (mean value: 24.8 cc). 3, 6 and 9 months postoperatively, the tumour volumes were reassesed on CT-scans. The patients were classified into 3 groups: responders (volume reduction), stabilized disease (no volume reduction) and nonresponders (uninhibited tumour growth). The response rates according to this classification are summarized in Table 2.

Figures 1 and 2 show survival curves estimated by the method of Kaplan-Meier[4]. With a mean follow-up of 55.8 months and 51 months the estimated mean survival probability is 105 months for patients with a glioma grade I and 102 months for those with a glioma grade II respectively. Patients with a glioma grade III have an estimated mean survival probability of 65.7 months (mean follow up: 59.6 months). After combined

Table 2. *Response to Interstitial Irradiation 3–9 Months Postoperatively*

	Responder (no. pat.)	Stabilized disease (no. pat.)	Nonresponder (no. pat.)
Grade I	16/16 (100%)	—	—
Grade II	24/28 (86%)	1/28 (3%)	3/28 (11%)
Grade III	20/26 (77%)	2/26 (8%)	4/26 (15%)
Grade IV (perm.)	5/14 (36%)	2/14 (14%)	7/14 (50%)
Grade IV (temp.)	10/13 (77%)	1/13 (8%)	2/13 (15%)

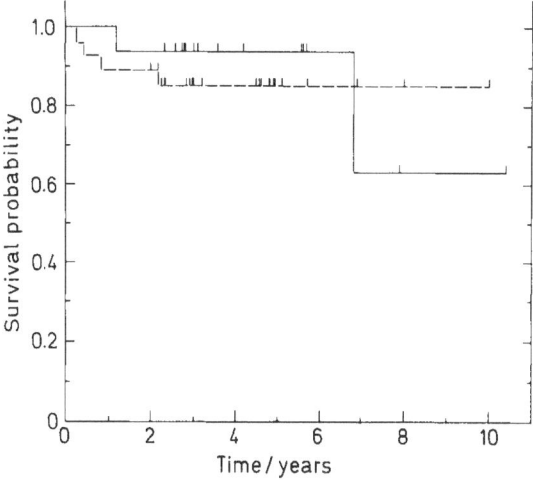

Fig. 1. Kaplan-Meier representation of the probability of survival after Iodine-125 implantation. Solid line: Patients with a pilocytic astrocytoma grade I (16 cases). Broken line: Patients with a glioma grade II (28 cases). *Ticks*: Censored patients

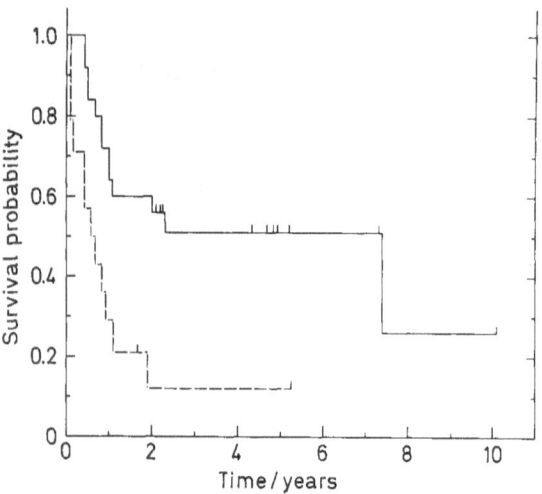

Fig. 2. Kaplan-Meier representation of the probability of survival after Iodine-125 implantation. Solid line: Patients with a glioma grade III (26 cases). Broken line: Patients with a glioma grade IV (14 cases, continuous interstitial irradiation). *Ticks*: Censored patients

continuous interstitial and fractionated external irradiation of a glioma grade IV the mean survival time is 15.6 months. The mean relapse free survival probabilities, estimated only for those patients, who responded to therapy, are 90.2 months (grade I), 89.9 months (grade II) and 79.7 months (grade III) respectively.

There was neither operative or peri-operative morbidity nor mortality. Pre-operatively the Karnofsky performance status[5] of all patients ranged from 50 to 100% (mean: 84.4%). The Karnofsky scale was reassesed 9 months postoperatively. No patient, who responded to therapy or had a stabilized disease as documented by CT-examinations showed clinical worsening at this time. 9 to 24 months after seed implantation in 17 patients with low grade gliomas and in 15 patients with a glioma grade III transient radiation induced vasogenic oedema was evident on CT-scans. Concommittantly in 2 of these patients the Karnofsky performance status decreased reversibly. In another 2 patients the worsening of the clinical status has been irreversible. 6 years after treatment one other patient (brain stem astrocytoma, grade III) developed a severe radiation necrosis with a progressive midbrain syndrome.

Discussion

With mean follow-up times of 55.8 months and 51 months in this study the mean estimated survival probability is 105 months for patients with an astrocytoma grade I and 102 months for patients with a grade II glioma. These results are roughly comparable to the data of Ostertag *et al.*, who reported a 5-years survival rate of 95% for pilocytic astrocytomas (28 patient) and 78% for patients with an astrocytoma grade II (69 cases)[9]. Mundinger and Weigel estimated a 5-years survival rate of 55% for pilocytic astrocytomas and of 23% for astrocytomas grade II, following stereotactic Iodine-125 implantation[8].

In our series in 85% of the patients with a glioma grade III and in 50% of the patients with a glioma grade IV, the combination of continuous interstitial and fractionated external irradiation caused local tumour control in cases, where cytoreductive surgery and conventional radiotherapy have not been feasible. Tumour control has been achieved in 85% of patients with a grade IV glioma, who had been treated with reduced implantation times and thus with higher dose rates. Seed implantation as well as interstitial irradiation have been well tolerated in our series (no operative or peri-operative mortality nor morbidity). Radiation induced tumour regression improved the Karnofsky performance status in the majority of the patients by 5–15%. Severe radiation necrosis, which occured in one patient (malignant astrocytoma, midbrain) 6 years postoperatively, was most probably caused by the relatively high dose (Iodine-125: 70 Gy, boost dose: 20 Gy) given to this region.

The following facts make the interpretation of our favourable results difficult: The study is retrospective without a control group. Most of the patients were young, i.e. in a good general condition and showed a relatively high Karnofsky score. The latter could have influenced our results positively. On the other hand, the location of the tumours in functionally important areas of the brain must be considered unfavourable with regard to prognosis.

Despite the possible bias due to the above mentioned drawbacks of this study our data show the efficacy and benignity of low dose rate interstitial irradiation in circumscribed deep sited cerebral gliomas. An indispensable prerequisite is the use of modern stereotactic treatment-planning and implantation techniques, which enable the application of any radiation dose to an intracranial target volume with the highest precision and an optimal sparing of the surrounding tissue[1,11–13].

References

1. Bauer-Kirpes B, Sturm V, Schlegel W, Lorenz WJ (1988) Computerized optimization of 125-I implants in brain tumors. Int J Radiat Oncol Biol Phys 14: 1013–1023

2. Ehrlich SS, Davis RL (1978) Spinal subarachnoid metastasis from primary intracranial glioblastoma multiforme. Cancer 42: 2854–2864

3. Hochberg FA, Pruitt A (1980) Assumptions in the radiotherapy of glioblastomas. Neurology 30: 907–911

4. Kaplan EL, Meier P (1958) Nonparametric estimation from incomplete observations. J Am Statist Assoc 53: 457–481

5. Karnofsky DA, Abelmann WH, Carver LF, Burchenal JH (1948) The use of nitrogen mustards in the palliative treatment of carcinoma with particular reference to bronchogenic carcinoma. Carcinoma: 634–656

6. Kernohan JW, Mabon RF, Svien HJ, Adson AW (1949) A simplified classification of the gliomas. Proc Mayo Clin 24: 71–75

7. Mundinger F (1966) The treatment of brain tumors with radioisotopes. Progr Neurol Surg 1: 220–257

8. Mundinger F, Weigl K (1988) Considerations in the usage and results of Curietherapy. In: Lunsford LD (ed) Modern stereotactic neurosurgery. Martinus Nijhoff, Boston, pp 245–258

9. Ostertag CB (1989) Stereotactic interstitial radiotherapy for brain tumors. J Neurosurg Sci 33: 83–89

10. Riechert T, Mundinger F (1955) Beschreibung und Anwendung eines Zielgerätes für stereotaktische Hirnoperationen. Acta Neurochir (Wien) [Suppl] 3: 308–337

11. Schlegel WJ, Scharfenberg H, Sturm V, Penzholz H, Lorenz WJ (1981) Direct visualization of intracranial tumours in stereotactic and angiographic films by computer calculation of longitudinal CT-sections: A new method for stereotactic localization of tumour outlines. Acta Neurochir (Wien) 58: 27–35

12. Schlegel W, Scharfenberg H, Doll J, Hartmann G, Sturm V, Lorenz WJ (1984) Three dimensional dose planning using tomographic data. In: IEEE Comp Society (eds) Proc of the 8th Int. Conf. on the Use of Computers in Radiation Therapy, Silver Spring, IEEE Comp. Soc. Press., pp 191–196

13. Sturm V, Pastyr O, Schlegel W, Scharfenberg H, Zabel HJ, Netzeband G, Schabbert S, Berberich W (1983) Stereotactic computer tomography with a modified Riechert Mundinger device as the basis for integrated stereotactic neuroradiological investigations. Acta Neurochir (Wien) 68: 11–17

14. Szikla G, Peragut JC (1984) Irradiation interstitielle des gliomes. Neurochirurgie 21 [Suppl 2]: 187–228

15. Zülch KJ (1979) Histological typing of tumors of the central nervous system. WHO, Geneva

Correspondence: J. Voges, M.D., Department of Stereotactic and Functional Neurosurgery, University of Köln, Josef-Stelzmann-Strasse 9, D-50931 Köln, Federal Republic of Germany.

Acta Neurochir (1993) [Suppl] 58: 112–114

Interstitial Implant Radiosurgery for Cerebral Metastases

F. W. Kreth, P. C. Warnke, and **Ch. B. Ostertag**

Abteilung Stereotaktische Neurochirurgie, Neurochirurgische Universitätsklinik, Freiburg, Federal Republic of Germany

Summary

The effectiveness of interstitial implant radiosurgery (IRS) as an alternative or adjuvant treatment to radiotherapy (WBRT) or surgery of cerebral metastases remains unclear.

In a retrospective study (1982–1991) we analysed four therapeutic regimes after stereotactic biopsy: IRS with a tumour dose of 60 Gy in combination with WBRT (40 Gy/5 × 2 Gy/week — 38 patients), IRS only (tumour dose 60 Gy — 22 patients), WBRT only (40 Gy/5 × 2 Gy/week — 49 patients), and IRS only for recurrent cerebral metastases (tumour dose 60 Gy — 21 patients). Low-activity iodine-125 seeds were used exclusively. IRS was performed in the case of circumscribed, mostly solitary metastases ≤ 4 cm in diameter.

Patients undergoing combined treatment had the best survival, with a median survival time of 17 months in comparison with 12 months after IRS alone and 7.7 months after WBRT. The median survival of patients with recurrent metastases after IRS was 6 months. A comparison of treatments in the multivariate analysis showed that IRS + WBRT was not superior to IRS alone. The metastases could be locally controlled in every case. There were no radionecroses requiring treatment. Most favourable determinants after IRS or IRS + WBRT were a solitary metastasis and a long time interval between diagnosis of the primary and diagnosis of the cerebral metastases.

Our results demonstrate the effectiveness of IRS. For a single cerebral metastasis, IRS as a minimally invasive method offers major advantages.

Keywords: Cerebral metastases; interstitial implant radiosurgery; radiotherapy; radiosurgery.

Introduction

The value of radiosurgical methods (linear accelerator radiosurgery, interstitial implant radiosurgery) in the treatment of cerebral metastases alone or in combination with percutaneous radiotherapy (WBRT) has yet to be determined. In a retrospective study carried out between 1982 and 1991 the effectiveness of interstitial implant radiosurgery (IRS) and WBRT, either alone or in combination, was analysed.

Material and Methods

Between 1982 and 1991, the diagnosis of cerebral metastasis was confirmed by CT-guided biopsy in 410 patients. All patients after IRS or IRS + WBRT and all patients with available complete follow up after WBRT and an identical radiotherapeutic regime were included in this study. Patients with supportive therapy only were not analysed. The following treatment regimes were completed:

1. IRS alone with a tumour dose of 60 Gy (22 patients);
2. IRS and WBRT with a local tumour dose of 60 Gy and subsequent radiotherapy with 40 Gy (5 × 2 Gy/week) (38 patients);
3. WBRT alone for multiple or poorly circumscribed metastases with a diameter of > 4 cm (49 patients) (40 Gy/5 × 2 Gy/week);
4. IRS alone as treatment for recurrent disease confirmed by biopsy after previous operation and WBRT (tumour dose 60 Gy) (21 patients). IRS or IRS + WBRT were performed for patients with circumscribed mostly solitary metastases with a diameter ≤ 4 cm. With the increase in clinical experience and the regular use of improved imaging techniques (MRI), since 1988 there has been a growing tendency to perform IRS alone.

Sources and Dosimetry

Low-activity iodine-125 seeds* were used exclusively. IRS was carried out using temporary implants. For clinical use the calculated dose ranged between 60 Gy and 100 Gy (Table 1). This dose was calculated to accumulate at the outer rim of the tumour as defined by contrast enhancement of the CT scan and serial biopsy. In 18 patients two seed implantations were carried out due to two confirmed cerebral metastases.

Results

The present series includes 130 patients with multiple or single metastases with a mean follow up period of 17 months (Tables 2a and b). The median survival time was 17 months after IRS and WBRT (1 censored event),

---

* Model 6702 — Amersham Buchler GmbH.

Table 1. *Distribution of Characteristic Parameters for IRS*

Tumour-radius (mm)	
mean (median)	15(15)
SD	5.4
Tumour-volume (ml)	
mean (median)	16.5(13)
SD	15.6
Dose-rate (cGy/h)	
mean (median)	7.0(7.2)
SD	1.5
Total dose (GY)	
mean (median)	65(60)
SD	13.2

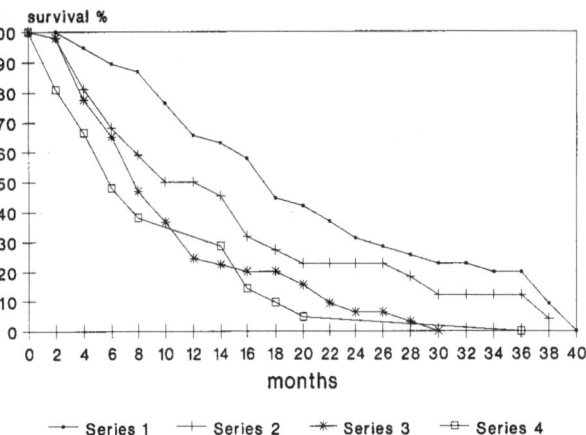

Fig. 1. Survival plots after IRS + WBRT (series 1), IRS alone (series 2), WBRT alone (series 3) and after IRS as treatment for recurrence (series 4). Significant differences between plot 1 and 3 (p < 0.01) and between plot 1 and 2 (p < 0.05) (Kaplan-Meier Test-Lee-Desu-Statistic)

12 months after IRS (2 censored events), and 7.7 months after WBRT (4 censored events). The median survival after local treatment of recurrent disease was 6 months with a total median survival of 17 months (Fig. 1). 12 patients with a metastasis from a malignant melanoma had a median survival of 16 months after IRS or IRS + WBRT, 15 patients with hypernephroma metastases survived 18.5 months.

Prognostically favourable factors in the multivariate analysis after IRS and IRS + WBRT were a latency period ⩾ 1 year (time interval between diagnosis of primary tumour and cerebral metastasis) p < 0.001, a solitary cerebral metastasis (p < 0.001), a Karnofsky score (KS) of ⩾ 70 (p < 0.05) and absence of disseminated disease (p < 0.05). The prognosis for lung cancer

Table 2b. *Primary Tumours*

Primary	No.
Primary unknown	7(5.4%)
Hypernehphroma	26(20.0%)
Bronchial-carcinoma (non small cell)	54(41.5%)
Melanoma	16(12.3%)
Gastro-intestinal	11(8.5%)
Uterus/ovary	2(1.5%)
Breast	11(8.5%)
Thyroid gland	3(2.3%)
Total	130(100%)

Table 2a. *Distribution of Important Clinical Variables in the Treatment Subgroups*

	IRS + WBRT	IRS	WBRT	IRS Recurrence treatment	No.
Age (years)					
mean (median)	51(55)	59(59)	57(59)	47(47)	57(57)
SD	11	10	11	9	11
KS					
mean (median)	72(80)	70(80)	68(70)	70(70)	70(70)
SD	18	18	15	17	16.7
Latency (months)					
mean (median)	34(17)	23(17)	13(1)	46(29)	26(9)
SD	42	28	25	47	37
Size (mm)					
mean (median)	32(30)	22(15)	30(25)	32(35)	29(30)
SD	13	11	14	13	13
Primary					No.
localized	12	13	24	15	64
disseminated	26	9	25	6	66
Metastases					
single	35	19	17	9	80
multiple	3	3	32	12	50
Localisation					
lobar	27	15	38	16	96
non-lobar	11	7	11	5	34

was worse than for other primary tumours (p < 0.01). IRS + WBRT was not superior to IRS alone in the multivariate model.

In the subgroup of patients receiving WBRT the presence of a solitary metastasis was the most favourable factor.

Follow-up

75% of all the patients showed improvement or stabilisation of their KS three months after the end of treatment. This was observed more frequently after WBRT + IRS than after WBRT (p < 0.05). There were no differences between IRS + WBRT and IRS alone with regard to the follow-up. No patient died from the locally irradiated metastasis. The treated metastasis could be locally controlled in every case (in eight patients not until re-implantation with a total tumour dose of 100 Gy.) No patients suffered radionecrosis requiring treatment. In 10 patients receiving IRS alone WBRT was performed after a latency period (median: 6 months) because of confirmed multiple cerebral metastases.

Discussion

The treatment of choice for patients with multiple or poorly circumscribed metastases is percutaneous radiotherapy. Sixty-seven percent of this subgroup showed an improvement or stabilization of the KS three months after completion of WBRT. This was in accordance with the results reported in the literature[2-5]. The results after IRS are not comparable with those after WBRT. With IRS a calculated radionecrosis is induced with consecutive removal by macrophage activity. This explains why IRS is considered a surgical rather than a radiotherapeutic procedure[6]. The indications for surgery and IRS are similar.

According to the results of regression analysis, the patients with brain metastases most likely to benefit from IRS are those with a single, well-circumscribed lesion, a latency period of ≥ 1 year, absence of disseminated disease and a KS $\geq 70\%$. Median survival of ≥ 12 months after IRS was in agreement with the results after surgery and WBRT reported in the literature[7,9]. IRS + WBRT was not superior to IRS only in

multivariate analysis. This finding was remarkable since most studies found an important influence of WBRT on survival after surgery[10]. Therefore, it would seem that local control of the metastasis is more likely to be achieved using IRS than surgical extirpation. In the present study low-activity IRS controlled the growth of brain metastases regardless of tumour histology, prior treatment, or resistance to WBRT without any radionecrotic complications. This was in contrast to other studies using high-activity IRS[1,8]. IRS alone as a minimally invasive method has strong advantages. Whether surgery or IRS is indicated for surgically accessible metastases is a controversial issue.

References

1. Bernstein M, Gutin PH (1981) Interstitial irradiation of brain tumors: A review. Neurosurgery 6: 741–750
2. Borgelt B, Gelber R, Kramer S, Brady L, Chang C, Davis L, Perez C, Hendrickson F (1980) The palliation of brain metastases: Final results of the first two studies by the Radiation Therapy Oncology Group. Int J Radiat Oncol Biol Phys 6: 1–19
3. Gelber RD, Larson M, Borgelt B, Kramer S (1981) Equivalence of radiation schedules for the palliative treatment of brain metastases in patients with favorable prognosis. Cancer 48: 1749–1753
4. Hoskin P, Crow J, Ford H (1990) The influence of extent and local management on the outcome of radiotherapy for brain metastases. Int J Radiat Oncol Biol Phys 19: 111–115
5. Kurtz JM, Gelber R, Brady LW, Carella RJ, Cooper JS (1981) The palliation of brain metastases in a favorable patient population: A randomized trial by the Radiation Therapy Oncology Group. Int J Radiat Oncol Biol Phys 7: 891–895
6. Ostertag CB, Weigel K, Warnke P, Lombeck G, Kleihues P (1983) Sequential morphological changes in the dog brain after interstitial Iodine-125 irradiation. Neurosurgery 13: 523–528
7. Patchell RA, Tibbs PA, Walsh JW, Dempsey RJ, Maruyama Y, Kryscio RJ, Markesbery WR, Mcdonald JS, Young B (1990) A randomized trial of surgery in the treatment of single metastasis to the brain. N Engl J Med 322: 494–500
8. Prados M, Leibel S, Barnett CM, Gutin P (1989) Interstitial brachytherapy for metastatic brain tumors. Cancer 63: 657–660
9. Sause WT, Crowley JJ, Morantz R, Rotman M, Mowry PA, Bouzaglou A, Borst JR, Selin H (1990) Solitary brain metastasis: Results of an RTOG/SWOG protocol evaluation. Surgery + RT versus RT alone. Am J Clin Oncol 13: 427–432
10. Smalley SR, Schray MF, Laws ER, O'Fallon JR (1987) Adjuvant radiation therapy after surgical resection of solitary brain metastasis: Association with pattern of failure and survival. Int J Radiat Oncol Biol Phys 13: 1611–1616

Correspondence: Prof. Dr. Ch. B. Ostertag, Abteilung für Stereotaktische Neurochirurgie, Neurochirurgische Universitätsklinik, Hugstetterstrasse 55, D-79106 Freiburg, Federal Republic of Germany.

Acta Neurochir (1993) [Suppl] 58: 115–118

Comparison Between Radiosurgery and Stereotactic Fractionated Radiation for the Treatment of Brain Metastases

A. A. F. De Salles[1], M. Hariz[2], C. L. Bajada[3], St. Goetsch[3], T. Bergenheim[2], M. Selch[3], F. E. Holly[3], T. Solberg[3], and D. P. Becker[1]

[1] Division of Neurosurgery, University of California, Los Angeles, [2] Department of Neurosurgery, Umeå University, Sweden, and [3] Department of Radiation Oncology, University of California, Los Angeles, U.S.A.

Summary

This study evaluates the treatment of intracerebral brain metastases with single dose stereotactic radiosurgery in comparison to stereotactic fractionated radiotherapy (SFR). Twenty six patients with 41 lesions were evaluated. Thirty four lesions in 19 patients were treated with radiosurgery, and 7 lesions in 7 patients were treated with SFR. The radiosurgery group was treated with an average number of isocenters of 1.4, and an average of 9 arcs. The average dose was 2140 cGy delivered to the 70% isodose line. The average volume of the lesions was 5.22 cc. The SFR group lesions received a mean dose to the indicated area delivered by 4 to 6 coplanar fields. The dose was 600 cGy per fraction, 2 to 3 fractions were given. The average volume of the treated lesions was 21.2 cc.

Follow-up extended from 2–18 months. Twenty five lesions of the radiosurgery group had image follow-up. The overall local control was seen in 92% of the patients. Six lesions of the SFR group had image follow-up, the local control was 83%. The small number in each group, the non-randomized nature of the study, and the relatively short follow-up preclude a definitive conclusion.

SFR may be the method of choice for large lesions surrounded by significant edema. The delivery of the dose in large fractions may obviate the transient acute reactions seen when radiosurgical dose is delivered to large lesions surrounded by edema. However, both forms of therapy have proven to be effective in the control of brain metastases.

Keywords: Stereotactic radiosurgery; radiation therapy; metastases.

Introduction

The natural history of a brain metastases gives a median survival of 1.2 months[11]. The addition of conventional whole brain radiation therapy can extend patients life to a median of six months[2,5,6]. Recurrence in the brain was the cause of death in 30–50% of the patients[8]. The surgical resection of a solitary brain metastasis followed by partial or whole brain radiation therapy has increased the patients survival to 9 months[4], and decreased the brain relapse rate from 85% to 21%[10]. These conclusions were further validated with a randomized trial in which surgery and radiation therapy were compared to radiation therapy alone[9]. This study revealed a significant increased survival in the group that received surgery and radiation therapy. The combined modality group also had an increase in the time spent in good functional condition[9]. However, many patients with brain metastatic disease do not undergo surgery because of medical infirmity, tumor location or tumor multiplicity. Therefore, radiosurgery has been applied for brain metastases[1,3]. Adler *et al.* showed 50% of the lesions to be reduced, 29% of the lesions to disappear, and 12% of the lesions to be stabilized in a mean follow-up of 5.5 months[1]. Coffey *et al.* also revealed good radiosurgical local control, 53% of the lesions disappeared or became smaller, and 29% of the lesions remained unchanged[3]. These encouraging results led to this cooperative study which reviews results of single dose radiosurgery in comparison to results of radiation therapy followed by fractionated stereotactic radiation boost.

Materials and Methods

Stereotactic radiosurgery was performed at the University of California, Los Angeles, U.S.A., and stereotactic fractionated radiation (SFR) was performed at the University of Umeå, Sweden. Twenty six patients were studied. There were 12 females and 14 males. Forty one lesions were treated, 34 with radiosurgery and 7 with SFR. There were 24 adenocarcinoma, 7 melanomas, 7 renal cell carcinomas, 2 squamous cell carcinomas, and one retinoblastoma. Both treatment groups underwent stereotactic CT-scan localization and sometimes

MRI was also obtained to further define the lesions. A Philips SRS 200 system and the BRW frame were used for the radiosurgery group. The SFR group was treated using the Laitinen stereoadapter guidance and a BBC Dynaray linear accelerator. The volume of the lesions in the radiosurgery group ranged from 0.09 cc to 51.84 cc with an average of 5.2 cc. Each lesion received an average of 2140 cGy with a range of 1500 Gy to 3000 cGy. This dose was delivered to the margin of the lesion at a mean isodose line of 70%, ranging from 50% to 95%. An average of 1.4 isocenters, (range 1–4), 9 arcs (range 6–20), and a 20mm diameter collimator (range 7–32) were applied (Table 1). The procedure was completed in one day and the patients went home following removal of the stereotactic frame. The patients were seen in follow-up in 2–4 weeks and every two months thereafter. Scans were obtained at three month intervals or when there was a question of recurrence, at which time positron emission tomography or a thallium scan was also obtained.

The SFR group lesions volume ranged from 11.5 cc to 42.3 cc with a mean of 21.2 cc. The dose to the target was defined as the mean dose to the indicated target area. Table 2 shows the characteristics of the patients. The patients were irradiated in 2 to 3 sessions. At each session the Laitinen stereoadaptor with lateral target plates was mounted to the patient's head and locked to the linear accelerator table. The target dose was 600 cGy delivered by 4 to 6 coplanar fields. The marginal dose was delivered to the 85% isodose line, ranging

from 80 to 90%. All patients in the SFR group received stereotactic radiation as a boost to conventional whole brain radiation therapy. This group of patients had a follow-up CT scan at two to six months (Table 2).

Results

The follow-up extended from 2 to 18 months. Radiosurgery controlled 92% of the lesions, i.e., 72% decreased in size and 20% remained unchanged. Lesion progression was seen in 8% of the patients. Four patients presented with metastases to other areas of the brain during the follow-up. Temporary side effects were headache in one patient, and nausea and vomiting in another. One patient with a prior history of seizures had one seizure episode following radiosurgery. One patient had enlargement of his lesion with central necrosis, a thallium scan was negative, his lesion was not scored as progression. Four patients had distant failure. Three of the patients

Table 1. *Treatment Characteristics of Radiosurgery Patients*

Patient number	Age	Sex	Diagnosis	Location of lesion	Irradiation history	Status	Maximum dose	Prescribed dose	Prescribed isodose line
1	58	F	renal	temporal	3000 cGy	alive	2105 cGy	2000 cGy	95%
				basal ganglia			4000 cGy	2000 cGy	50%
2	50	F	lung	temp./parietal	5140 cGy	dead	2500 cGy	1500 cGy	60%
3	45	F	melanoma	hippocampus	4000 cGy	dead	2100 cGy	2000 cGy	95%
4	45	M	melanoma	parietal	3750 cGy	dead	2778 cGy	2500 cGy	90%
				frontaltemporal			6000 cGy	3000 cGy	50%
				temporal			6000 cGy	3000 cGy	50%
5	60	M	lung	frontal	no	alive	2778 cGy	2500 cGy	90%
				temporal			2778 cGy	2500 cGy	90%
				frontal			2778 cGy	2500 cGy	90%
				occipital			4000 cGy	2000 cGy	50%
				temporal			4000 cGy	2000 cGy	50%
				temporal			5000 cGy	3000 cGy	60%
				temporal			4000 cGy	2000 cGy	50%
6	34	M	melanoma	frontal	4000 cGy	alive	2353 cGy	2000 cGy	60%
				frontal			2353 cGy	2000 cGy	60%
				temporal			2353 cGy	2000 cGy	60%
7	58	F	breast	frontal	4500 cGy	dead	2222 cGy	2000 cGy	90%
8	57	M	lung	occipital	7800 cGy	dead	2727 cGy	1500 cGy	55%
9	44	F	lung	post. fossa	3000 cGy	alive	2727 cGy	1500 cGy	55%
10	75	M	base of tongue	cavernous sinus	no	alive	3600 cGy	1800 cGy	50%
11	65	M	lung	cavernous sinus	4000 cGy	alive	4400 cGy	2200 cGy	50%
12	51	F	breast	frontal	4400 cGy	dead	3600 cGy	1800 cGy	50%
				frontal			3600 cGy	1800 cGy	50%
				cerebellar			3600 cGy	1800 cGy	50%
13	71	F	lung	basalganglia	4400 cGy	dead	5000 cGy	2500 cGy	50%
14	50	M	renal	frontal	no	alive	6000 cGy	3000 cGy	50%
15	73	M	prostate	petrous apex	no	alive	3600 cGy	1800 cGy	50%
16	79	M	lung	cerebellar	4000 cGy	dead	4000 cGy	2000 cGy	50%
17	60	M	parotid	cavernous sinus	neutrons	alive	4000 cGy	2000 cGy	50%
18	49	F	renal	occipital	5000 cGy	alive	2667 cGy	2000 cGy	75%
				cerebellum			2667 cGy	2000 cGy	75%
				temporal			2667 cGy	2000 cGy	75%
19	60	M	lung	posterior fossa	no	dead	4000 cGy	2000 cGy	50%

Table 2. *Treatment Characteristics of Stereotactic Fractionated Radiation Patients*

Patient number	Age	Sex	Diagnosis	Location of lesion	Irradiation history	Status	Dose (Gy)	IDL* (%)
1	68	M	lung	parietal	3400 cGy	dead	3 × 6	80
2	62	F	lung	temp-par	3300 cGy	dead	2 × 6	90
3	68	F	breast	parietal	2700 cGy	dead	3 × 6	90
4	39	M	renal	caudatus	2700 cGy	dead	2 × 6	80
5	63	F	retinoblastoma	parietal	2900 cGy	dead	2 × 6	80
6	61	F	breast	frontal	1800 cGy	dead	3 × 6	80
7	60	M	esophageal	putamen	2700 cGy	dead	3 × 6	80

* Isodose line.

had previous whole brain radiation therapy. One patient was referred for conventional whole brain radiation therapy following his failure. The survival to date in the radiosurgery group is eight months, ranging from 2 to 18 months. Eleven patients are alive and eight have died. Three of the patients died secondary to central nervous system (CNS) disease and five died of systemic disease.

Total regression was seen in three of the six patients who had imaging studies in the SFR group, one lesion decreased in size, one lesion had central necrosis, one lesion progressed. Two of the patients died secondary to their CNS disease and five died secondary to systemic disease. The average survival from time of treatment was seven months. Summary of the results is shown in Table 3.

Discussion

Radiosurgery and stereotactic fractionated radiation play an important role in the treatment of brain metastases, either as a boost after whole brain conventional radiation therapy or at the relapse after surgical resection followed by whole brain irradiation. Patients that cannot benefit from surgical resection due to medical infirmity, multiplicity of lesions or lesions located in areas of difficult surgical approach can be helped with

Table 3. *Imaging Follow-up*

	Radiosurgery[a]	SFR[b]
Decreased	18 (72%)	5 (83%)
Unchanged	5 (20%)	1 (17%)
Progression	2 (8%)	0

[a] In the radiosurgery group 25/34 lesions had post radiosurgery imaging.
[b] In the stereotactic fractionated radiation group 6/7 lesions had post SFR imaging.

radiosurgery of SFR. Radiosurgery has been shown to prolong patient's survival with a good quality of life[7].

This study shows that radiosurgery compares favorably to SFR. The survival in the radiosurgery group is longer than that of the SFR group, this may be secondary to the larger size of the lesions in the SFR group. The nature of the study and the number of patients preclude definitive conclusions. The two failures in the radiosurgery group occurred in patients with metastases of squamous cell carcinoma to the cavernous sinus: Prior to treatment, the lesions were tracking along the cranial nerves, and continued to track along the cranial nerves outside of the treatment area after radiosurgery. Therefore, radiosurgery may not be indicated in the treatment of metastatic disease to the cavernous sinus, unless there is no other medical option available. This preliminary study suggests that it may be beneficial to treat the large lesions with surrounding edema with SFR to avoid transient acute reactions frequently seen when radiosurgery is performed to a large volume.

References

1. Adler JR, Cox RS, Kaplan I, Martin DP (1992) Stereotactic radiosurgical treatment of brain metastases. J Neurosurg 76: 444–449
2. Cairncross JG, Kim JH, Posner JB (1980) Radiation therapy for brain metastases. Ann Neurol 7: 529–541
3. Coffey RJ, Lunsford LD, Flickinger JC (1992) The role of radiosurgery in the treatment of malignant brain tumors. Stereotact Radiosurg V3N1: 231–244
4. Galicich JH (1981) Surgery of malignant brain tumors: Seminars in Neuro VI. Thieme-Stratton, New York, pp 159–168
5. Hendrickson FR (1977) The optimum schedule for palliative radiotherapy for metastatic brain cancer. Int J Radiat Onc Bio Phys 2: 165–168
6. Kurtz JM, Gelber R, Brady LW (1981) The palliation of brain metastases in a favorable patient population. A randomized clinical trial by the Radiation Therapy Oncology Group. Int J Radiat Onc Bio Phys 7: 891–895
7. Loeffler JS, Kooy HM, Wen PY, Fine HA, Mannarino EG, Tsai

JS, Alexander E (1990) The treatment of recurrent brain metastases with stereotactic radiosurgery. J Clin Onc 8: 576–582

8. Order SE, Hellman S, Von Essen CF (1968) Improvement in quality of survival following whole brain irradiation for brain metastases. Radiology 91: 149–153

9. Patchell RA, Tibbs PA, Walsh JW (1990) A randomized trial of surgery in the treatment of single metastases to the brain. N Engl J Med 322: 494–500

10. Smalley SR, Schray MF, Laws ER, O'Gallon JR (1987) Adjuvant radiation therapy after surgical resection of solitary brain metastases: Association with pattern of failure and survival. Int J Radiat Onc Bio Phys 13: 1611–1616

11. Takakura K, Sano K, Hojo S, Hirano A (1982) Metastatic tumors of the central nervous system. Igaku-Shoin, Tokyo

Correspondence: Antonio A. F. De Salles, M.D., PhD., Division of Neursurgery, 300 UCLA Medical Plaza, Suite B-212, Los Angeles, California 90024-6975, U.S.A.

Acta Neurochir (1993) [Suppl] 58: 119–122

Interstitial Irradiation for Newly Diagnosed or Recurrent Malignant Gliomas: Preliminary Results

M. Scerrati, R. Roselli, P. Montemaggi[1], M. Iacoangeli, A. Prezioso, and **G. F. Rossi**

Institute of Neurosurgery and [1]Institute of Radiology, Catholic University, Rome, Italy

Summary

The preliminary results obtained in 19 patients treated with interstitial irradiation for malignant gliomas are reported. Three different groups are included in the study: I Newly diagnosed tumours not suitable for surgery: 13 cases (10 anaplastic astrocytomas (AA) and 3 glioblastomas (GBM), mean volume 46.56 cc, source Ir 192) were implanted permanently (n = 11, mean peripheral dose 93.54 Gy) or temporarily (n = 2, 50 Gy = 0.5 Gy/hr). External beam irradiation was additionally applied in all cases. II Residual or recurrent tumours: 5 patients (2 AA and 3 GBM, mean volume 7.2 cc, source Ir 192) received temporary implants (150 Gy peripheral dose = 1.5 Gy/hr) after surgery and conventional radiotherapy. III Newly diagnosed surgically removable tumours: only one patient with AA (15 cc volume, source Ir 192) received temporary implantation with the same dose regimen used in Group II before surgery and external beam irradiation.

A median survival time of 26.75 mos (34.62 mos for AA, with 3 long-term survivors) was observed in the patients of Group I. Three patients of Group II are still alive after 8, 12 and 12 mos after brachytherapy, the other 2 (GBM) survived 7 and 12 mos. The single patient so far included in Group III is still alive after 6 mos.

Although the study is still in progress, these preliminary data seem to indicate that interstitial radiotherapy can be effective in prolonging survival of patients with malignant gliomas.

Keywords: Malignant gliomas; interstitial irradiation; brachytherapy.

Introduction

The advantage of interstitial irradiation for malignant gliomas is that a high radiation dose is delivered to the hypoxic as well as to the infiltrating peripheral cells, i.e. to the elements which lead to fatally local recurrence or progression[1,2]. Accordingly, it has been proposed either as an alternative to surgery in nonresectable tumours[7–10,12,15], as irradiation of residual/recurrent neoplasms[3–5,8,16] or, more recently, as a neo-adjuvant "up-front" procedure[6,14].

We report here the preliminary results obtained in a series of 19 patients harbouring cerebral malignant gliomas who have been treated with brachytherapy.

Clinical Material and Methods

Brachytherapy as Alternative Treatment to Surgery (Group I)

Thirteen patients with malignant brain tumours which, because of their location, were not considered for surgical removal, received interstitial irradiation followed with external beam irradiation. The main characteristics of the treated patients are outlined in Table 1. In all cases the histological diagnosis was obtained by means of serial stereotactic biopsies, also aimed at verifying the tumour borders[13]. The sources were implanted permanently in 11 patients and temporarily in the other two. Multiple evenly spaced implants were used whenever possible. The Talairach stereotactic system, upgraded with personal accessories for polar and orthogonal approaches[11], was used for both biopsy and implant. Minimum peripheral dose ranged from 70 to 110 Gy (m = 93.54) in the permanent implants and was 50 Gy (0.5 Gy/hr) in the two patients treated with temporary irradiation. External beam irradiation (35–50 Gy, 1.8 Gy/fr) was applied to a second target volume extending 2 cm beyond the tumour borders

Table 1. *Group I (n = 13)*

Histology	
Anapl. astrocyt.	10
Glioblastoma	3
Sites	
Hemisphere	8
Thalamus/basal ganglia	5
Target volume (cc)	14.1–113
	(m = 46.56)
Age (yrs)	26–68
	(m = 43.35)
Karnofsky score	0.70–0.80
	(m = 0.77)

treated with brachytherapy: it was begun 60–90 days after permanent implants and 2–3 weeks after the removal of temporary implants. Position of the implanted isotopes and isodose curves was tridimensionally checked both intraoperatively (stereotactic radiograms) and postoperatively (CT scan) to obtain the final dosimetry.

Brachytherapy for Residual or Recurrent Tumours (Group II)

Five patients previously subjected to surgical removal and external beam irradiation for malignant gliomas (2 anaplastic astrocytomas and 3 glioblastomas) underwent interstitial radiotherapy for residual or recurrent disease. All tumours were well outlined on CT or MRI, their volume ranged from 4.2 to 14 cc (m = 7.2 cc). All patients received temporary irradiation with multiple, parallel and evenly spaced implants (Iridium-192 wires). A tumour dose of 150 Gy (1.5 Gy/hr) was applied to the peripherally enhanced zone, *corresponding* to 50 Gy (0.5 Gy/hr) at 1 cm distance from it.

Neo-Adjuvant ("Up-Front") Brachytherapy (Group III)

A new protocol has recently been developed for the treatment of malignant cerebral tumours, in which brachytherapy is scheduled before surgical removal and external beam irradiation. The aim of this protocol is to deliver high levels of radiation dose to the tumour and to the infiltrated periphery (BAT). Only one patient with anaplastic astrocytoma (volume 15 cc) has been entered so far in this new protocol. Brachytherapy was given with the same dose regimen and methodology used for Group II. Subsequent surgery was aimed at the removal of the entire volume included by the 50 Gy isodose delivered with brachytherapy and was followed by external beam irradiation (50 Gy, 1.8 Gy/fr) on an additional area extending 2 cm from the resected borders.

Results

Group I

The follow-up of the 13 cases of this group ranged from 7 to 80 months (m = 35.40). Three patients with

anaplastic astrocytomas are still alive 81, 84 and 92 months after brachytherapy. The survival estimates for the entire group show a survival probability of 72.73% (± 13.43) at 11.5 months, of 54.55% (± 15.01) at 22.75 months and of 27.27% (± 13.43) at 35.25 months. The median survival time is 26.75 months (Fig. 1a). If only the 10 anaplastic astrocytomas are considered, the survival probability rises to 75.00% (± 15.31) at 22.75 months and to 37.50% (± 17.12) at 35.25 months. The median survival time of this subgroup is 34.62 months (Fig. 1b). The mean score of Karnofsky performance status before the treatment was 0.77, it increased to 0.86 during the first year, ranging thereafter from a maximum of 0.73 at 24 and 36 months from treatment to a minimum of 0.63 in the patients still alive at 48 and 60 months (Fig. 2). Six patients of this group developed radionecrosis within 1 year of treatment: four of them required surgical decompression, while the other two were well controlled with steroids.

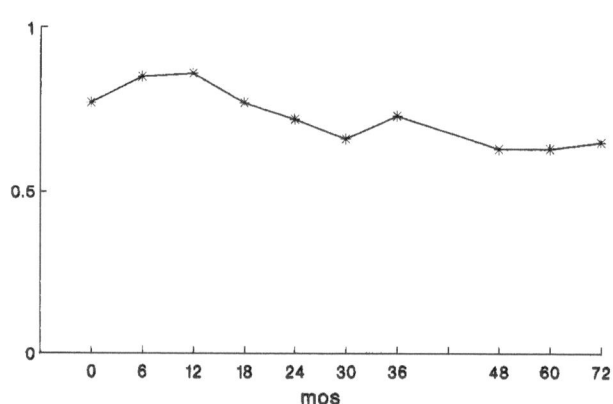

Fig. 2. Group I: Trend of the KPS (mean value) after brachytherapy

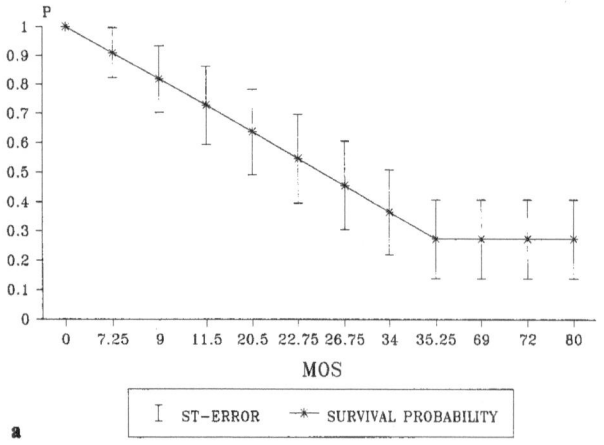

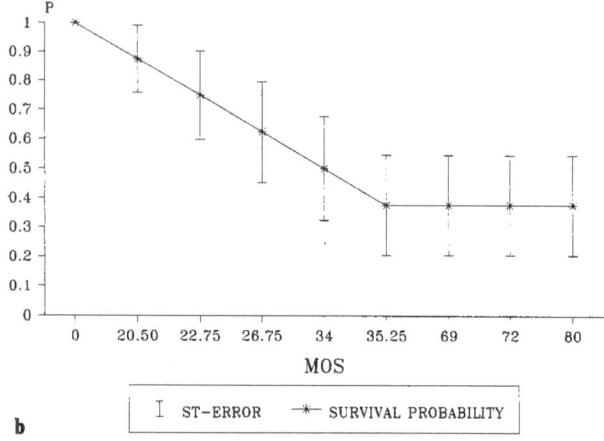

Fig. 1. Group I: Actuarial survival curve of the whole series (a), and of anaplastic astrocytomas (b) after brachytherapy. (a) Median survival time 26.75 months, (b) 34.625

Group II

The few cases so far included in this group do not allow any actuarial estimate of survival. Two patients, both with glioblastomas, survived 7 and 12 months after brachytherapy (12 and 24 months after surgical removal); the other three (1 glioblastoma and 2 anaplastic astrocytomas) are still alive 12, 8 and 12 months after interstitial irradiation (18, 28 and 40 months after surgical resection). Clinicoradiological signs of radionecrosis did not appear in any patient, all of them maintained a Karnofsy score between 0.70 and 0.80.

Group III

The single patient treated so far is still surviving after 6 months and is in excellent clinical condition.

Discussion

The use of interstitial radiotherapy for malignant cerebral tumours was first proposed by European neurosurgical groups as *alternative treatment to surgery in nonresectable tumours*[7-10,12,15]. In the series reported by Mundinger[7] the mean survival was 248 ($\pm$ 205) days for glioblastomas (n = 62), 724 ($\pm$ 502) days for anaplastic astrocytomas (n = 43) and 904 ($\pm$ 554) days for anaplastic oligodendrogliomas (n = 25). Survival rates reported by Szikla[15] indicate a 5-year probability of 55% in 7 grade III and 19% in 9 grade IV gliomas. These data are consistent with those of Rougier[9], who quotes 5-year median survival time in 12 anaplastic astrocytomas. However, the personal classification criteria followed by the French school make their results hardly comparable with those of the other groups, including ours. A survival probability of 37.5% at 3 years (m.s.t. 34.6 mos) for anaplastic astrocytomas with three long survivals seems noteworthy, despite the small number of patients so far included in our series.

Much attention has to be paid, however, to the risk of radionecrosis, which developed in the 46% of our cases. Smaller volumes should be considered for brachytherapy (< 15 cc), especially when surgical removal is problematic.

A median survival time of 18 mos has been reported by Mundinger[8] in 42 *recurrent malignant gliomas*, with 78% of patients alive after 2 years. The use of interstitial radiotherapy in the treatment of recurrences has recently gained remarkable consideration, particularly in the United States. In the 41 recurrent malignant gliomas treated by Gutin[4,5] a median survival time of 35 weeks

for glioblastomas and 153 weeks for anaplastic astrocytomas was observed; survival longer than 3 and 4 years was reached in 10 and 4 anaplastic astrocytomas respectively. More recently, a survival rate of 60% at 12 months and of 25% at 24 months postimplantation was reported by Willis[16] in 17 cases, and a median survival time of 44 weeks in the 18 patients treated by Bernstein[3]. Our experience in the treatment of recurrences with brachytherapy is only in its beginning stage: a cumulative survival of 18 and 24 months and of 28 and 41 months has been achieved in 2 glioblastomas and 2 anaplastic astrocytomas, respectively, with three patients still surviving. Radionecrosis was not observed in any case, probably due to the small volumes treated (m = 7.2 cc).

No comments are possible at the present time on the use of brachytherapy as *neoadjuvant (up-front) treatment*, due to the scarcity of data in the literature[6,14] and to the single case to date included in our protocol.

In conclusion, the results so far reported, though from single series, seem to indicate that interstitial radiotherapy can play a useful role in prolonging survival of patients harbouring malignant gliomas. However, a rigid selection of patients entering the different protocols together with the development of randomized and multicenter studies are still required to validate its efficacy.

Acknowledgements

This study is partially supported by a grant of Italian Ministry of University and of Scientific and Technologic Research.

References

1. Bashir R, Hochberg F, Oot R (1988) Regrowth patterns of glioblastoma multiforme related to planning of interstitial brachytherapy radiation fields. Neurosurgery 23: 27–30
2. Bernstein M, Gutin PH (1981) Interstitial irradiation of brain tumours. A review. Neurosurgery 9: 741–750
3. Bernstein M, Laperriere N, Leung P, McKenzie S (1990) Interstitial brachytherapy for malignant brain tumors: Preliminary results. Neurosurgery 26: 371–380
4. Gutin PH, Philips TL, Wara WM, Leibel SA, Hosobuchi Y, Levin VA, Weaver KA, Lamb S (1984) Brachytherapy of recurrent malignant brain tumors with removable high-activity iodine-125 sources. J Neurosurg 60: 61–68
5. Gutin PH, Leibel SA, Wara WM, Choucair A, Levin VA, Philips TL, Silver P, Da Silva V, Edwards MSB, Davis RL, Weaver KA, Lamb S (1987) Recurrent malignant gliomas: Survival following interstitial brachytherapy with high-activity iodine-125 sources. J Neurosurg 67: 864–873
6. Karlsson U, Black P, Nair S, Yablon JS, Brady LW (1991) Radical proposal for the treatment of malignant astrocytoma. Am J Clin Oncol 14: 75–79

7. Mundinger F, Busam B, Birg W, Schildge J (1979) Results of interstitial Iridium-192 brachy-curie therapy and Iridium-192 protracted long-term irradiation. In: Szikla G (ed) Stereotactic cerebral irradiation. Elsevier/North Holland, Amsterdam, pp 303–319

8. Mundinger F (1987) Stereotactic biopsy and technique of implantation (instillation) of radionuclides. In: Jellinger K (ed) Therapy of malignant brain tumors. Springer, Wien New York, pp 134–194

9. Rougier A, Pigneux J, Cohadon F (1984) Combined interstitial and external irradiation of gliomas. Acta Neurochir (Wien) [Suppl] 33: 345–353

10. Scerrati M, Arcovito G, D'Abramo G, Montemaggi P, Pastore G, Piermattei A, Romanini A, Rossi GF (1982) Stereotactic interstitial irradiation of brain tumors: Preliminary report. RAYS 7: 93–99

11. Scerrati M, Fiorentino A, Fiorentino M, Pola P (1984) Stereotaxic device for polar approaches in orthogonal systems. Technical note. J Neurosurg 61: 1146–1147

12. Scerrati M, Roselli R, Iacoangeli M, Montemaggi P, Cellini N, Falcinelli R, Rossi GF (1989) Comments on brachycurie therapy of cerebral tumours. Acta Neurochir (Wien) [Suppl] 46: 94–96

13. Scerrati M, Rossi GF, Roselli R (1987) The spatial and morphological assessment of cerebral neuroectodermal tumours through stereotactic biopsy. Acta Neurochir (Wien) [Suppl] 39: 28–33

14. Selker RG, Eddy MS (1989) Results of 125-I implants in newly diagnosed and "failed" glioblastoma (and other) patients. Proceedings 8th International Conference on Brain Tumor Research and Therapy. J Neuro Oncol 7 [Suppl] 95: S26

15. Szikla G, Schlienger M, Blond S, Daumas-Duport C, Missir O, Miyahara S, Musolino A, Schaub C (1984) Interstitial and combined external irradiation of supratentorial gliomas. Results in 61 cases treated 1973–1981. Acta Neurochir (Wien) [Suppl] 33: 355–362

16. Willis BK, Heilbrun MP, Sapozink MD, McDonald PR (1988) Stereotactic interstitial brachytherapy of malignant astrocytomas with remarks on postimplantation computed tomographic appearance. Neurosurgery 23: 348–354

Correspondence: Massimo Scerrati, M.D., Instituto di Neurochirurgia, Università Cattolica S. Cuore, Largo A. Gemelli 8, I-00168 Roma, Italia.

Pain

Acta Neurochir (1993) [Suppl] 58: 125–130

The Pathophysiology of Peripheral Neuropathic Pain—Abnormal Peripheral Input and Abnormal Central Processing

C. J. Woolf

Department of Anatomy and Developmental Biology, University College London, London, U.K.

Summary

A review is given on the pathogenesis of peripheral neuropathic pain. Central neuropathic pain resulting from damage of the spinal cord or brain is not covered.

The following conclusions are proposed. At the time of peripheral injury, an abnormal injury discharge may be sufficient to produce long term changes in the excitability of the spinal cord and/or an excitoxic death of dorsal horn neurons. These acute changes might set the scene for the maintenance of sensory disorders both as a result of an ongoing ectopic input, which might persistently induce a state of central sensitization and for a structural reorganization of the synaptic connections of the dorsal horn. The implications of these findings is that it may be possible to prevent some of the long term consequences of nerve damage, that the treatment at the time of injury may need to be quite different for that required later, and finally that treatment directed only at the periphery may be insufficient to eliminate the sensory disturbance of chronic neuropathic pain.

Keywords: Peripheral nerve lesion; neuropathic pain; pathophysiology; prevention.

Neuropathic or neurogenic pain is that pain deriving from a lesion to or disorder of the nervous system. Although patients with neuropathic pain typically present with a characteristic set of sensory disorders independent of the cause; a constant scalding or burning pain, a partial loss of sensitivity, tactile or cold allodynia and hyperpathia to repeated stimulation, it is necessary from a pathophysiological perspective to classify the pain anatomically. This article will review the pathogenesis of peripheral neuropathic pain. Central neuropathic pain resulting from damage to the spinal cord or brain will not be dealt with.

Peripheral neuropathic pain includes a number of diverse conditions, the commonest of which are; trigeminal neuralgia, postherpetic neuralgia, painful diabetic neuropathy, the reflex sympathetic dystrophies including causalgia, mononeuropathies and peripheral nerve injury. Dorsal root avulsion injuries represent a transitional condition between peripheral and central neuropathic pain. The time course, pattern and nature of the sensory disorders differ between the different conditions, but even within a single condition, the presentation varies from patient to patient. Although progress is being made in understanding the pathophysiological processes involved, there remain problems with laboratory investigations into a syndrome, the susceptibility of which in individual patients is low. Only an extremely small number of patients with soft tissue injury develop a full-blown reflex sympathetic dystrophy, the number of patients who have causalgia after traumatic peripheral nerve injury is also very low and there are no accurate predictors for establishing which patients with herpes zoster are likely to develop intractable postherpetic neuralgia.

It is widely accepted that the treatment for peripheral neuropathic pain is generally inadequate[5,6]. For those patients with a demonstrable sympathetic component, i.e. sympathetic maintain pain[9], interruption of the sympathetic nervous system, surgically, with a local anaesthetic or pharmacologically, remains the major approach. However, in some patients at least, sympathetic involvement may be transient[5]. Non-surgical treatment including anticonvulsants, antidepressants, calcium channel blockers, steroids, topical aspirin or capsaicin, physiotherapy and transcutaneous nerve stimulation, have potential benefit for a period[6]. Surgical intervention in the periphery is more controversial, in many cases it is ineffective or may worsen the situation[21,35,36]. Dorsal root entry zone lesions, anterolateral cordotomy and central stimulation have been considered, but seem to be appropriate only for patients with a disabling pain, refractory to all other treatments and usually associated with a terminal disease or limited life expectancy.

Improvements in our understanding of the pathogenesis of peripheral neuropathic pain may lead to improvements in its management. Unfortunately multiple mechanisms may operate with different time courses and treatment might have to be directed at different sites at different times. Two general mechanisms appear to operate to produce neuropathic pain; abnormal peripheral input and abnormal sensory processing, although the latter may in some situations be contingent on the former. The two mechanisms are likely to co-exist.

Abnormal Peripheral Input

1) Acute Injury Discharge

Damage to a peripheral nerve initiates an acute injury discharge in the axotomized afferent fibres[33]. This acute injury discharge lasts for tens of seconds and is due to a sudden depolarization of the peripheral membrane consequent on ionic shifts. Because the discharge will involve many, if not all of the axotomized afferents, an enormous and highly abnormal input will enter the CNS. Apart from producing intense and excruciating pain, such input appears to produce long-lasting changes in the dorsal horn. The evidence for this is that animal models of neuropathic pain are prevented if a local anaesthetic is administered topically prior to damage to a nerve[18,24].

Brief afferent inputs now have the capacity to produce relatively long lasting changes in excitability[34,37] as a result of the summation of C-fibre mediated slow synaptic potentials in the spinal cord[36]. The summation operates via glutamate acting on the N-methyl-D-aspartate and (NMDA) receptors[36] although neuropeptides may also be involved. Activation of the NMDA receptor results in calcium entry into the cell which can directly activate protein kinases, although substance P and other neuropeptides can also activate kinases through other second messenger systems. The activated protein kinases produce a number of changes in the neurone, including phosphorylating ion channels, such as the NMDA receptor ion channel[11]. Blocking the NMDA receptor with specific antagonists eliminates the summation of slow potentials[36], the excitability increase produced by afferent input[40] and neuropathic pain[14,25]. The excitability increases produced by brief

electrical stimulation of a peripheral nerve, however[34] only lasts for an hour or so. How could such changes contribute to persistent neuropathic pain. One possibility is that the NMDA receptor mediated change in second messengers may have consequences other than just the short-lived phosphorylation of receptors orion channels. Nerve injury results, for example in the persistent increase in the expression of the immediate-early proto-oncogene C-fos[12,26], which may control the expression of a variety of late effector genes. This could result in a maintained increase in excitability way beyond the initiating input, in a manner analogous to the synaptic plasticity in the hippocampus that is thought to determine memory and learning. Alternatively NMDA receptor activation might lead to an excitotoxic effect on dorsal horn neurones leading to cell death[4]. If a selective death of inhibitory interneurones occurred, this would result in a permanent disinhibition, or excitability increase. Indirect evidence from this idea has been produced[29] but whether actual degeneration of specific subpopulations of neurons occurs has not been established yet.

What is clear though, is that sensory input produced at the time of a nerve injury may be sufficient to establish a modifiability in the CNS which may contribute to the persistance of neuropathic pain. The clinical implication of this relates to any elective surgery on peripheral nerves where, preventing the injury discharge, may influence the pain outcome. Such a claim has been made for phantom limb pain[3] but further prospective, randomized trials are required. Figure 1 summarizes the long term consequence of an acute nerve injury.

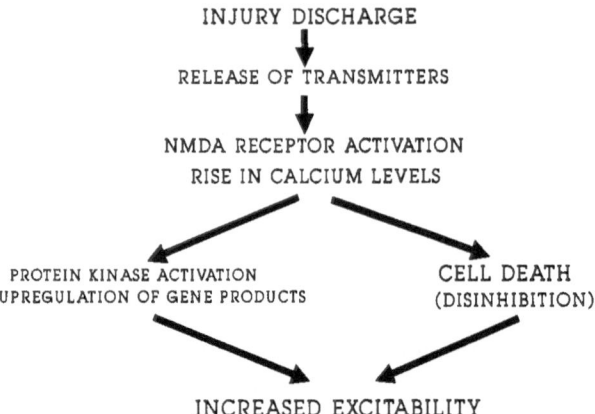

Fig. 1. Acute nerve injury. The injury discharge produced at the time of a nerve injury may lead to long term alteration in the excitability of the spinal cord as a result either of changes in gene expression in dorsal horn neurones or as a result of a selective loss of inhibitory interneurones. In both cases an NMDA receptor mediated changes seems to be involved

2) Chronic Ectopic Discharge

Several days after the acute injury discharge, an ectopic afferent input begins to be generated, first in myelinated and later in unmyelinated injured axons, at the proximal stump (neuroma) along the axon and also from cell bodies in the dorsal root ganglion[15,20,32]. Ectopic activity appears to be the consequence of the development of abnormal pacemaker properties, which have been explained on the basis of the accumulation of sodium channels in the injured neuron[7]. While the pacemaker may become autonomous, running as a rhythmic oscillator, a feature of injured neurons is the development of abnormal sensitivity to mechanical, thermal and chemical stimuli[7]. Sensitivity to circulating and locally released catecholamines in axotomized[32] and unaxotomized fibres following nerve injury[23] is of particular importance because of its possible role in the aetiology of sympathetic maintained pains.

One further aspect of the chronic ectopic activity is of interest. That is the development of cross-excitation. This may occur between sensory fibres, so that abnormal activity in one fibre might initiate firing in other fibres due to abnormal proximity of the fibres, loss of Schwann cells, accumulation of potassium or the formation of junctions between adjacent membranes. It also might develop between postganglionic sympathetic neurons and primary sensory neurons so that normal sympathetic reflex output might generate an input in sensory fibres.

A fundamental issue concerning the ectopic input resulting from peripheral nerve injury is whether it is an epiphenomenon or whether it represents the actual mechanism responsible for the generation of neuropathic pain. This has considerable bearing on treatment strategies. The ectopic input is, for example, susceptible to blockade by systemic local anaesthetics at doses well below those that produce conduction block[16] and this could explain the effectiveness of such drugs in treating neuropathic pain, although central actions[41] have to be considered. The simplest explanation for how ectopic afferent input produced neuropathic pain would be that such activity in nociceptive A delta and C fibres would initiate sensory signals in the neural pathways in the central nervous system that ultimately lead to the sensation of pain (Fig. 2). In other words the CNS remains normal, it's just an ongoing input in "pain" fibres that produces the chronic pain. That this is definitely not the case for the tactile hyperalgesia or mechanical allodynia associated with neurogenic pain has now been clearly established[8,22]. Such pain is in-

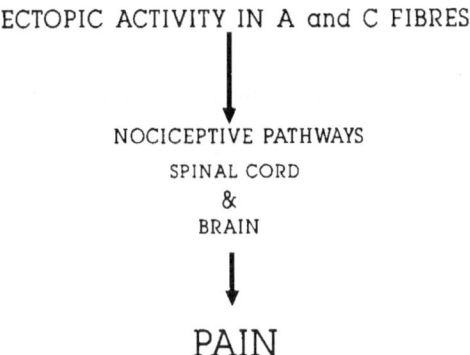

ECTOPIC ACTIVITY IN A and C FIBRES

NOCICEPTIVE PATHWAYS
SPINAL CORD
&
BRAIN

PAIN

Fig. 2. Chronic nerve injury. The simplest model of neuropathic pain is that ectopic input in axotomized primary afferents activate normal transmission pathways in the central nervous system leading to the sensation of pain. This model cannot explain however, how mechanical allodynia is mediated by low threshold mechanoreceptors, fibres which never normally produce pain

itiated by activity in low threshold mechanoreceptor primary afferents. Blockade of conduction in large myelinated fibres eliminates the touch evoked allodynia, while leaving C-fibre mediated thermal and pain sensitivity intact. How does this occur?—The answer lies in abnormal central processing.

Abnormal Central Processing

Two general categories of abnormal central processing can occur; an increase in excitability (central sensitization, a form of functional plasticity) or a re-wiring of synaptic connections (structural reorganization). Both may occur together.

1) Central Sensitization

This is the phenomenon whereby the response of dorsal horn neurons to normal afferent input is augmented or facilitated. The hypersensitivity state can be produced by brief bursts of activity in C afferents[38] as a result of an NMDA and tachykinin mediated action on spinal neurons[40,45]. Experimentally 20 second stimulation of cutaneous C-fibres produce an excitability increase that lasts up to 10 minutes while muscle afferents produce an effect that lasts up to an hour[34,42]. Recently this phenomenon has been demonstrated directly in human subjects[31]. Central sensitization results from the modification of the receptive field properties of dorsal horn neurons including a reduction in threshold, an increase in responsiveness, recruitment of novel inputs and an increase in size of the receptive field[13,27,39].

Any ongoing input in C-fibres would be expected to produce an ongoing but afferent-dependent state of central sensitization. The maintenance of an altered central processing by an abnormal peripheral input has recently been demonstrated both for experimental allodynia in human subjects[19] and in a small group of patients with painful neuropathy[17]. In this respect some component of the pain of peripheral neuropathy in certain patients may be similar to the pain resulting from acute tissue injury which also depends on induction of central sensitization[38]. The difference being that the inducing stimulus is transient, as opposed to the persistent ectopic C input.

The pain of reflex sympathetic dystrophy can also be explained by this model. Reflex activity in postganglionic sympathetic neurons may begin to drive C-afferents, which have developed an adrenergic sensitivity as a result of axotomy[9]. This C-afferent input would then produce central sensitization. Once central sensitization is present, either as a result of ectopic pacemaker activity in C-fibres or secondary to sympathetic drive, it will result in abnormal responses to all other inputs including A beta afferents (Fig. 3).

Whether a state of central sensitization can be induced that becomes independent of any afferent input is not known. If this were so then ectopic input would not necessarily produce the pain but provide the peripheral input that is interpreted as being painful. Clinically there are many examples where local anaesthetic block results in prolonged relief of pain[2] but this could be because the input that produces central sensitization is removed, or because the input that generates pain as a result of an autonomous state of central sensitization, is no longer present. Experimentally at least, central sensitization only lasts for relatively short periods (up to several hours after a C-fibre input) so that eliminating an ongoing peripheral C-fibre drive would be expected to produce prolonged pain relief. The reason why this may not be so relates to the second type of central change that occurs after nerve injury— structural reorganization.

a) Structural reorganization. Peripheral nerve injury results in a complex series of changes in the cell bodies of the axotomized neurons in the dorsal root ganglion. The changes include alterations in morphology (the chromatylytic reaction), reductions in the levels of transmitters and changes that enable regeneration to occur. The latter includes decreases in the levels of cytoskeletal proteins and the upregulation of growth-associated proteins that are normally expressed only during development[28].

Although regeneration of the injured axon in the periphery has been extensively investigated and thought to contribute to post nerve-injury sensory disorders by either a failure of regeneration or the inappropriate peripheral innervation, recently it has become clear that structured changes also occur in the central terminals of primary sensory neurons after damage to the peripheral axon. Two major alterations have been detected. The first is an apparent degeneration, atrophy or withdrawal of central terminals from their normal synaptic connections on dorsal horn neurons[1,10]. This would leave a large number of evoked or vacant synaptic sites. The second is that the growth-associated proteins that seem necessary for peripheral regeneration are also transported to the central terminals of the injured neurons, in the spinal cord[43]. The combination of central terminals with the molecular machinery for growth and vacant synapses provides the mechanism and opportunity for growth to occur. There is now evidence that this in fact happens[44]. In terms of the production of pain, what is of particular interest is that the growth involves low threshold mechanoreceptive fibres growing from lamina III into lamina II, the site of termination of C-fibres. If these sprouting fibres form new synaptic connections, then a re-wiring of the dorsal horn might contribute to mechanical allodynia. Activation of low threshold mechanoreceptors would drive cells that normally receive nociceptive inputs (Fig. 4).

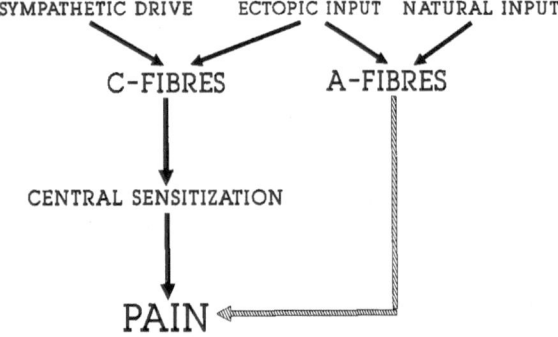

Fig. 3. Chronic nerve injury. C-fibre input has the capacity to produce increased excitability in dorsal horn neurons, the phenomenon of central sensitization. Ongoing abnormal C-fibre input could be generated either by the development of abnormal pacemaker properties in these afferents leading to an ectopic input or by the development of a sensitivity to catecholamines such that sympathetic activity leads to C-fibre input. Once central sensitization was established A-fibre input would begin to produce pain

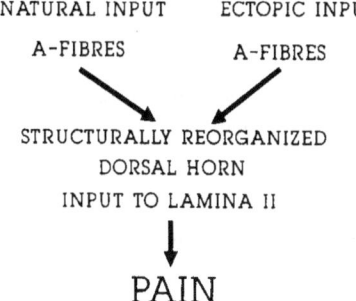

NATURAL INPUT   ECTOPIC INPUT
   A-FIBRES   A-FIBRES

       STRUCTURALLY REORGANIZED
              DORSAL HORN
         INPUT TO LAMINA II

                PAIN

Fig. 4. Chronic nerve injury. Structured activation in the dorsal horn resulting from peripheral nerve injury results in the central terminal of large myelinated, low threshold mechanoreceptor fibres sprouting from lamina III, their normal site of termination into lamina II, the normal site of C-fibre terminals. If functional synaptic contacts were established this might lead to pain as a result of A-fibre input, whether ectopic or as a result of natural stimulus

Conclusion

At the time of peripheral injury, an abnormal injury discharge may be sufficient to produce long term changes in the excitability of the spinal cord and/or an excitoxic death of dorsal horn neurons. These acute changes might set the scene for the maintenance of sensory disorders both as a result of an ongoing ectopic input, which might persistently induce a state of central sensitization and for a structural reorganization of the synaptic connections of the dorsal horn. The implications of these findings is that it may be possible to prevent some of the long term consequences of nerve damage, that the treatment at the time of injury may need to be quite different for that required later, and finally that treatment directed only at the periphery may be insufficient to eliminate the sensory disturbance of chronic neuropathic pain.

References

1. Aldskogius H, Arvidsson J, Grant G (1985) The reaction of primary sensory neurons to peripheral nerve injury with particular emphasis on transganglionic changes. Brain Res 373: 15–21
2. Arner S, Meyerson BA (1988) Lack of analgesic effect of opioids on neuropathic and idiopathic forms of pain. Pain 33: 11–23
3. Bach FW, Jenson TS, Kastrup J, Stigdby B, Dejgard A (1990) The effect of intravenous lidocaine on nociceptive processing in diabetic neuropathy. Pain 40: 29–34
4. Bennett GJ (1991) Evidence from animal models on the pathogenesis of painful peripheral neuropathy: relevance for pharmacotherapy. In: Basbaum AI, Besson JM (eds) Towards a new pharmacotherapy of pain. Wiley, Chichester, pp 365–379
5. Bowsher D (1991) Neurogenic pain syndromes and their management. Brit Med Bull 47: 644–666
6. Bullitt E (1992) The treatment of hyperalgesia following nerve injury. In: Willis WD Jr (ed) Hyperalgesia and allodynia. Raven, New York, pp 345–361
7. Burchiel KB (1984) Effects of electrical and mechanical stimulation on two foci of spontaneous activity which develop in primary afferent neurons after peripheral axotomy. Pain 18: 249–265
8. Cambell JN, Raja SN, Meyer RA, McKinnon SE (1988) Myelinated afferents signal the hyperalgesia associated with nerve injury. Pain 32: 89–94
9. Cambell JN, Meyer RA, Davis KD, Raja SN (1992) Sympathetically maintained pain. A unifying hypothesis. In: Willis WD Jr (ed) Hyperalgesia and allodynia. Raven, New York, pp 141–149
10. Castro-Lopes JM, Coimbra A, Grant G, Arvidsson J (1990) Ultrastructural changes of the central scalloped (C_1) primary afferent endings of synaptic glomeruli in the substantia gelatinosa Rolandi of the rat after peripheral neurotomy. J Neurocytol 19: 329–337
11. Chen L, Huang L-Y M (1992) Protein kinase C reduces MG_{2+} block of NMDA-receptor channels as a mechanism of modulation. Nature 356: 521–523
12. Chi S-J, Levine JD, Basbaum AI (1989) Time course of peripheral neuroma-induced expression of Fos protein immunoreactivity in spinal cord of rats and effects of local anesthetics. Neurosci Abst 15: 155
13. Cook AJ, Woolf CJ, Wall PD, McMahon SB (1987) Dynamic receptive field plasticity in rat spinal cord dorsal horn following C-primary afferent input. Nature 325: 151–153
14. Davar G, Hama A, Deykin A, Vos B, Maciewicz R (1991) MK-801 blocks the development of thermal hyperalgesia in a rat model of experimental painful neuropathy. Brain Res 553: 327–330
15. Devor M (1991) Neuropathic pain and injured nerve: peripheral mechanisms. Brit Med Bull 47: 619–630
16. Devor M, Wall PD, Catalan N (1992) Systemic lidocaine silences ectopic neuroma and DRG discharge without blocking nerve conduction. Pain 48 (1992) 261–268
17. Gracely RH, Lynch SA, Bennett GJ (1992) Painful neuropathy: Altered central processing, maintained dynamically by peripheral input. Pain (in press)
18. González-Darder JM, Barberá J, Abellán MJ (1986) Effects of prior anaesthesia on autotomy following sciatic transection in rats. Pain 24: 87–91
19. Koltzenburg M, Wahren LK, Torebjörk HE (1992) Dynamic changes of mechanical hyperalgesia in neuropathic pain states and healthy subjects depend on the ongoing activity of unmyelinated nociceptive afferents. Pflügers Archiv 420: R52
20. Nordin M, Nystrom B, Wallin U, Hagbarth KE (1984) Ectopic sensory discharges and paraesthesias in patients with disorders of peripheral nerves, dorsal roota and dorsal columns. Pain 20: 231–245
21. Noordenbos W, Wall PD (1981) Implications of the failure of nerve resection and graft to cure chronic pain produced by nerve lesions. J Neurol Neurosurg Psychiatry 44: 1068–1073
22. Price DD, Bennett GJ, Raffii M (1989) Psychological observations on patients with neuropathic pain relieved by a sympathetic block. Pain 36: 273–288
23. Sato J, Perl ER (1991) Adrenergic excitation of cutaneous pain receptors induced by peripheral nerve injury. Science 251: 1608–1610
24. Seltzer Z, Beilin BZ, Ginzburg R, Paran Y, Shimko T (1991) The role of injury discharge in the induction of neuropathic pain behaviour in rats. Pain 46: 327–336
25. Seltzer Z, Cohn S, Ginzburg R, Bellin BZ (1991) Modulation of neuropathic pain behaviour in rats by spinal disinhibition and NMDA receptor blockade of injury discharge. Pain 45: 69–75
26. Sharp FR, Griffith J, Gonzalez MF, Sagar SM (1989) Trigeminal

nerve section induces Fos-like immunoreactivity (FLI) in brainstem and decreases FLI in sensory cortex. Mol Brain Res 6: 217–220

27. Simone DA, Sorkin LS, Oh UT, Chung JM, Owens C, LaMotte RH, Willis WD (1991) Neurogenic hyperalgesia: Central neural correlates in responses of spinothalamic tract neurons. J Neurophysiol 66: 228–246

28. Skene HJP (1989) Axonal growth associated proteins. Ann Rev Neurosci 12: 127–156

29. Sugimoto T, Bennett GJ, Kajander KC (1990) Transsynaptic degeneration in the superficial dorsal horn after sciatic nerve injury: effects of a chronic constriction injury, transection and strychnine. Pain 42: 205–213

30. Thompson SWN, King AE, Woolf CJ (1990) Activity-dependent changes in rat ventral horn neurons in vitro; summation of prolonged afferent evoked postsynaptic depolarizations produce a D-2-amino-5-phosphonovaleric acid sensitive windup. Eur J Neurosci 2: 638–649

31. Torebjörk HE, Lundberg LER, LaMotte RH (1992) Central changes in processing of mechanoreceptor input in capsaicin-induced sensory hyperalgesia in humans. J Physiol 448: 765–780

32. Wall PD, Gutnick M (1974) Ongoing activity in peripheral nerves: The physiology and pharmacology of impulses originating from a neuroma. Exp Neurol 43: 580–593

33. Wall PD, Waxman S, Basbaum AI (1974) Ongoing activity in peripheral nerve: Injury discharge. Exp Neurol 45: 576–589

34. Wall PD, Woolf CJ (1984) Muscle but not cutaneous C-afferent input produces prolonged increases in the excitability of the flexion reflex in the rat. J Physiol (Lond) 356: 443–458

35. White JC, Sweet WH (1969) Pain and the neurosurgeon. A forty-year experience. Thomas, Springfield, Ill

36. Wirth FR, Rutherford RB (1970) A civilian experience with causalgia. Arch Surg 100: 633–638

37. Woolf CJ (1983) Evidence for a central component of post-injury pain hypersensitivity. Nature 306: 686–688

38. Woolf CJ (1991) Generation of acute pain: Central mechanisms. Brit Med Bull 47: 523–533

39. Woolf CJ, King AE (1990) Dynamic alterations in the cutaneous mechanoreceptive fields of dorsal horn neurons in the rat spinal cord. J Neurosci 10: 2717–2726

40. Woolf CJ, Thompson SWN (1991) The induction and maintenance of central sensitization is dependent on N-methyl-D-aspartic acid receptor activation; implications for the treatment of post-injury pain hypersensitivity states. Pain 44: 293–299

41. Woolf CJ, Wiesenfeld-Hallin Z (1985) The systemic administration of local anaesthetics produces a selective depression of C-afferent fibre evoked activity in the spinal cord. Pain 23: 361–374

42. Woolf CJ, Wall PD (1986) The relative effectiveness of C primary afferent fibres of different origins in evoking a prolonged facilitation of the flexor reflex in the rat. J Neurosci 6: 1433–1443

43. Woolf CJ, Reynolds ML, Molander C, O'Brien C, Lindsay RM, Benowitz LI (1990) GAP-43, a growth associated protein, appears in dorsal root ganglion cells and in the dorsal horn of the rat spinal cord following peripheral nerve injury. Neuroscience 34: 465–478

44. Woolf CJ, Shortland P, Coggeshall RE (1992) Peripheral nerve injury triggers central sprouting of myelinated afferents. Nature 355: 75–77

45. Xu X-J, Maggi CA, Wiesenfeld-Hallin Z (1991) On the role of NK-2 tachykinin receptors in the mediation of spinal reflex excitability in the rat. Neuroscience 44: 483–490

Correspondence: Clifford J. Woolf, M.D., Department of Anatomy and Developmental Biology, University College London, Gower Street, London WC1E 6BT, U.K.

Acta Neurochir (1993) [Suppl] 58: 131–135

Painful Nerve Injuries: Bridging the Gap Between Basic Neuroscience and Neurosurgical Treatment

K. J. Burchiel, T. J. Johans, and **J. Ochoa**[1]

Division of Neurosurgery, Oregon Health Sciences University, and [1] Department of Neurology, Neuromuscular Division, Good Samaritan Hospital, Portland, Oregon, U.S.A.

Summary

Pain which followed suspected nerve injury was comprehensively evaluated with detailed examination including history, neurologic exam, electrodiagnostic studies, quantitative sensory testing, thermography, anesthetic and sympathetic nerve blocks. Forty two patients treated surgically fell into four discrete groups: 1) Distal sensory neuromas treated by excision of the neuroma and reimplantation of the proximal nerve into muscle or bone marrow, 2) Suspected distal sensory neuromas in which the involved nerve was sectioned proximal to the injury site and reimplanted, 3) Proximal neuromas-incontinuity of major sensorimotor nerves treated by external neurolysis, and 4) Proximal major sensorimotor nerve injuries at points of anatomic entrapment treated by external neurolysis and transposition, if possible. Patient follow up was possible in 40/42 patients (95%) from 2–32 months (average F/U = 11 mo.). Surgical success was defined as: ≥ 50% improvement in pain (VAS) or pain relief subjectively rated as good or excellent, and no postoperative narcotic usage. Overall, 40% (16/40) of patients met those criteria. Success rates varied as follows: Group 1 (n = 18) 44%, Group 2 (n = 10) 40%, Group 3 (n = 5) 0%, and Group 4 (n = 7) 57%. A total of 12 of 40 patients (30%) were employed both pre- and postoperatively.

We conclude that: 1) Neuroma excision, neurectomy and nerve release for injury-related pain of peripheral nerve origin yields substantial subjective improvement in a minority of patients, 2) External neurolysis of proximal mixed nerves is ineffective in relieving pain, 3) Surgically proving the existence of a neuroma, with confirmed excision may be preferable, 4) Traumatic neuroma pain is only partly due to a peripheral source, 5) Demographic and neurologic variables do not predict success, 6) The presence of a discrete nerve syndrome and mechanical hyperalgesia do modestly predict pain relief, 7) Ongoing litigation is the strongest predictor of failure, 8) Change in work status is not a likely outcome.

Keywords: Peripheral nerve lesion; neuropathic pain; surgical treatment; outcome.

Introduction

Neuropathic pain can be defined as pain secondary to a change in the normal function of the peripheral or central nervous system, such that pain is perceived in the absence of any noxious stimulus. Numerous converging lines of evidence indicate that there are at least five potential mechanisms to explain neuropathic pain after nerve injury: 1) Sensitization of nociceptors, 2) Spontaneous ectopic impulse generation in nociceptive and non-nociceptive axons, 3) Abnormal chemo- and mechanosensitivity of axons within the neuroma, 4) A breakdown of the normal isolation between nociceptive and non-nociceptive axons (ephapses), and 5) Changes in the responsiveness of CNS neurons to inputs that are otherwise innocuous ("centralization").

Four of these mechanisms invoke the role of peripheral nervous system abnormalities which might be alleviated by removal of the offending structure, the "neuroma", i.e., incontinuity nerve lesions, true end-bulb neuromas or a combination. Our thesis has been that after a thorough evaluation in the neuromuscular laboratory, patients may be logically grouped, and that some patients may improve after peripheral nerve surgery intended to eliminate the abnormal nerve segment.

Methods and Materials

Preoperative Assessment

An extensive preoperative evaluation (Table 1) provided a large base of information for post-operative outcome measures on 42 patients. One would expect that poorly localized pain and ill defined signs and symptoms would be poor prognostic factors. To study this in a more scientific fashion, the concept of a discrete nerve syndrome was developed. A discrete nerve syndrome (DNS) was considered to be a condition in which a single nerve could account for all the neurological findings and pain distribution. The presence and distribution of mechanical and thermal hyperalgesia were considered very important in the determination of a discrete nerve

Table 1. *Assessment of Painful Nerve Injuries*

History (videotaped)
Physical/neurologic examination
Sympathetic function tests
Quantitative sensory testing
  Mechanical algometry
  Thermal stimuli (cold, warm, and pain)
Electrodiagnosis (routine)
  EMG/NCV
Thermography (baseline/provocative)
  Video telethermography
  Contact thermography
Nerve block (with placebo block)
  Peripheral nerve block
  Sympathetic block
Microneurography (elective)

EMG electromyography; *NCV* nerve conduction
velocity.

syndrome. Symptoms of "reflex sympathetic dystrophy" (RSD), were
determined by assessing for trophic changes, vasomotor, color, temperature changes, bone scan, and the results of sympathetic blocks.

Intraoperative Methods

Operations were performed under local anesthesia with intravenous sedation and analgesia. The presence of Tinel's sign can be
taken as a reliable sign of the location of the neuroma. Prior scars
or incisions are also helpful indicators of the region for exploration.
Dissection was carried out using loupe or microscopic magnification.
Mechanosensitivity of the injured nerve segment with production of
Tinel's sign was often an important guide to the dissection. Once the
nerve with the suspected neuroma were isolated electrical stimulation
of the nerve (0.5–1.5 mA, 100 microsecond pulse width, 50–100 HZ,
balanced biphasic constant current square wave impulses) was carried
out using electrified nerve hooks. This method was very useful for
identify the neural structure in surrounding scar, and in replicating
the patients pain distribution with stimulation-induced paresthesias.

In groups 1 and 2, once mapping of the neuroma was completed,
the proximal nerve was locally anesthetized and sharply divided.
The nerve was then implanted retrograde into an intramuscular
pocket using 6–0 prolene suture, or into bone marrow if no muscle
was available, e.g., in digital neuromas. In group 3 an external
neurolysis was performed, i.e., dissection in the plane of the external
epineurium, with translocation of the injured segment into a virgin
tissue bed, if possible. In group 4, external neurolysis was performed
and the nerve transposed using standard techniques appropriate to
the site of entrapment.

Postoperative Follow-up

Patients were assessed both by clinical evaluation and phone
conversation with a mean follow up of 11 months (2–32 mo.) (Table 2).
The Visual Analog Scale (VAS) for pain was determined preoperatively
and at each subsequent contact. The VAS is the patient's subjective
pain rating from 0 (no pain) to 10 (maximum pain). Narcotic use
was also serially determined. The patients were also asked to give
an overall assessment of the results of surgery, (excellent, good, poor
or failure). The surgical outcome was considered successful if there
was a 50% improvement based on the VAS (0–10 point scale), *or*
a subjective rating of good to excellent, and the patient was off of
narcotics.

Results

Of the initial 42 patients evaluated and surgically
treated, adequate information was available on 40 patients (95% follow-up). The surgical treatment rendered
varied according to the nerve involved, providing four
discrete subgroups; 1) distal sensory neuromas treated
by excision of the neuroma and reimplantation of the
proximal nerve into muscle or bone marrow, 2) suspected distal sensory neuromas in which the involved
nerve was sectioned proximal to the injury site and
re-implanted, 3) proximal neuromas incontinuity of

Table 2. *Pre- and Post-Operative Variables Analyzed for Neurectomy/Neurolysis Patients*

History
Nerve involved   Post-operative employment status (EMPO)
Pre-operative narcotic use (NARP)   Litigation (LIT)
Post-operative narcotic use (NRPO)   Workman's compensation (WC)
Pre-operative employment status (EMPP)   Prior surgical procedures (PP)

Physical
Tinel's sign (TIN)   Thermal hyperalgesia (THER)
Reflex sympathetic dystrophy (RSD)   Discrete nerve syndrome (DNS)
Mechanical hyperalgesia (MECH)

Patient assessment
Pre-operative visual analog scale (VASP)
Post-operative visual analog scale (VAPO)
Subjective overall result (E, G, P, F)

Calculated
Percent improvement (change in visual analog scale)

$$\frac{VASP - VAPO}{VASP} \times 100$$

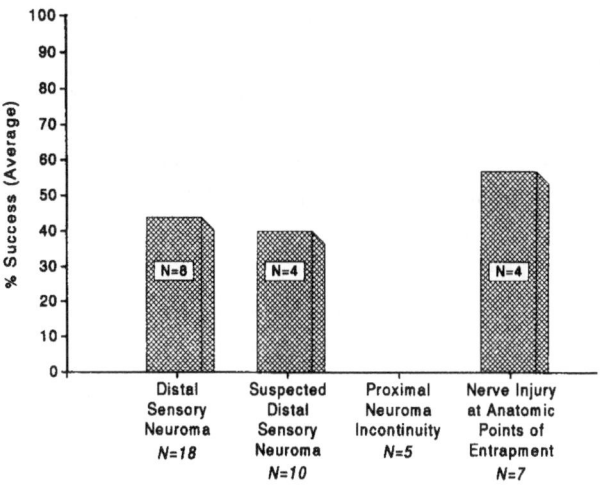

Fig. 1. Relationship of treatment groups to outcome. Graph showing the successful outcomes by groups. There were no successful outcomes in group 3, proximal incontinuity neuromas treated by external neurolysis only

Table 3. *Relationship of Potential Predictors and Outcome*

Predictor	Incidence	Success rate	P-Value (X_z)
Tinel sign	77.5%	45%	0.22
R.S.D	27.5%	27%	0.31
Mechanical hyperalgesia	79%	37%	0.96
Thermal hyperalgesia	52%	41%	0.83
Discrete nerve syndrome	57.5%	48%	0.23
Narcotic use, pre-op.	50%	50%	0.5
Litigation	17.5%	14%	0.12
Workman's compensation	53%	38%	0.8
Employment, pre-op.	33%	54%	0.2
Prior procedures	43%	41%	0.9

major sensorimotor nerves treated by external neurolysis, 4) proximal major sensorimotor nerve injuries at points of anatomic entrapment treated by external neurolysis and nerve transposition if possible.

Figure 1 depicts the relationship of successful outcome for these four subgroups. Those patients with proximal incontinuity neuromas of major sensorimotor nerves (group 3) had the worst surgical outcomes, since there were no successes in this small subgroup of five patients. In contrast, those patients with traumatic neuromas-incontinuity at areas of anatomic entrapment did quite well, as four of seven (57%) of the patients had successful surgical outcomes. The most common neuroma treated involved distal sensory nerves, either proven at surgery (group 1) or strongly suspected by physical examination (group 2). These subgroups had a 44% (8/18) and 40% (4/10) success rate, respectively.

Ten potential predictors obtained through history and physical examinations were correlated with successful surgical outcome, and the results and incidence of each are listed in Table 3. Although none of the predictive factors reached statistical significance by chi-square analysis, litigation, discrete nerve syndrome, preoperative employment and the presence of Tinel's sign approached statistical significance. The presence of a discrete nerve syndrome was moderately correlated to pain improvement on the VAS. Patients with a DNS had an average reduction of 3.0 points on the postoperative VAS for pain, compared to the 1.7 point drop for those patients without DNS (Pearson's R analysis,

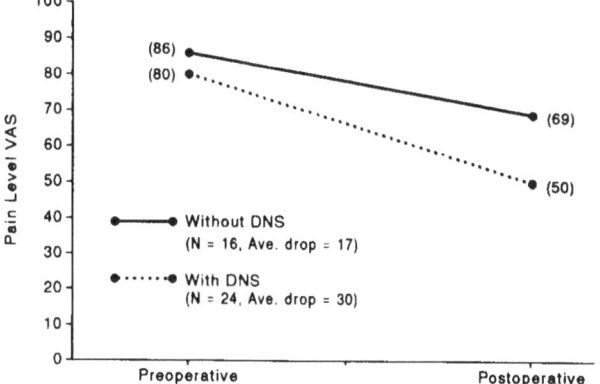

Fig. 2. Relationship of discrete nerve syndrome to outcomes pre- and postoperative pain ratings by VAS in patients with (n = 16, avg. drop in VAS = 17) and without (n = 24, avg. drop in VAS = 30) a discrete nerve syndrome (DNS). Of the 14 potential variables analyzed, DNS was the most correlated to pain improvement, but this was not statistically significant (Pearson's R analysis, r = − 0.27, p = 0.10). Numbers in parentheses are average VAS scores

Fig. 2). Mechanical hyperalgesia (allodynia), either dynamic or static, as an independent variable, was not significantly related to success as assessed by chi-square analysis. When both DNS and mechanical hyperalgesia were considered together, there was a moderate correlation with pain improvement (Fig. 3). The group of patients with both DNS and mechanical hyperalgesia had an average pain score reduction of 4.4 points, which was greater than the patients with either factor alone. Many of the patients in this group had dramatic reductions of their preoperative pain scores (Fig. 4). Overall, among the four subgroups, the average pain level drop was 30%, while the overall success rate was 40%.

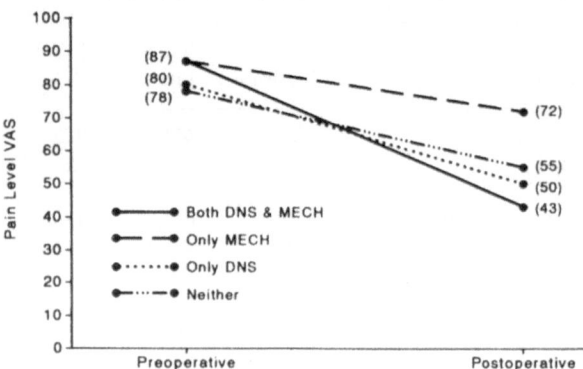

Fig. 3. The relationship of discrete nerve syndrome (*DNS*) and mechanical hyperalgesia (*MECH*). Pre- and postoperative pain ratings in patients with a discrete nerve syndrome (*DNS*) or mechanical hyperalgesia (*MECH*), both or neither. The 15 patients with both DNS and MECH had greater reduction in pain than did the 21 patients with only DNS, only MECH, or neither (ANOVA interaction $p = 0.08$). Numbers in parentheses are average VAS scores

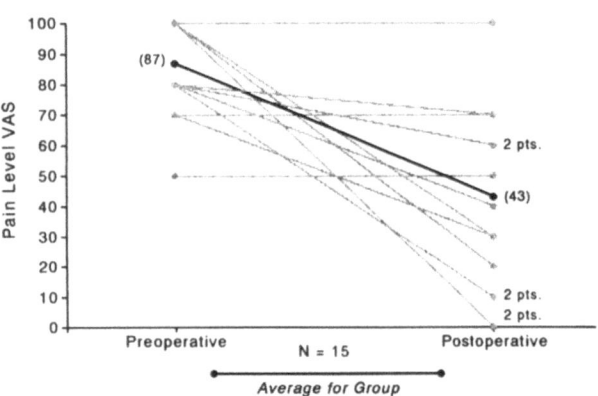

Fig. 4. The relationship of discrete nerve syndrome/mechanical hyperalgesia and pain improvement. Both DNS and mechanical hyperalgesia. Pre- and postoperative VAS data on individual patients with both a discrete nerve syndrome (DNS) and mechanical hyperalgesia (MECH). Note that a number of patients had dramatic resolution in pain complaints. Numbers in parentheses are average VAS scores

Discussion

Central nervous system effects of peripheral nerve injury are profound and their relevance to painful nerve injuries cannot be ignored. Nevertheless, we have adopted the thesis that in many patients the majority of the pathophysiological abnormalities are peripheral, rather than central. The physiologic consequences of nerve injury have been extensively studied in the laboratory in recent years. The basic properties of axons are altered by injury such that rather than high-fidelity *conductors* of neuronal signals, they become *generators* of abnormal activity. The pathophysiology of injured

axons has been extensively reviewed[2]. Suffice it to say that injured axons develop several properties: 1) Ectopic discharge, 2) mechanosensitivity, and 3) chemosensitivity, particularly to alpha-adrenergic agents or neurotransmitter. These properties would logically lead to the conclusion that removal of the generator of abnormal neuronal activity, or transposition of the injured nerve segment to a less mechanically stimulated area should be beneficial. It is on this basis that an algorithm for the management of these patients has been described[1,3].

Treatment Recommendations

It is disappointing, after such a thorough investigation, to be unable to identify prognostic factors to help influence future surgical outcomes for neuroma patients. However, these results are promising and do tend to question widespread pessimism that is associated with the treatment of these patients. For example, patients who were on workman's compensation did no worse than the group as a whole. Neither preoperative narcotic use nor employment status were predictive of outcome. Of particular note, no patient unemployed preoperatively became employed post-operatively, even if the surgery was a success by our criteria. All patients employed preoperatively, remained employed postoperatively, even if the surgery was a failure. The physical findings sometimes associated with "reflex sympathetic dystrophy", findings of mechanical or thermal hyperalgesia, or the presence of Tinel's sign did not independently reach predictive significance. However, when combined into the collective term of "discrete nerve syndrome", these factors were moderately correlated with pain improvement. Finally, patients with discrete nerve syndrome in combination with mechanical hyperalgesia had greater reduction of pain than with either factor alone.

Although painful nerve injuries are fairly common, suprisingly little recent clinical investigative work on neuromas has been produced. The numerous non-surgical treatment options for neuroma reflect that no one approach has shown uniformly superior or effectie results. As in other chronic pain syndrome the use of chronic narcotics should be avoided. Furthermore, preoperative evaluation by a psychologist with particular experience in treating patients with chronic pain should be completed to screen out patients with major psychopathology or functional overlay. Finally, we feel that it is imperative that treating specialists avoid the tendency to lump patients with painful nerve injuries into vague and possibly pejorative diagnostic categories, such as "reflex sympathetic dystrophy", "causalgia", and "neuroma pain". Only by careful clinical study can we differentiate patients that we may palliate by surgical means.

References

1. Burchiel KJ, Ochoa J (1991) Pathophysiology of injured axons. In: Burchiel KJ (ed) Surgical management of peripheral nerve injury and entrapment. Neurosurg Clin North Am 2: 105–116
2. Burchiel KJ, Ochoa J (1991) Surgical management of post-traumatic neuropathic pain. In: Burchiel KJ (Ed) Surgical management of peripheral nerve injury and entrapment. Neurosurg Clin North Am 2: 117–126
3. Burchiel KJ, Zimmerman CG (1991) Surgical management of post-traumatic neuropathic pain. Eur J Pain 12(3): 69–76

Correspondence: Prof. Kim J. Burchiel, M.D., Division of Neurosurgery, 3181 S.W. Sam Jackson Park Road, L-472, Portland, OR 97201, U.S.A.

Acta Neurochir (1993) [Suppl] 58: 136–140

What Can the Neurosurgeon Offer in Peripheral Neuropathic Pain?

J. Gybels, R. Kupers, and **B. Nuttin**

Department of Neurology and Neurosurgery, University of Leuven, Leuven, Belgium

Summary

Neurosurgery has much to offer in the treatment of peripheral neuropathic pain but selection of the best procedure for a given patient remains problematic: planning of the treatment must be based on an analysis of the pathophysiological mechanism in the given case but the identification of this mechanism is often difficult. Available procedures are: 1) Nerve repair, neurolysis and nerve relocation; 2) Interventions on the sympathetic nervous system; 3) Neurostimulation; 4) Intraspinal morphine; 5) Ablative lesions.

Neurosurgeons have, or should have, the necessary neuroscience background and microsurgical skills to be important partners of the team caring for patients with peripheral neuropathic pain.

Keywords: Neurosurgery; peripheral neuropathic pain; man; rat.

Introduction

When discussing neurosurgical treatment of neuropathic pain one can choose different approaches.

A first and most logical possibility is to treat the subject according to the mechanisms for the genesis of pain in peripheral nerve pathology. Although much has been learned about mechanisms of peripheral neuropathic pain in the last few years (see Woolf, this volume, pp.   ) it remains difficult for the clinician to take this approach, because in a given patient, several of these mechanisms may be acting in varying degrees and their identification on a routine clinical basis remains somewhat problematic.

A second possibility is to develop surgical treatment according to disease: entrapment neuropathy, neuroma, plexus avulsion, post-herpetic neuralgia, and so on. But a syndrome oriented classification is inadequate, because for the planning of the treatment analysing the pathophysiological mechanism is mandatory[7], although on the basis of the clinical reports it can easily be understood how complex the physiopathology may be.

We are then left to use the classical and straight forward but rather artificial framework of developing the subject according to techniques. Nerve repair and nerve relocation, interventions on the sympathetic nervous system, stimulation at different sites of the nervous system, drugs injected into the spinal fluid.

Nerve Repair, Neurolysis and Nerve Relocation

This is a vast subject, and space allows only for a few general remarks. Much useful up-to-date information can be found in a textbook by Mackinnon and Dellon[6] and a review by Burchiel and Ochoa[1].

After appropriate conservative treatment and expert psychological evaluation, *nerve block* is an important step towards a potential surgical approach to a painful neuroma. This clinical test has its scientific logic since it has been experimentally shown that both in the animal and humans there is in the presence of a neuroma spontaneous ongoing and mechanically evoked activity in both A and C fibre afferents. Testing for *placebo* response is an important step in a diagnostic block, but the interpretation of the block may remain difficult, since a placebo response does not necessarily mean psychopathology or malingering. The clinical syndrome of neuroma may be due to an anatomical neuroma, but is also observed when there is pain in a region of a scar with altered sensation in the distribution of the nerve when the region of the scar is palpated. *Nerve repair* is probably one of the most important means of preventing a neuroma. If there is no hope of restoring continuity, two main surgical strategies have been proposed. One is to remove the neuroma and prevent the formation of a new neuroma. This has been an elusive goal and many ingenious methods of restricting regeneration of axons have been studied and used ranging from the placement of mechanical barriers, to expose in experiments the painful nerve to ricin, an agent that is transported back in the nerve body where it initiates cell death, to placing the cut end of the nerve in a new environment such as innervated muscle. A second surgical strategy is to move the cut nerve or the neuroma into a position where mechanical stimulation is less likely such as for instance the bone marrow. According to a recent review by Burchiel and Ochoa[1], whether the nerve is burried in muscle or bone or the neuroma is removed or simply relocated, success with surgical manipulation of the painful neuroma now ranges from 65% to 82% and in many instances, patients who have persistent pain after excision of a neuroma can still be operated upon successfully.

For a neuroma in continuity, the surgical strategy is different. In the larger mixed nerves of the extremities, resection of the site of the nerve injury may not be feasable. Here a choice may have to be made between *external neurolysis* and *nerve relocation* and *internal neurolysis* with intra-operative neurophysiological testing, neurolysis of the functional fascicles and nerve graft and muscle or bone implant or the non-functional fascicles. Burchiel *et al.* (this volume, pp. 131) in their follow-up of the outcome after surgical treatment of post-traumatic pain of peripheral nerve origin in 42 surgically treated patients concluded that: 1) neuroma excision, neurectomy and nerve release for injury-related pain of peripheral nerve origin yielded substantial functional relief in a minority of patients; 2) external neurolysis of proximal mixed nerves was ineffective in relieving pain; 3) demographic and neurological variables did not predict success and 4) return to work was not a likely outcome in these patients.

Surgery of traumatized nerves for painful sequelae is an extra-ordinarily difficult field asking for particular expertise; it seems to me that the conclusions of an influential paper by Noordenbos and Wall[8] have perhaps led in the neurosurgical community to a too negative view of the possibilities of surgery.

Interventions on the Sympathetic Nervous System

Although the pathogenesis of disorders like causalgia, algoneurodystrophy, reflex sympathetic dystrophy (R.S.D.) etc... remains little understood, much has been learned about the chemosensitivity of injured nerve fibres to norepinephrine and the role of the sympathetic nerve system in these disorders.

R.S.D., in contrast to causalgia, includes those patients whose pain is initiated by soft tissue or bone injury without direct nerve injury. Roberts[12] coined the term sympathetically maintained pain (S.M.P.). This syndrome requires that the pain responds to sympathetic blockade, and this criterion is particularly useful from a therapeutic point of view. The different clinical syndromes, which are grouped together under the label S.M.P. have in common that initially the pain may be confined to a nerve distribution as in neuralgia, but progressively will spread beyond this territory. Frequent symptoms and signs are persisting pain with a burning character, allodynia, dysaesthesia, hyperalgesia, hyperpathia and social and psychological distress.

Sympathetic blockade requires some comment. An absence of proper *placebo* control destroys the meaning of the test. False positive and false negative results are possible as a result of over- or underanaesthetizing the region, with a loss of somatic nerve function or inadequate sympathetic blockade respectively. Because of these difficulties, additional tests are used to diagnose S.M.P. These include systemic administration of L-adrenergic antagonists such as phentolamine and provocative tests with adrenergic agonists such as intradermal injection of phenylephrine. Moreover, the demonstration of hyperalgesia to cooling stimuli, such as placing gently a drop of acetone or ice water onto the area that is hyperalgesic to mechanical stimuli, may represent a simple and sensitive, but not specific test for S.M.P. Campbell *et al.*[2] have given a vivid account with the help of an illustrative case as to how S.M.P. can be diagnosed.

In their review of the sympathetic nervous system in painful nerve injury, O'Neill and Burchiel[10] note that in S.M.P. success rate with chemical or surgical sympathetic block have ranged from 12 to 90% and go on to say that this most probably reflects the variance in diagnostic criteria, the effectiveness of sympathectomy and the natural history of S.M.P. Recurrence of pain may occur in as many as one-third of patients following sympathectomy in follow-up periods ranging from 3–8 years[11]. Even full relief of pain after several technically successful sympathetic blocks is not necessarily an adequate basis for recommending operation[4]. An extreme position is taken by Ochoa[9] when he said: "Regardless of conventional wisdom in current textbooks, the acknowledged reality by academic neurologists capable of evaluating patients with chronic neurological symptoms fully diagnosable as "Causalgia", "R.S.D.", or "S.M.P."

(since they do respond transiently to diagnostic sympathetic blocks) is that such patients are not cured by sympathectomy. They often respond for weeks or a few months, on bases that may be speculated upon. However, the painful syndrome regularly recurs".

Neurostimulation

Many patients with a peripheral neuropathy will not be well treated by nerve surgery and sympathectomy. Long-term follow-up studies of large series carefully analysed now allow the opportunity to evaluate the results of neurostimulation in these conditions.

Stimulation of the *peripheral nerves* (PNS) for pain relief, using implanted devices, has been used for over 20 years. In a review of the reference[4] 46% of the 299 described cases can be considered as a success. In a personal series of 11 cases, mostly with traumatic lesions of small nerves of the hand there was total pain relief in 8. None of these patients needed analgesics, and all were able to perform their daily life and professional activities. As illustrated in Fig. 1, in which the duration of stimulation at latest follow-up compared to the initial duration is plotted logarithmically against time, a most remarkable finding was that in all patients, except one, pain relief was obtained with diminishing frequency of use. This was usually described as an "after effect", a pain-free period stopping the stimulation, which increased with time. This resulted in some patients in a PNS use for less

than half an hour every month, although at first they had to use it continuously. We conclude that in view of the rather favourable long-term results of peripheral nerve stimulation and the simple low-risk technique it is surprising that peripheral nerve implantation is rarely performed.

On the contrary *dorsal cord stimulation* (DCS) and dorsal root stimulation (DRS) has been applied to many patients. In a recent review, Siegfried[14] reported his results of a series of 669 consecutive patients. In 398 the stimulator was internalized. 127 of these 398 suffered from a peripheral nerve lesion. Follow-up ranged from 1–16 years. A five-grade rating score including pain duration, pain intensity on a visual analogue scale, influence on mood and behaviour, and intake of drugs was used. Of the 127 implanted patients 84 were scored as very good and 32 as good. This was by far the most successful score (90%) as compared to other pathological conditions and this is indeed very impressive taking into account the difficulty in treating post-traumatic neuropathic pain.

In certain conditions, *V.P.M.–V.P.L. stimulation* (D.B.S.) may be indicated, f.i. in cranial nerve neuropathy or when DCS gives inconsistent paraesthesiae due to movement at the level of the cervical spine.

Although clinicians have for a long time performed V.P.M.–V.P.L. stimulation partly on theoretical considerations but mainly on empirical grounds, we have now preliminary data, which we think are the first experimental evidence showing that V.P.L. stimulation reduces allodynia. We used our own modified version

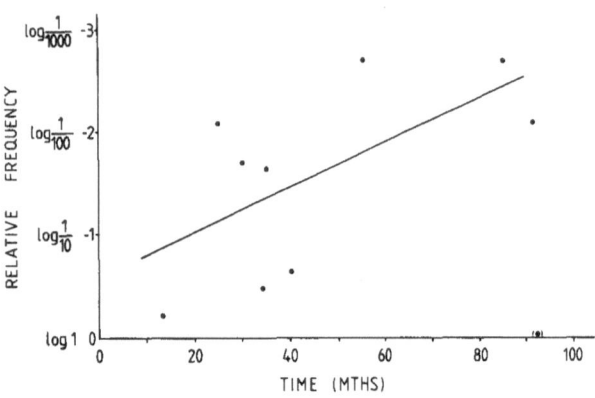

Fig. 1. Relative frequency of use of PNS plotted against time. On the axis, the follow-up period for the 10 patients with good or moderate results is shown. On the y-axis, a logarithmic scale is used for the relative frequency of use at latest follow-up, compared with the initial use. With the least squares method, a linear regression was performed with the 9 patients who reported diminishing frequency of use. The equation of this straight line is: $y = -0.022x - 0.55$. The correlation coefficient is 0.63. From Gybels *et al.*, 1990[3]

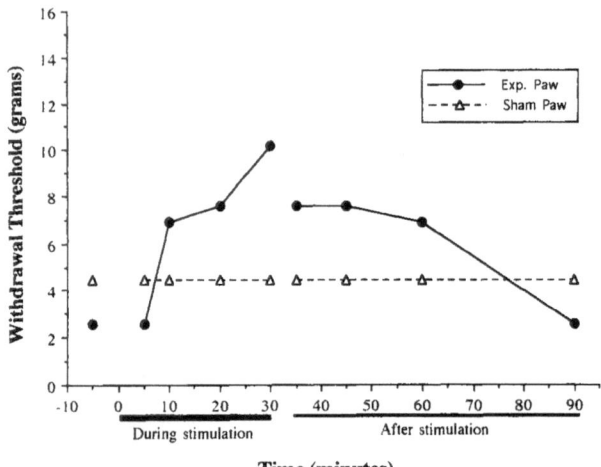

Fig. 2. Influence of contralateral electrical stimulation of V.P.L. on mechanical allodynia produced in a rat by partial entrapment of the sciatic nerve. On x-axis: time in minutes, on y-axis intensity of mechanical stimulation in grams

of the model of peripheral neuropathic pain proposed by Seltzer[13] producing a high percent of rats with mechanical allodynia and to a lesser degree thermal allodynia and signs of spontaneous pain (paw elevation and scratching). Electrodes were implanted into the V.P.L. and their position verified electrophysiologically. Figure 2 shows how in an experimental rat V.P.L. stimulation antagonizes the observed mechanical allodynia.

Intraspinal Infusion of Opioids

The therapeutic efficacy of spinally administered opioids in the treatment of postoperative pain and cancer pain is well established now. However, there is much less consensus on its efficacy, and not many reports have documented experience in patients with chronic non-cancer pain. We have in "available therapy resistent" cases of failed back surgery (N 16) in whom about half had neuropathic "root" pain, peripheral deafferentation (N 4) and reflex sympathetic dystrophy (N 2), with median pain duration of 6 years, implanted a drug pump (Medtronic Synchromed system) delivering intrathecal morphine after epidural placebo and morphine test administration. During the screening procedure, patients received in a blind manner epidural morphine and placebo. About 50% of the patients tested showed a good analgesic response to epidural morphine but not to placebo administration. At the moment of the evaluation of the treatment outcome measures, mean follow-up time was 11 ± 5 months. Pain relief, use of analgesic medication, change in sleep pattern, change in daily activities and therapy satisfaction were evaluated. Verbal descriptors were converted into a numerical score. Good to very good results were obtained in the majority of the patients; the median initial dose of morphine was 0.3 mg/24 h, complications were rare. Long-term side effects occurred in about half of the patients and mainly concerned body oedema, excessive transpiration and reduced libido.

In a double blind prospective study (N 8) 5 patients suffered the failed back surgery syndrome, 2 with neuropathic "root" pain, 1 patient post-herpetic neuralgia pain (acute phase) and 1 surgical trauma to the root. The patients received courses of 4 weeks of morphine or a placebo (Fig. 3). Of the 8 patients tested, 2 responded favourably, 1 patient with failed back surgery syndrome and root pain (patient 4 in Fig. 3), and the patient with post-herpetic neuralgia (patient 2 in Fig. 3)[5]. These data indicate that for some patients with peripheral neuropathic pain there may be room for this form of treatment, but much more work remains to be done, both from the experimental and clinical point of view before it can be recommended.

Conclusion

My leitmotiv in writing this lecture which I was invited to give was to show that contrary to what is often said, neurosurgery has much to offer in peripheral neuropathic pain. This is an extraordinarily difficult field in which physiopathological diagnosis, basic knowledge, clinical acumen and craftmanship are all of prime importance. Many of the papers on this subject state that "in carefully selected patients, etc..." and a major issue now is how to select the right patient for the right procedure. It seems to me that neurosurgeons are those professionals who have, or should have, the necessary neuroscience background and microsurgical skills to be important partners of the team caring for patients with peripheral neuropathic pain.

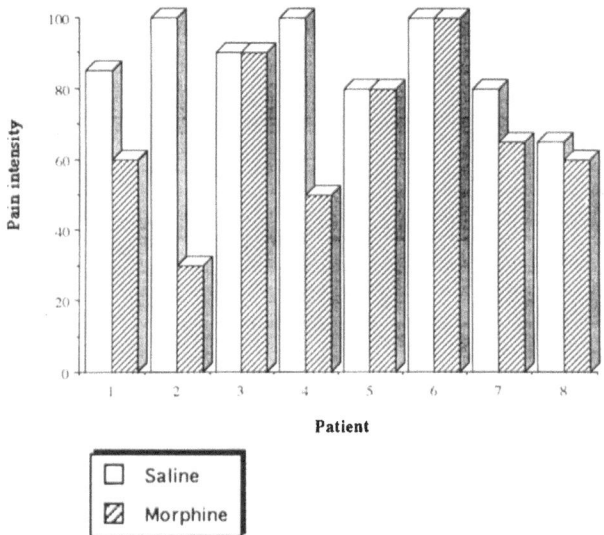

Fig. 3. Pain intensity ratings in a double blind prospective study of 8 patients during saline and morphine treatment. Pain intensity was assessed by means of the 101-point rating scale, with "0" indicating no pain and "100" the worst imaginable pain. Morphine or saline was epidurally administered by means of a PCA pump. The maximal dose was set at 0.3 mg/h (= 7.2 mg/day)

Acknowledgements

The authors are indebted to Mrs. M. Feytons-Heeren for expert technical assistance. Part of this work was supported by the N.F.W.O. of Belgium (grant no. 3.0031.87) and by Medtronic.

References

1. Burchiel KJ, Ochoa JL (1991) Surgical management of post-traumatic neuropathic pain. Neurosurg Clin North Am 2: 117–126
2. Campbell JN, Raja SN, Meyer RA (1991) How should sympathetically maintained pain be diagnosed: a case study. In: Besson JM, Guilbaud G (eds) Lesions of the primary afferent fibers as a tool for the study of clinical pain. Elsevier, Amsterdam, pp 45–51
3. Gybels J, Kupers R, Van Calenbergh F (1990) Physiological approach to management of pain. In: Dimitrijevic MR, Wall PD, Lindblom U (eds) Recent achievements in restorative neurology, Vol 3: Altered sensation and pain. Karger, Basel, pp 79–86
4. Gybels JM, Sweet WH (1989) Neurosurgical treatment of persistent pain. Karger, Basel, pp 442
5. Kupers R, Mayaert J, Gybels J (1992) The efficacy of intraspinal morphine in patients with non-cancer pain. In: Van Zundert A (ed) Highlights in regional anesthesia and pain therapy. Third Joint ESRA-ASRA Congress, pp 236–246
6. Mackinnon SE, Dellon AL (1988) Surgery of the peripheral nerve. Thieme, New York, pp 638
7. Meyerson BA (1990) Neuropathic pain: An overview. Adv Pain Res Ther 13: 193–199
8. Noordenbos W, Wall PD (1981) Implications of the failure of nerve resection and graft to cure chronic pain produced by nerve lesions. J Neurol Neurosurg Psychiatry 44: 1068–1073
9. Ochoa J (1991) Afferent and sympathetic roles in chronic neuropathic pains: Confessions on misconceptions. In: Besson JM, Guilbaud G (eds) Lesions of primary afferent fibers as a tool for the study of clinical pain. Elsevier, Amsterdam, pp 25–43
10. O'Neill OR, Burchiel KJ (1991) Role of the sympathetic nervous system in painful nerve injury. Neurosurg Clin North Am 2: 127–136
11. Payne R (1986) Neuropathic pain syndromes, with special reference to causalgia and reflex sympathetic dystrophy. Clin J Pain 2: 59–73
12. Roberts WJ (1986) A hypothesis on the physiological basis for causalgia and related pains. Pain 24: 297–311
13. Seltzer R, Dubner R, Shir Y (1990) A novel behavioural model of neuropathic pain disorders produced in rats by partial sciatic nerve injury. Pain 43: 205–218
14. Siegfried J (1991) Therapeutical neurostimulation—Indications reconsidered. Acta Neurochir (Wien) [Suppl] 52: 112–117

Correspondence: Prof. J. Gybels, Department of Neurology and Neurosurgery, UZ Gasthuisberg, Herestraat 49, B-3000 Leuven, Belgium.

Acta Neurochir (1993) [Suppl] 58: 141–144

The Possible Role of the Cerebral Cortex Adjacent to the Central Sulcus for the Genesis of Central (Thalamic) Pain—a Metabolic Study

M. Hirato, S. Horikoshi, Y. Kawashima, K. Satake, T. Shibasaki, and **C. Ohye**

Department of Neurosurgery, Gunma University, Gunma, Japan

Summary

In nine patients with central (thalamic) pain after stroke, X-CT, MRI, PET scan and intraoperative thalamic microrecordings were performed. The PET studies made use of Sokoloff's method with [18]FDG and a steady-state method with $C^{15}O_2$-$^{15}O_2$. CT scan and MRI revealed definite thalamic damage (Th) in 3 cases, putaminal damage (Put) in 3 cases, combined damage (Th + Put) in one case, and cortical (parietal) damage in 2 cases.

In patients with a subcortical lesion, the greater the severity of superficial pain, the higher was the relative value of regional cerebral glucose metabolism (rCMRGlu) as compared to oxygen metabolism (rCMRO$_2$) in the cerebral cortex around the central sulcus on the damaged side. Also, in a case with combined (Th + Put) lesion, regional oxygen extraction ratio (rOEF) was increased in this area. Moreover, in another case, central pain disappeared after a small subcortical haemorrhage in the same structure. In all patients including those with a cortical lesion, rCMRGlu was decreased in the postero-lateral (sensory) thalamus on the invalued side. The possible role of the cerebral cortex around the central sulcus for the genesis of central pain is discussed.

Keywords: Central (thalamic) pain; cerebral cortex around central sulcus; PET scan; depth microrecording.

Introduction

We have previously presented results related to the thalamic pathophysiology of central pain based on the findings on X-CT, PET scan and intraoperative thalamic microrecordings. These results suggested that deep pain was associated with functional changes in the thalamic Vim (Ventralis intermedius) nucleus after a restricted damage of the VC (Ventralis Caudalis) nucleus. Superficial pain was associated with the maintained or increased activity of thalamic intralaminar nuclei and medial thalamus after partial destruction of the VC nucleus[3]. In this paper we discussed the possible role of the cerebral cortex around the central sulcus for the genesis of central pain.

Patients and Methods (Table 1)

According to CT and MRI findings, patients were classified in four groups: thalamic lesion group, 3 cases; putaminal lesion group, 3 cases; combined lesion (putaminal with thalamic lesion) group, 1 case; and cortical (parietal) lesion group, 2 cases. Clinical features of central pain were referred to as superficial pain and deep pain. It is noted that superficial pain is more marked in cases in which the CT scan and MRI reveals a definite thalamic lesion (Table 1).

[18]FDG studies were performed in all cases of central pain with Sokoloff's method[4]. Meanwhile, gas studies were performed in 6 cases of central pain with subcortical lesions, using the steady-state method with $C^{15}O_2$-$^{15}O_2$[2]. In our PET laboratory, the X-CT and the PET scanner (Hitachi; PCT-H1, full width half maximum = 8 mm) are installed in parallel, and a single-patient bed slides between them. Thus, the centers of each slice in X-CT and PET are adjusted automatically, making precise comparisons possible. Regional CMRGlu and blood flow, rOEF and rCMRO$_2$ in central pain patients were determined in the thalamus, the caudate nucleus, putamen, globus pallidus, and the cerebral cortex. Oxygen-glucose molar utilization ratios (OGMUR) were also calculated. The cortical area was divided into three parts; anterior (premotor area), middle (primary sensorimotor area) and posterior portion (inferior parietal lobule) of the cerebral cortex on the basis of the medullary pattern of cerebral white matter. The central sulcus was included in the middle portion of the cerebral cortex.

Stereotactic VIM-Vcpc (Ventralis caudalis parvocellularis) thalamotomy was carried out in 4 cases, and VIM thalamotomy in 3 cases. Patients with cortical CVD lesion were not operated upon. During the course of stereotactic thalamotomy using Leksell's stereotactic apparatus microrecording was performed. We analyzed spontaneous thalamic unitary neural activities. Thalamic sensory responses to natural peripheral stimulation and to electrical thalamic stimulation were also studied.

Results

In the patient with a thalamic lesion (case O.H.), the electrophysiological study showed that there were many irregular burst discharges in the Vop-VIM area and a kinesthetic response to stretch of contralateral flexor muscle (Fig. 1). In this case, rCMRGlu decreased in the

Table 1. *A List of Patients*

Case	Age	Sex	Side	Stroke	Sensory disturbance	Operation
A. Thalamic lesion						
1. K.Sa.	44	F	R	inf	S+++,D+,H',dst	Vim-Vcpc
2. K.Su.	48	F	R	hem	S+++,D+,H',pst, bp	Vim-Vcpc
3. O.H.	53	M	L	hem	S+++,D+,H',dst	Vim
B. Putaminal lesion						
1. H.T.	54	F	R	hem	S+,D++ +,H',dst	Vim-Vcpc
2. M.T.	53	M	L	hem	S+ +,D++,H',bp	Vim-Vcpc
3. K.S.	43	F	L	hem	S+ +,D++,H',pst, bp	(Vim)
C. Combined (Th + Put) lesion						
1. K.Y.	67	M	L	hem	S+++,D++,H',bp	Vim
D. Cortical lesion						
1. G.O.	60	M	R	inf	S+ +,D+,H',pst	/
2. I.S.	68	F	L	inf	S+ +,H'	/

S superficial pain; *D* deep pain; *H'* hyperpathia; *dst* dysesthesia; *pst* paresthesia; *bp* burning sensation.

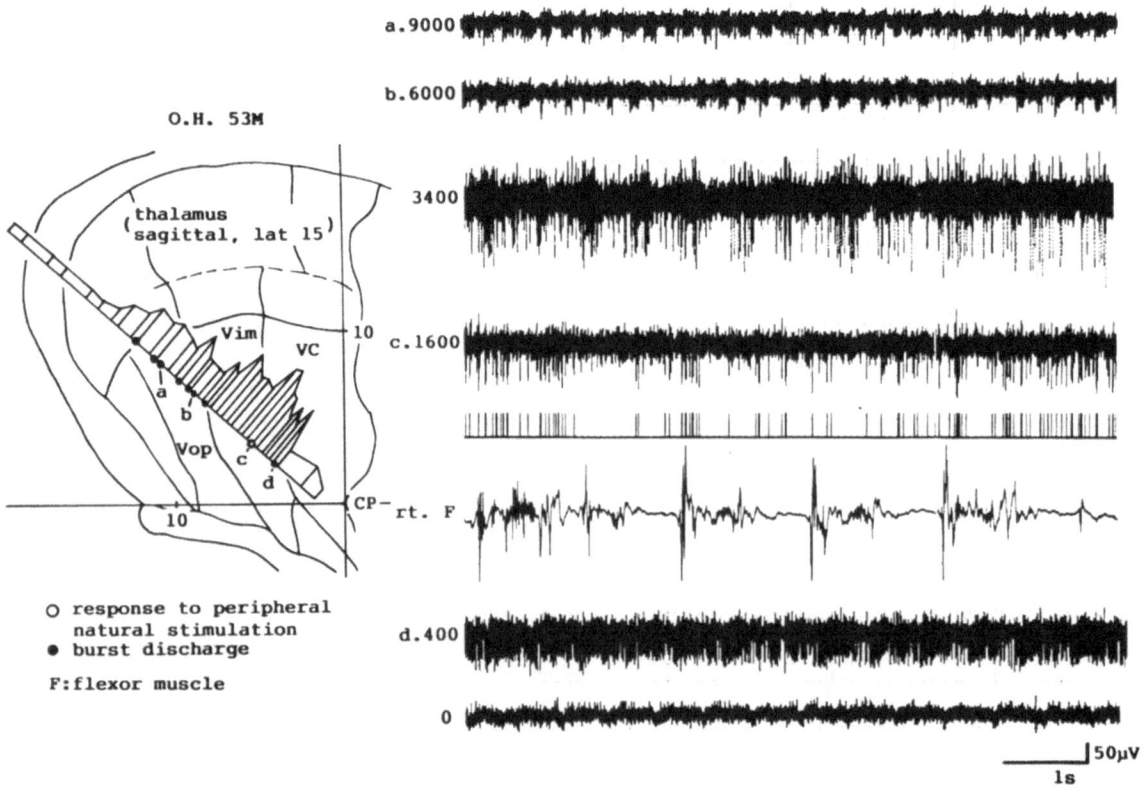

Fig. 1. Electrophysiological findings during the course of stereotactic thalamotomy (case O.H.). Right: Neural activity recorded along the trajectory shown in the figure at the left at several positions on the way toward the ventral thalamus from the prefrontal cortex. From top to bottom, sites of recording were located at 9000 micron (*a*), 6000 micron (*b*), 3400 micron, 1600 micron (*c*), 400 micron (*d*) and 0 micron from the target point. Spike activity is shown below the original recording (c. 1600 micron). EMG (right flexor muscle) was recorded simultaneously with neural activity and is shown below the pulse recording. Left: Sequential change of the background neural activity along the trajectory, represented by electrically integrated value (arbitrary unit). Open circle shows the site at which responses to peripheral natural stimulation were recorded. Closed circle denotes point where irregular burst discharges were identified. Figure shows the sagittal plane in the thalamus 15 mm lateral to the midline. *Vop* N. Ventrooralis posterior; *Vim* N. Ventralis intermedius; *VC* N. Ventralis Caudalis; *CP* commissure posterior

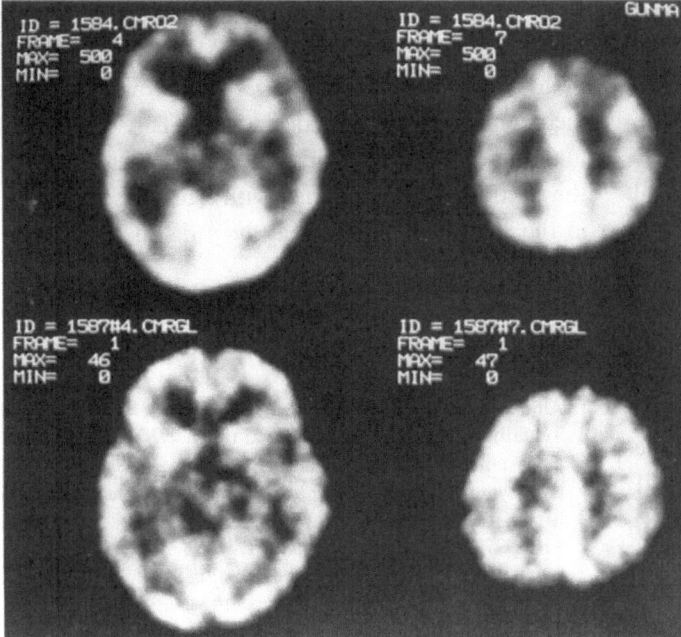

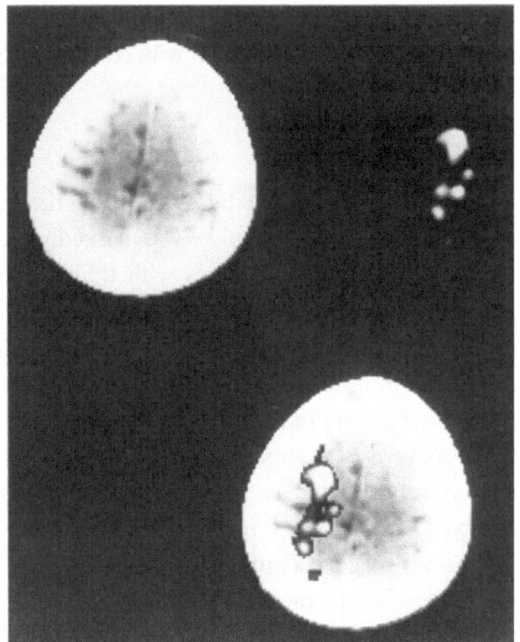

Fig. 2. PET images of cerebral oxygen metabolism (rCMRO₂) and glucose metabolism (rCMRGlu) (case O.H.). Top: Images of rCMRO₂ (0.8–3.0 ml/100 g/min); Bottom: Images of rCMRGlu (8–41 µmol/100 g/min); Left: This cut is from 48 mm above the orbitomeatal (OM) line and includes the basal ganglia and thalamus. Right: This cut is from 96 mm above the OM line and includes the cerebral cortex. Intensity of white colour is proportional to the intensity of oxygen consumption or glucose accumulation. Note that in the cerebral cortex around central sulcus rCMRGlu was increased whereas rCMRO₂ was unchanged, illustrating the dissociation between oxygen compared to glucose metabolism

Fig. 3. Superposition of XCT and PET images of oxygen extraction ratio (rOEF) (case K.Y.). Top left: CT scan at the level of cerebral cortex. Central sulcus is easily identified in the left hemisphere. Top right: Images of regional oxygen extraction ratio (0.67–0.80); Bottom: Superposition of the XCT and rOEF images. Note that rOEF is increased in the cerebral cortex around the central sulcus

lesioned thalamus, but increased in the cerebral cortex around central sulcus on the lesioned side. However, rCMRO₂ did not increase, resulting in the dissociation between the regional cerebral glucose compared to oxygen metabolism in the same area (Fig. 2). In patients in whom both the glucose and oxygen studies were performed, the value of OGMUR on the involved side cerebral cortex around the central sulcus was compared: mean value was 3.15 (± 0.94) in the premotor area and 3.45 (± 0.92) in the primary sensori-motor area of the cerebral cortex in cases with thalamic lesion, 4.50 and 5.24, respectively in the case with putaminal lesion (case M.T.), and 6.63 and 7.15 in other case with putaminal lesion (case H.T.). In the patient with a combined lesion (case K.Y.), the corresponding figures 3.91 and 4.16 were obtained. Therefore, OGMUR decreased more in the cases with a thalamic lesion than in those with a putaminal lesion.

In a patient with a combined putaminal and thalamic lesion (case K.Y.), neural activity was reduced in the VIM-VC area. Peripheral receptive fields to electrical thalamic stimulation (0.6–0.8 mA, 100 Hz) were predominant in the face, hand, and sometimes in the foot area. In this case, rOEF was markedly increased in the cerebral cortex around the central sulcus on the side with the lesion (side of lesion = 0.67–0.80, contralateral side = 0.52–0.53), despite the chronic stage of CVD (Fig. 3). Moreover, in the patient with a putaminal lesion, in whom many irregular burst discharges were encountered in the VIM-VC area, a small subcortical haemorrhage had accidentally occurred near the cerebral cortex around the central sulcus during the operation. Fortunately, however, central pain disappeared after the haemorrhage, probably due to the secondary cortical damage.

In two patients with cortical lesions, who showed mild superficial pain with or without deep pain (Table 1), rCMRGlu was decreased in the lesioned cerebral cortex. Though no CVD lesion could be demonstrated by CT scan, rCMRGlu was reduced in the postero-lateral (sensory) thalamus on the lesioned side (23.1 µmol/100 g/min to 28.2 in case O.G., and 28.9 to 42.2 in case I.S.). In patients with a subcortical lesion, rCMRGlu commonly decreased in this same area. Therefore, rCMRGlu in this area was decreased in all cases with central pain including those who presented with cortical lesions.

Discussion

This study showed that OGMUR in the cerebral cortex around the central sulcus was markedly decreased on the damaged side in cases with thalamic lesions. However, in patients with a putaminal lesion, it was moderately decreased (more in the rostral part). Clinically, superficial pain is more marked in cases with definite thalamic damage. In patients with subcortical lesions, the more severe the superficial pain, the higher was the relative value of glucose metabolism compared to that of oxygen (which was a reciprocal value of OGMUR) in the cerebral cortex around central sulcus on the side involved. Also, in the patient with combined lesions, rOEF was increased in the same area. In an other patient, an accidental CVD lesion in this area suppressed the central pain. Therefore, it is suggested that the abnormal (increased) neural activity present in the cerebral cortex around the central sulcus is related to central (thalamic) pain, especially when associated with definite thalamic damage. The Primary sensory cortex plays an important role in pain processing, whereas the motor and premotor cortex may exert a facilitatory influence on neurons in the spinal dorsal horn, and the thalamic intralaminar nuclei, respectively[1]. If abnormal (increased) activity develops in this area, it may affect pain processing. Furthermore, rCMRGlu is decreased in the sensory thalamus in all patients with central pain including those with a cortical lesion. Therefore, abnormal (increased) activity in the cerebral cortex around the central sulcus combined with a decreased activity in the sensory thalamus may be a distinctive feature related to central (thalamic) pain.

References

1. Albe-Fessard D, Condes-Lara M, Sanderson P, Levante A (1984) Tentative explanation of the special role played by the areas of paleospinothalamic projection in patients with deafferentation pain syndromes. In: Kruger L, Liebeskind JC (eds) Advances in pain research and therapy, Vol 6. Raven, New York, pp 167–182
2. Frackowiak RSJ, Lenzi GL, Jones T, Heather JD (1980) Quantitative measurement of regional cerebral blood flow and oxygen metabolism in man using 15 and positron emission tomography: Theory, procedure, and normal values. J Comput Assist Tomogr 4: 727–736
3. Hirato M, Kawashima Y, Shibazaki T, Shibasaki T, Ohye C (1991) Pathophysiology of central (thalamic) pain: A possible role of the intralaminar nuclei in superficial pain. Acta Neurochir (Wien) [Suppl] 52: 133–136
4. Sokoloff L, Reivich M, Kennedy C, *et al* (1977) The (^{14}C)-deoxyglucose method for the measurement of local cerebral glucose utilization: Theory, procedure, and normal values in the conscious and anesthetized albino rat. J Neurochem 28: 897–916

Correspondence: M. Hirato, M.D., Department of Neurosurgery, Gunma University, 3-39-15, Showa-machi, Maebashi, Gunma, 371, Japan.

Acta Neurochir (1993) [Suppl] 58: 145–149

Trigeminal Neuropathic Pain

K. J. Burchiel

Division of Neurosurgery, Oregon Health Sciences University, Portland, Oregon, U.S.A.

Summary

Typical trigeminal neuralgia is characterized by episodic, unilateral, lancinating, triggerable, often shock-like facial pains, and pain-free intervals. Trigeminal neuropathic pain could be defined as constant unilateral facial pain of variable intensity, is non-triggerable, and unremitting. Atypical trigeminal neuralgia is an overlap syndrome with both episodic and constant pain. Patients with clear pathology of the trigeminal system often have sensory loss and atypical pains. The leading theory of causation of typical trigeminal neuralgia points to minimal compression and demyelination at the root entry zone (REZ) of the trigeminal nerve. In the experience of many neurosurgeons, atypical pains result from lesions or injuries of the trigeminal nerve root *distal* to the REZ.

In a review of 122 patients, distal trigeminal pathology correlated with a clinical syndrome of trigeminal neuropathic pain.

An hypothesis is presented that peripheral lesions of the trigeminal nerve behave *as do other peripheral nerve lesions* in that they are associated with sensory loss, deafferentiation and non "tic-like" pains. This may be due to the *asynchronous* spatial temporal dispersion of abnormal centrally-propagating axonal activity in nociceptive and non-nociceptive fibers from the region of pathology. This is opposed to the putative *synchronous* multifiber volleys which may emanate from the proximal trigeminal REZ in patients with typical trigeminal neuralgia. It appears that classical trigeminal neuralgia may be an exception to the rule that nerve injuries typically produce symptoms like constant pain and allodynia. Trigeminal neuropathic pain is, in some ways, a more general syndrome in that it is a painful nerve injury of the distal trigeminal nerve. Only when patients with this defined syndrome are identified, we will be in position to determine the effectiveness of therapy.

Keywords: Trigeminal neuralgia; trigeminal neuropathic pain; pathophysiology.

Introduction

Trigeminal neuralgia is a fascinating disorder that has tantalized theorists for many decades. What pathophysiologic mechanism can account for the classical paroxysmal lancinating pains which are triggerable by non-noxious tactile stimulation even well outside of the painful division, pain-free intervals, and other characteristic features of trigeminal neuralgia? Despite the fact that we do not yet understand the mechanism of trigeminal neuralgia, it is probably the most tractable chronically painful condition known. Several drugs, and numerous types of surgical approaches, can produce complete alleviation of the pain with little morbidity.

The uniqueness of typical trigeminal neuralgia and its favorable response to therapy have tended to obscure the fact that the trigeminal nerve is commonly subject to trauma along its intra- and extracranial course from, for example, head injury, tumor, sinus disease, facial trauma, dental procedures, and other causes. Like other peripheral nerves, the trigeminal nerve *must* have a spectrum of associated syndromes of post-traumatic neuropathic pain. However, trigeminal neuropathic pain is given little attention in most textbook reviews of facial pain[29]. Perhaps this is a consequence of the relative difficulty of both diagnosis and treatment of facial pains other than trigeminal neuralgia.

Facial pain that is not classical trigeminal neuralgia is often lumped into a non-specific category of "a typical facial pain". While this may be useful from a therapeutic standpoint, it does not clarify the etiology of the pain. Furthermore, the term atypical facial pain is essentially synomomous with a functional or psychogenic pain disorder. My thesis herein is that between the realms of classical trigeminal neuralgia and clearly psychogenic facial pains, lies a group of patients that have discrete facial pain, minimal or no psychopathology, and clearcut structural pathology of the trigeminal nerve. The purpose of this report is to emphasize that trigeminal neuropathic pain exists as a distinct entity, and can be diagnosed by clinical and neurophysiological criteria. Furthermore, an hypothesis to explain trigeminal neuropathic pain is presented.

Materials and Methods

The medical record library of Oregon Health Sciences University was searched for the years 1977–1987 for patients with a diagnosis of "facial pain". Patients fulfilling the following criteria were diagnosed as having *trigeminal neuropathic pain*:

1) Constant unilateral localized facial pain
2) Abnormal or unpleasant dysesthesiae, e.g., described as burning, aching, or throbbing
3) Delayed onset after a precipitating injury
4) Pain in an area of a clinically detectable sensory deficit
5) Presence of allodynia (mechanical hyperalgesia)
6) No history or evidence of significant psychopathology.

Results

From this record review 122 patients were identified. Eighty-one of these patients were found to have typical trigeminal neuralgia by virtue of a typical history. Thirty-four patients of this group (28%) were diagnosed as having trigeminal neuropathic pain.

The diagnoses associated with these 34 cases could be categorized as follows: 12 cases of dental or temporomandibular joint pathology, 10 cases of head and neck cancer, 4 cases of chronic sinus disease, 2 cases following facial trauma, and one case of a cerebellopontine angle tumor. In five cases no definite precipitating injury or pathology could be identified, but these cases were included in the analysis since they fulfilled the remainder of the above criteria.

Discussion

Typical Trigeminal Neuralgia

It is not usually characterized by overt facial sensory loss. If "idiopathic" trigeminal neuralgia can be associated with subclinical, but demonstrable, hypesthesia in the affected region, this would be evidence of minor deafferentation. Lewery and Grant[28], and Dott[9] found hypesthesia in 25% and 40% of cases, respectively. More recently Jannetta[20] has indicated that sensory loss can be found in about 25% of patients with trigeminal neuralgia. However, the data is not consistent. Hampf et al.[18] studied 18 "virgin" trigeminal neuralgia patients, and found no change in the quantitative sensory exam. This raises the question of patient selection for previous studies. That is, it is possible that prior results indicating the presence sensory loss in patients were contaminated by numbers of patients with more overt trigeminal neuropathy, or even structural lesions that were undetectable given the neuroimaging modalities available at the time of the study. Suffice it to say that classical trigeminal neuralgia is typified by little or no clinical deafferentation. On the contrary, trigeminal neuralgia seems to be more of an "irritative" phenomenon, rather than a deafferentation pain.

Despite the absence of obvious facial sensory loss, trigeminal root entry zone (REZ) demyelination may, in part, underlie the etiology of trigeminal neuralgia[17]. Numerous neuropathological studies have, in fact, found evidence of demyelination in the region of the REZ[1,22–25]. Jannetta[21] has hypothesized that focal injury and demyelination of the trigeminal root at the REZ both incites and maintains classical trigeminal neuralgia. A number of evoked potential studies now support the notion that there is an area of demyelination in the retrogasserian root. Leanori and Favale[27] found somatosensory evoked potential (SSEP) evidence of damage at the root entry zone or slightly distal in 33 of 68 cases of trigeminal neuralgia. Using trigeminal SSEPs, Cruccu and colleagues[6] have concluded that the primary lesion in classical trigeminal neuralgia affects afferent fibers in the proximal portion of the root or the intrinsic portion of the pontine tract. Stohr et al.[35] and Bennett and Jannetta[3] have documented SSEP delays in 41% and 83% of trigeminal neuralgia patients, respectively.

Patients with multiple sclerosis, but otherwise clinical "symptomatic" trigeminal neuralgia, have also been studied by SSEPs. Iragui et al.[19] found that the evoked potential abnormality in these patients localized to the trigeminal sensory root and lateral lemniscus. Garcin et al.[16] found similar postmortem changes at the REZ in patients with both MS and trigeminal neuralgia, although the demyelination became more extensive as the disease progressed and the pain became more continuous and atypical. Demyelination has also been detected in biopsies of the trigeminal nerve in patients with trigeminal neuralgia. In the case presented by Lazar and Kirkpatrick[26], the trigeminal root at the REZ was biopsied and a demyelinating plaque was found in the central portion of the root. Interestingly, the "peripheral nerve" portion of the biopsy was entirely normal.

There have been numerous reports to substantiate the proposal by Jannetta that trigeminal neuralgia is caused by microvascular compression of the trigeminal root at the REZ[21]. This area, the approximate location of the transition of peripheral and central myelin, or the Obersteiner-Redlich zone, may well be a site that is prone to ectopic impulse generation given the potential impedance mismatch between differently myelinated axon segments[30]. It has also been shown that minimal experimental trigeminal denervation can also produce hyperactivity in trigeminal WDR neurons[33,34], and the activity of these cells can be diminished by agents such as baclofen, phenytoin and carbamazepine which facilitate segmental inhibition as Fromm has pointed out[11-15].

With the exception of two rare conditions, the pain of trigeminal neuralgia is virtually unique. Glossopharyngeal neuralgia and nervus intermedius neuralgia can produce similar pains in the tonsillar area and ear, respectively. Interestingly, both of these disorders are said to be associated with vascular compression of the nerve root[10,21], as is the motor equivalent of trigeminal neuralgia, hemifacial spasm[31,32]. Other nerves rarely, if ever, generate a syndrome akin to trigeminal neuralgia. The pains of tabes dorsalis are said to be "lighting like", but in the last decade of the twentieth century there are many neurologists and neurosurgeons who have never seen a patient with this disorder. In postherpetic neuralgia fleeting lancinating pains are seen but are essentially always superimposed on a background of constant pain described as "burning", "aching", "clawing", etc.

What about the trigeminal nerve predisposes it to the development of the syndrome of trigeminal neuralgia? If we accept that the origin of the syndrome is, at least initially, related to vascular compression, the size and location of the nerve in the posterior fossa may be simple, but important, factors. Whatever the cause, a minimal and exquisitely localized demyelinating neuropathy seems to be among the requisite elements.

One fascinating difference between trigeminal neuralgia and more typical neuropathic pains is that trigeminal neuralgia can be alleviated, sometimes for years, by relatively minor damage to the peripheral nerve, ganglion, or retrogasserian root. In contradistinction to trigeminal neuralgia, further neural injury is usually ineffective in alleviating neuropathic pain associated with clinical deafferentiation. In fact, ablative therapy often *worsens* neuropathic pain, and a destructive approach is usually avoided by experienced clinicians.

Neuropathic pain occurs in many areas of the body, and certainly there is a spectrum of injury severity. The failure to recognize such overlap syndromes in other nerves, even rarely, argues strongly that differences between trigeminal neuralgia and other neuropathic disorders are qualitative, rather than quantitative.

Trigeminal Neuropathic Pain

One fascinating aspect of the trigeminal sensory system is that it *can* produce a variety of pains. Several distinct syndromes are recognizable: Trigeminal neu-

ralgia, postherpetic neuralgia, post-traumatic neuropathic pain, and anesthesia dolorosa. For this reason, the trigeminal nerve deserves much greater attention in the study of neuropathic pain. Although I would disagree with drawing a direct parallel between typical neuropathic pain and classical trigeminal neuralgia, I believe that we can learn much about neuropathic pain by studying the trigeminal system.

Compression or injury to the trigeminal nerve distal to the root entry zone, most commonly due to a structural lesion like a tumor, aneurysm, or vascular malformation, produces "trigeminal neuropathy". This neuropathy is associated with a more constant pain *highly reminiscent of the pain of peripheral nerve injury*. These disorders are more comparable to other neuropathic pains, in that the pain is much more continuous, and that sensory loss is a prominent feature. This syndrome could be termed appropriately "*trigeminal neuropathic pain*".

Surgical observations indicate that trigeminal root compression *distal to the REZ* results in a more atypical facial neuralgia. For example, Bullitt and her associates[4] have made the following observation:

"Peripherally placed tumors tend to produce atypical facial pain associated with sensory deficits. Tumors encoraching upon the gasserian ganglion may produce either atypical facial pain or trigeminal neuralgia, and tumors located within the posterior fossa are usually associated with trigeminal neuralgia, although atypical facial pain may also occur".

In this regard, Cusick[8] has also observed that:

"Increasing evidence, however, indicates that mass lesions affecting the trigeminal nerve proximal to the gasserian ganglion may be frequently associated with atypical trigeminal neuralgia and that a high degree of clinical curiosity should be maintained regarding such a relationship".

Evoked potential studies also support the concept that, in comparison to trigeminal neuralgia, more atypical trigeminal neuralgia secondary to neuropathy is correlated with electrophysiologic evidence of more damage to the nerve[7,37]. Finally, the treatment of trigeminal neuropathic pain is, like the treatment of other peripheral neuropathic pains, much more difficult than the management of trigeminal neuralgia. Szapiro et al.[36] reported that in patients with purely paroxysmal trigeminal neuralgia, 95% were relieved by trigeminal microvascular decompression, while in patients with baseline on-going pain and paroxysmal pain, only 58% were successfully treated by trigeminal decompression.

Evaluation of trigeminal neuropathic pain: All patients with a suspected diagnosis of trigeminal neuropathic pain must have a thorough historial review, general physical and detailed neurologic exam. Facial pain that is clearly not trigeminal neuralgia must also be evaluated with psychological testing and interview. Oral surgery evaluation is also recommended to rule out treatable pathology. Neuroimaging using CT and/ or MRI scanning with contrast enhancement is essential. To detect subtle sensory loss, quantitative sensory testing using Von Frey hairs and quantitative thermo-testing are very sensitive and specific techniques. The role of trigeminal evoked potentials in the diagnosis of this disorder has yet to be determined, but also may provide objective evidence and localization of nerve pathology.

Pharmacological therapy of neuropathic pain: We can learn a great deal from the pharmacological and physiological properties of drugs that are effective in the treatment of trigeminal neuralgia and other more typical neuropathic pains. Carbamazepine, phenytoin and baclofen can all be *highly effective* for the treatment of trigeminal neuralgia. However, in a very real sense, all the agents which have been tried for neuropathic pains have been "borrowed" from other disorders, such as epilepsy, spasticity, and depression. The effectiveness of these agents for neuropathic pain is *highly questionable*. This is not surprising since neuropathic pain does not resemble these entities clinically. We should not suppose that they would necessarily be similar to neuro-pathic pain mechanistically. To develop more effective pharmacotherapy, we simply need further insights from laboratory models of neuropathic pain such as that developed in the rat by Bennett[2]. What the disparity between the medical treatment of trigeminal neuralgia and trigeminal neuropathic pain should tell us is that, at the very least, the mechanisms of these two disorders are quite different.

Mechanism of Trigeminal Pain: a Theory

The mechanism of classical trigeminal neuralgia is still a matter of considerable debate. Perhaps the best synthesis of the available data has been presented by Fromm[12]. He postulates that bursts of impulses originate from the region of nerve compression/demyelein-ation at the trigeminal REZ. The mechanism of the synchronous generation of these REZ discharges may involve ectopic spike generation, ephapses, spike reflection, or other unknown mechanisms. This afferent barrage causes sensitization of wide dynamic range (WDR) neurons in the trigeminal subnucleus caudalis. These neurons are further sensitized by loss of descending inhibition related to minimal deafferentation. By this hypothesis, abnormal trigeminal WDR excitability and activity leads to the central propagation of a noci-

ceptive signal and the perception of a paroxysm of lancinating pain.

Patients with trigeminal neuropathic pain describe ongoing burning, or aching pain, and little or no episodic pain. It is reasonable to hypothesize that significant distal trigeminal nerve injury produces these atypical pain syndromes by a mechanism much like what has been posited for other painful incontinuity nerve injuries[6]. An animal model of trigeminal neuropathic pain has been produced which mimicks both the structural pathology and clinical phenomena of trigeminal neuropathic pain[5]. By this hypothesis, *asynchronous* summation of on-going ectopic impulse generation, chemosensitivity, mechanosensitivity, taken together with central reorganization in response to deafferentation leads to the constant but varying pains of which are the hallmark of this syndrome. Any generator of abnormal neural activity in the periphery would be subject to substantial spatiotemporal dispersion due to the distance from the central terminals, the variety of nerve fibers involved in the area of injury, interruption of nerve continuity or myelination, and areas of nerve ischemia, to name a few possibilities. The asynchronous nature of the centrally propagating impulses would make it *unlikely* that critical summation at the WDR neuron would occur, resulting in brief, lancinating pain. On the other hand, if this theory is correct, *synchronous* discharge of an impulse generating mechanism in the periphery should be able to produce brief, lancinating pain. In fact, this can be demonstrated in the clinical ability to produce a Tinel's sign from mechanical stimulation of the area of nerve injury in patients with facial or dental trauma. A graphic schema of this theory is presented in Fig. 1.

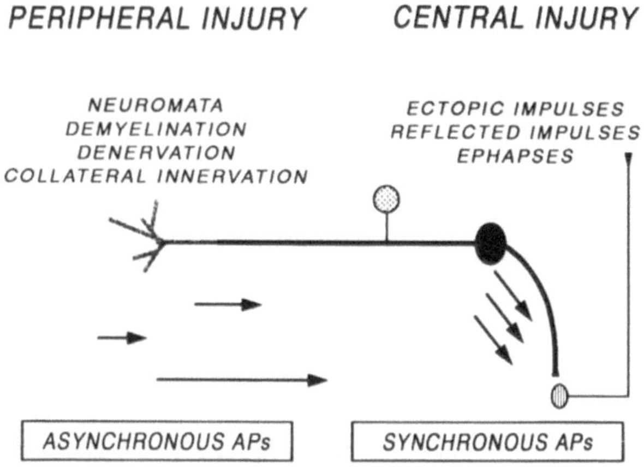

Fig. 1. Schema of an hypothesis for the pathophysiology of trigeminal neuralgia and trigeminal neuropathic pain

In conclusion, I have presented an argument that trigeminal neuropathic pain is similar in origin and clinical presentation to other peripheral neuropathic pains. Typical trigeminal is unique to disorders of compression and/or demylelination of the proximal trigeminal REZ. In contrast, damage to the distal trigeminal nerve is more associated with trigeminal neuropathic pain. A combination of REZ and more distal trigeminal injury produces an overlap syndrome, atypical trigeminal neuralgia. Only when patients with the defined syndrome of trigeminal neuropathic pain are identified, will we be in position to determine the effectiveness of therapy for this disorder.

References

1. Beaver DL, Moses HL, Ganote CE (1965) Electron microscopy of the trigeminal ganglion. III. Trigeminal neuralgia. Arch Pathol 79: 571–582
2. Bennett GJ, Xie YK (1989) A peripheral mononeuropathy in rat that produces disorders of pain sensation like those seen in man. Pain 33: 87–108
3. Bennett MH, Jannetta PJ (1983) Evoked potentials in trigeminal neuralgia. Neurosurgery 13: 242–247
4. Bullitt E, Tew JM, Boyd J (1986) Intracranial tumor in patients with facial pain. J Neurosurg 64: 865–871
5. Burchiel KJ (1980) Abnormal impulse generation in focally demyelinated trigeminal roots. J Neurosurg 53: 674–683
6. Burchiel KJ, Ochoa J (1991) Pathophysiology of injured axons. In: Burchiel KJ (ed) Surgical management of peripheral nerve injury and entrapment. Neurosurg Clin North Am 2: 105–116
7. Cruccu G, Leanori M, Feliciani M, Manfredi M (1990) Idiopathic and symptomatic trigeminal pain. J Neurol Neurosurg Psychiatry 53: 1034–1042
8. Cusick JF (1981) Atypical trigeminal neuralgia. JAMA 245: 2328–2329
9. Dott NM (1935) Paroxysmal neuralgia and other pains in the jaws and face. Br Dent J 58: 616
10. Fraioli B, Esposito V, Ferrante L, Trubiani L, Lunardi P (1989) Microsurgical treatment of glossopharyngeal neuralgia: Case reports. Neurosurgery 25: 630–632
11. Fromm GH (1991) Medical treatment of patients with trigeminal neuralgia. In: Fromm GH, Sessle BJ (eds) Trigeminal neuralgia: Current concepts regarding pathogenesis and treatment. Butterworth-Heinemann, Boston, pp 131–144
12. Fromm GH (1991) Pathophysiology of trigeminal neuralgia. In: Fromm GH, Sessle BJ (eds) Trigeminal neuralgia: Current concepts regarding pathogenesis and treatment. Butterworth-Heinemann, Boston, pp 105–130
13. Fromm GH, Chattha AS, Terrence CF, Glass JD (1981) Role of inhibitory mechanisms in trigeminal neuralgia. Neurology 31: 683–687
14. Fromm GH, Sato K, Nakata M (1992) The action of GABAB antagonists in the trigeminal nucleus of the rat. Neuropharmacology 31: 475–480
15. Fromm GH, Terrence CF, Chattha AS, Glass JD (1980) Baclofen in trigeminal neuralgia: Its effect on the spinal trigeminal nucleus: A pilot study. Arch Neurol 378: 768–771
16. Garcin R, Godlewski S, Lapresle J (1960) Nevalgie du trijumeau et sclerose en plaques (a propos d'une observation anatomo-clinique). Rev Neurol 102: 441–451
17. Gardner WJ (1970) Causation of trigeminal neuralgia. In: Hassler R, Walker AE (eds) Trigeminal neuralgia: Pathogenesis and pathophysiology. Saunders, Philadelphia, p 153
18. Hampf G, Bowsher D, Wells C, Miles J (1990) Sensory and autonomic measurements in idiopathic trigeminal neuralgia before and after radiofrequency thermocoagulation: Differentiation from other causes of facial pain. Pain 40: 241–248
19. Iragui VS, Wiederholt WC, Romine JS (1986) Evoked potentials in trigeminal neuralgia associated with multiple sclerosis. Arch Neurol 43: 444–446
20. Jannetta PJ (1976) Microsurgical approach to the trigeminal nerve for tic douloureux. Prog Neurol Surg 7: 180–200
21. Jannetta PJ (1980) Neurovascular compression in cranial nerve and systemic disease. Ann Surg 192: 518–525
22. Kerr FWL (1967) Correlated light and electron microscopic observations of the normal trigeminal ganglion and sensory root in man. J Neurosurg 26: 132–137
23. Kerr FWL (1967) Pathology of trigeminal neuralgia: Light and electron microscopic observations. J Neurosurg 26: 151–156
24. Kerr FWL, Miller RH (1966) The pathology of trigeminal neuralgia. Electron microscopic studies. Arch Neurol 15 308–319
25. Kumagami H (1974) Neuropathological finding of hemifacial spasm and trigeminal neuralgia. Arch Otolaryngol 99: 160–164
26. Lazar ML, Kirkpatrick JB (1979) Trigeminal neuralgia and multiple sclerosis. Demonstration of the plaque in an operative case. Neurosurgery 5: 711–717
27. Leanori M, Favale E (1981) Diagnostic relevance of trigeminal evoked potentials following intraorbital nerve stimulation. J Neurosurg 75: 244–250
28. Lewery F, Grant F (1938) Physiopathologic and pathoanatomic aspects of major trigeminal neuralgia. AMA Arch Neurol Psychiatr 46: 1126
29. Loeser JD (1990) Cranial neuralgias. In: Bonica JJ (ed) The management of pain, Vol 1. Lea and Febiger, Philadelphia, pp 676–686
30. Loeser JD, Calvin WH, Howe JF (1977) Pathophysiology of trigeminal neuralgia. Clin Neurosurg 24: 527–537
31. Loeser JD, Chen J (1983) Hemifacial spasm: Treatment by microsurgical facial nerve decompression. Neurosurgery 13: 141–146
32. Piatt JH, Wilkins RH (1984) Treatment of tic douloureux and hemifacial spasm by posterior fossa exploration: Therapeutic implications of various neurovascular relationships. Neurosurgery 14: 462–471
33. Sessle BJ (1987) The neurobiology of facial and dental pain: Present knowledge, future directions. J Dent Res 66: 962–981
34. Sessle BJ (1991) Physiology of the trigeminal system. In: Fromm GH, Sessle BJ (eds) Trigeminal neuralgia. Current concepts regarding pathogenesis and treatment. Butterworth-Heinemann, Boston, 71–104
35. Stohr M, Petruch F, Schleglmann K (1981) Somatosensory evoked potentials following trigeminal nerve stimulation in trigeminal neuralgia. Ann Neurol 9: 63–66
36. Szapiro J, Sindou M, Szapiro J (1985) Prognostic factors in microvascular decompression for trigeminal neuralgia. Neurosurgery 17: 920–929
37. Tanaka A, Takaki T, Maruta Y (1987) Neurinoma of the trigeminal root presenting as atypical trigeminal neuralgia: Diagnostic values of orbicularis oculi reflex and magnetic resonance imaging. Neurosurgery 21: 733–736

Correspondence: Kim J. Burchiel, M.D., John Raaf Professor and Head, Division of Neurosurgery, Oregon Health Sciences University L-472, Portland, OR 97201, U.S.A.

Acta Neurochir (1993) [Suppl] 58: 150–153

Motor Cortex Stimulation as Treatment of Trigeminal Neuropathic Pain

B. A. Meyerson, U. Lindblom[1], B. Linderoth, G. Lind, and P. Herregodts[2]

Departments of Neurosurgery and Neurology[1], Karolinska Hospital, Stockholm, Sweden, and the Department of Neurosurgery[2], Academisch Ziekenhuis, Brussels, Belgium

Summary

A report is given on first experiences with motor cortex stimulation in 10 patients with different forms of neuropathic pain.

Three of them had central pain as sequelae of cerebrovascular disease. In none of them did the stimulation provide pain relief.

Two patients had pain from peripheral nerve injuries. One did not respond, but the other obtained about 50% pain relief.

The remaining 5 patients with trigeminal neuropathy experienced definite pain relief varying between 60 and 90%.

During test stimulation most patients had one or two short-lasting generalized seizures. But no one had any motor effects after permanent implantation.

Motor cortex stimulation appears to be a new and promising possibility of pain treatment, especially in cases with trigeminal neuropathy, but many problems have yet to be solved, before a clear indication could be given.

Keywords: Pain; neuropathic pain; trigeminal system; brain stimulation; cerebral cortex.

Introduction

In 1990 Tsubokawa first reported that electric stimulation of the motor cortex may be effective for neuropathic pain of central, supraspinal origin. The outcome appeared very promising in view of the fact that such pain is notoriously difficult to manage. Apart from pharmacological treatment, generally with tricyclic drugs, stimulation of the sensory thalamus or the sensory limb of the internal capsule has been tried. However, only a limited number of patients will respond to such treatment and late failures are common (Meyerson 1990). Although experimental data are available on motor cortex stimulation in cats subjected to deafferentation (Namba and Nishimoto 1988; Tsubokawa 1991) as well as on corticofugal modulation of sensory processing in the spinal dorsal horn, the mechanisms

behind the suppression of clinical pain with this form of stimulation are poorly understood.

Our first experience with motor cortex stimulation was in June 1990 when we had a patient with severe trigeminal pain originating from surgery for a parotid tumour. Trigeminal ganglion stimulation via a percutaneously implanted electrode, high cervical stimulation of the spinal trigeminal tract as well as sensory thalamic stimulation had been tried but failed. As a last resort, treatment with motor cortex stimulation was initiated and proved to be effective. The patient still has good pain relief after two and a half years.

Here we report our first experiences with motor cortex stimulation in ten patients with different forms of neuropathic pain.

Patients

Three of the patients had central pain as sequelae of cerebrovascular disease: two had had a haemorrhage in the thalamus and one a brain stem infarction.

Two patients had pain due to peripheral nerve injury: one as a result of minor surgery in the shoulder region which had produced excruciating pain and extreme allodynia covering most of the upper quadrant of the trunk; the other patient had phantom pain in the right index finger combined with allodynia and hyperalgesia in the radial aspect of the hand. Both patients had failed to respond to spinal cord stimulation and the patient with the finger phantom had also been subjected to trial stimulation in the sensory thalamus (VPM).

The remaining five patients, all females, had trigeminal neuropathy, four of them following surgery in the trigeminal territory but not for tic douloureux. It was a characteristic that they all presented with abnormalities of facial sensation, dominated by dysaesthesia or allodynia to mechanical stimuli and hyperalgesia to pin prick. In two of them the disturbance of cutaneous sensibility had spread outside the territory of the trigeminal nerve into the ipsilateral half of the head and neck where they also reported the presence of spontaneous pain. Four of these patients had previously been subjected to other forms of neurostimulation (Table 1).

Table 1. *Results of Motor Cortex Stimulation.* Pain relief refers to relief of ongoing pain assessed with VAS

	Diagnosis	Prev. stim. treatment	History years	Sensibility disturb	Follow-up months	Pain relief
F/48	trigeminal	Gasseri stim				
	neuropathy	VPM stim	11	allodynia	11	> 50%
F/44	–"–	Gasseri stim	5	allodynia hyperalgesia	8	> 50%
F/51	–"–	—	9	hyperalgesia	11	> 75%
F/46	–"–	cervical DCS	3	dysaethesia	14	> 75%
F/46	–"–	Gasseri stim cervical DCS	4	allodynia	28	> 50%
SM/71	peripheral posttraum	DCS		hyperalgesia		
	neuropathy	VPM	4	allodynia	4	< 50%
F/44	–"–	DCS	4	allodynia		0
M/61	post-stroke	—	2	hyperaesthesia		0
M/48	–"–	—	14	hyperaesthesia		0
F/53	–"–	cervical DCS	7	hyperaesthesia		0

Surgery

In order to locate the central sulcus—primary motor cortex, somatosensory evoked responses to median nerve stimulation were recorded using an array of recording electrodes orientated anterioposteriorly about 6 cm from the vertex (cf. Tsubokawa 1991).

The location where the so-called N20 component disappeared was marked on the skin as it conceivably represented the transitional zone between the sensory and motor cortex. Under local anaesthetic a burr hole was made just posteriorly to this skin marking and a 4-polar electrode strip (Resume®, Medtronic Inc.) was introduced epidurally. The electrode was positioned anteriorly or ventrally. Local muscle twitches in the contralateral hand or arm were induced by low frequency, high intensity stimulation to make sure that the electrode was located on the motor cortex.

Test stimulation via a percutaneous extension lead was subsequently performed. Great care was taken to try various coupling combinations of the stimulating poles. Ten different couplings, comprising all 4 poles, are available, and in several patients it was found that the coupling configuration was critical. Therefore, the stimulation test period generally lasted for about four weeks.

In a few patients a multipolar electrode grid, comprising 24 or 32 stimulating poles, was instead applied via a small craniotomy. This method enabled a more detailed search for an electrode polar array which could provide stimulation-produced pain relief. In a subsequent intervention, the row of stimulating poles providing the best result could be identified and the multipolar electrode grid was exchanged to a permanently implanted 4-polar electrode.

Stimulation was applied with pulses of 0.3 ms duration at 50 Hz. The intensity was set about 20–30% less than that required to evoke local muscle twitches with 0.6–0.8 ms pulses at 1–2 Hz. Stimulation was not accompanied by any subjective sensations, and therefore, the condition of the battery had to be carefully controlled. In the test stimulation phase the patients were instructed to use the stimulator for 20–30 min, 3–5 times daily. Each day a new combination of polar couplings was tried. The effect on pain (Table 1) was assessed by using visual analogue scaling (VAS), and besides, the patients were instructed to observe and make notes of any other effects. Effects on evoked pain (allodynia, dysaesthesia) were assessed by means of quantitative sensory testing (Lindblom 1985).

Results

In none of the patients with central pain did the stimulation provide relief in spite of the usage of a multipolar electrode grid in one and relocation of the 4-polar electrode in another patient. The outcome was equally negative in the patient with peripheral neuropathy in the shoulder region whereas the patient with a finger phantom experienced about 50% pain reduction. In contrast, all patients with trigeminal neuropathy had definite pain relief varying between 60 and 90%. Three of them had to stimulate 3–6 times daily and two patients 1–2 times. In general, the stimulator had to be on for 20–30 min to obtain maximal effect. The follow-up ranges from 4 to 28 months and in no case has there so far been any tendency to a decreasing effect of stimulation with regard to the degree of pain reduction and/or duration of post-stimulatory relief. One of the patients (case F/48) with trigeminal neuralgia had previously been treated first with Gasserian ganglion stimulation via an implanted electrode (Meyerson and Håkanson, 1986) and subsequently with VPM-stimulation. She reported that cortical stimulation provided more effective pain relief than any of the other treatments.

In one patient an electrode was initially located on the motor cortex ipsilateral to the trigeminal pain but stimulation proved ineffective (cf. Parrent and Tasker 1992).

When different couplings of the stimulating poles were tested, it was often found that one or two of the ten available combinations were superior. Moreover, two of the patients with trigeminal neuropathy spontaneously reported that different electrode couplings selectively influenced the pain in different parts of the face and neck. This unexpected finding was repeatedly controlled with the patient not knowing which coupling was used. The spatial specificity of the stimulation was also evident in the patient in whom a multipolar (24

Effect on tactile sensation

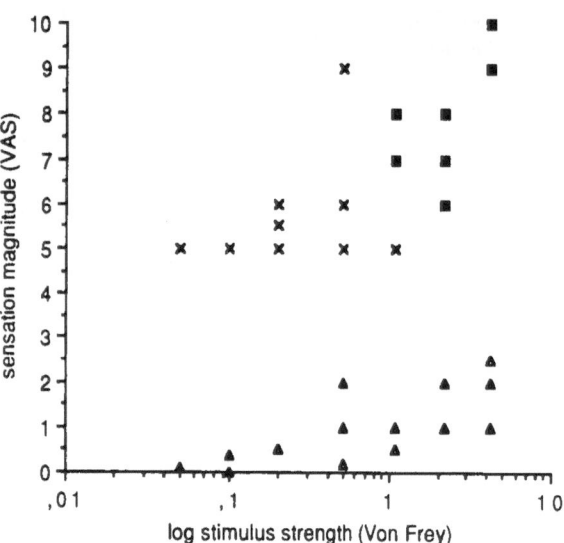

Fig. 1. Motor cortex stimulation. Facial dysaesthesia and allodynia assessed by von Frey hairs (stiffness expressed in gms of bending pressure on the abscissa) before and after motor cortex stimulation. Note that the cortical stimulation per se did not evoke any sensations. Evoked pain: × dysaesthesia, ■ allodynia before stimulation; ▲ sensation after stimulation. Spontaneous pain: Before stimulation: VAS 80/100; after stimulation: VAS 20/100

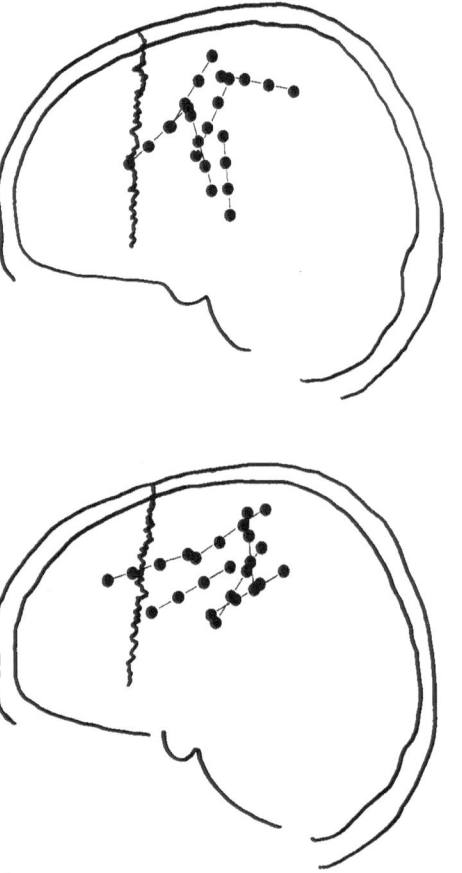

Fig. 2. Composite representation of location of 4-polar stimulating electrodes in patients who had good (upper) and no (lower) pain relief

poles) electrode grid was temporarily implanted.

All the patients with allodynia, dysaesthesia and hyperalgesia in the territory of the painful trigeminal region, as well as on the ear and the neck, reported that as a result of stimulation the unpleasant or painful sensations from touching or stroking the skin were markedly reduced. This effect of stimulation was controlled in one patient with the internal controls of the stimulator in off-position without the patient's knowing. There was no reduction of dysaesthesia and allodynia following 20 min of "stimulation" as assessed by quantitative testing with von Frey hairs (Fig. 1).

In Fig. 2 the effective and non-effective electrode locations, respectively, are shown. It should be emphasized that at all electrode sites but one, whether being effective for pain relief or not, it was possible to induce muscle twitches with low frequency, high intensity stimulation. The figure does not include the location of the multipolar electrode grids.

Complications

During the test stimulation period most patients had one or two short-lasting generalized seizures but in no case have there been any motor effects after permanent implantation. In two patients stimulation produced painful local sensations at the site of the implanted electrode. This necessitated a small craniotomy and denervation of the dura which was cut and sutured around the electrode. In one of the patients this led to an epidural clot which was at first undetected. The patient gradually developed a marked expressive aphasia which after evacuation of the clot persisted for several months. There is still a residual, slight degree of dysphasia.

In two patients it was necessary to re-operate at the site of the electrode cable connection located behind the ear because of threatening ulceration. There have been no equipment failures.

Discussion

Our results differ from those reported by Tsubokawa (1990) in that none of our patients with central pain enjoyed pain relief. Moreover, Tsubokawa (personal communication) has one case with trigeminal neuropathy who has failed to respond. All our patients had a well established symptomatic diagnosis of neuropathic pain and in none was there any co-existent nociceptive pain component. One common characteristic of the patients with trigeminal neuropathic pain was the disturbance of the facial sensibility, predominantly of a hypersensitive type with allodynia to touch and cold and hyperalgesia. In two cases this sensory disturbance extended outside the territory of the trigeminal nerve. Tsubokawa has reported experiments performed on cats in which transection of the spinothalamic tract resulted in hyperactivity of thalamic neurons. Spontaneous firing of such neurons could be inhibited by stimulation of the motor cortex whereas stimulation of the sensory cortex was ineffective. Namba and Nishimoto (1988) performed trigeminal denervation in cats resulting in deafferentation hyperactivity in the spinal trigeminal nucleus. Here, hyperactive WDR neurons could be inhibited by stimulation both of the sensory and of the motor cerebral cortex. It was further found that the activated corticofugal pathways passed by the sensory limb of the internal capsule and the sensory thalamus. It is in this context of interest that stimulation at both these locations may block pain due to trigeminal neuropathy in patients.

It is well known that electrical stimulation of the sensory-motor cortex in rats, cats and monkeys, not subjected to previous deafferentation, may produce presynaptic inhibition of spinal primary afferents (e.g. Lindblom and Ottosson 1957; Andersen *et al.* 1962; Carpenter 1963) and also of spinothalamic tract neurons (Coulter *et al.* 1974).

In the monkey there exists a strict somatotopic organization of the corticofugal inhibition of spinal primary afferents (Abdelmoumène *et al.* 1970) and this finding is of interest with regard to the unexpected finding that in our patients particular couplings of the stimulating poles sometimes had a regionally selective effect both on spontaneous and evoked pain (allodynia, dysaesthesia) in different parts of the face. Moreover, the critical location of the stimulating electrode is illustrated by the finding that in one of the patients with trigeminal neuropathy and a multipolar electrode grid, only a few stimulating poles provided pain relief. This may explain why the composite locations of effective and non-effective electrode locations illustrated in Fig. 2 show a considerable overlap. The possibility of inducing peripheral muscular activation is apparently not a reliable predictor of a pain relieving effect. This conclusion is further substantiated by the fact that in one patient with central pain, stimulation via a multipolar electrode grid using a large variety of electrode couplings no pain relief was obtained in spite of the induction of motor responses.

Although motor cortex stimulation appears to be a new, promising form of central nervous stimulation as pain treatment in selected patients, many problems have to be solved. As already discussed, there is no reliable method by which the ideal and effective electrode site can be defined; the optimal stimulus parameters have not been established; it is not known whether stimulation for long periods of time may induce kindling or have other harmful effects. As illustrated not least by this study and by the unexplicable discordance with Tsubokawa's results, indications are far from being established. A particular advantage over other forms of nervous system stimulation procedures is the lack of subjective sensations in this type of stimulation. This makes it possible for the first time to assess the effect of neurostimulation for pain, using a double-blind procedure. This advantage should be regularly explored in future studies with cortical stimulation.

References

1. Abdelmoumène M, Besson JM, Aléonard P (1970) Cortical areas exerting presynaptic inhibitory action on the spinal cord in cat and monkey. Brain Res 20: 327–329
2. Andersen P, Eccles JC, Sears TA (1962) Presynaptic inhibitory action of cerebral cortex on the spinal cord. Nature (London) 194: 740–741
3. Carpenter D, Lundberg A, Norrsell U (1963) Primary afferent depolarization evoked from the sensorimotor cortex. Acta Physiol Scand 59: 126–142
4. Coulter JD, Maunz RA, Willis WD (1974) Effects of stimulation of sensorimotor cortex on primate spinothalamic neurons. Brain Res 65: 351–356
5. Lindblom U (1985) Assessment of abnormal evoked pain in neurological pain patients and its relation to spontaneous pain: A descriptive and conceptual model with some analytical results. In: Fields, *et al* (eds) Advances in pain research and therapy, Vol 9. Raven, New York, pp 409–423
6. Lindblom U, Ottosson JO (1957) Influence of pyramidal stimulation upon the relay of coarse cutaneous afferents in the dorsal horn. Acta Physiol Scand 38: 309–318
7. Meyerson BA (1990) Electric stimulation of the spinal cord and brain. In: Bonica JJ (ed) The management of pain, 2nd Ed. Lea and Febiger, Philadelphia, pp 1862–1877
8. Meyerson BA, Håkanson S (1986) Suppression of pain in trigeminal neuropathy by electric stimulation of the Gasserian ganglion. Neurosurgery 18: 59–66
9. Namba S, Nishimoto A (1988) Stimulation of internal capsule, thalamic sensory nucleus (VPM) and cerebral cortex inhibited deafferentation hyperactivity provoked after Gasserian ganglionectomy in cat. Acta Neurochir (Wien) [Suppl] 42: 243–247
10. Parrent AG, Tasker RR (1992) Can the ipsilateral hemisphere mediate pain in man? In: Meyerson BA (ed) European society for stereotactic and functional neurosurgery. Acta Neurochir (Wien) 117: 89
11. Tsubokawa T, Katayama Y, Yamamoto T, Hirayama T, Koyama S (1991) Chronic motor cortex stimulation for the treatment of central pain. Acta Neurochir (Wien) [Suppl] 52: 137–139

Correspondence: B. A. Meyerson M.D., Ph.D., Department of Neurosurgery, Karolinska Hospital, S-10401 Stockholm, Sweden.

Acta Neurochir (1993) [Suppl] 58: 154–155

Long-Term Results of Stimulation of the Septal Area for Relief of Neurogenic Pain

J. R. Schvarcz

Department of Neurosurgery, School of Medicine, University of Buenos Aires, Buenos Aires, Argentina

Summary

Chronic brain stimulation may be a useful method of treating chronic neurogenic pain. However, the knowledge about the basic mechanisms responsible for pain relief is still fragmentary, and the clinical results have often been inconsistent even contradictory.

In an attempt to explore the possibility of stimulating other cerebral targets, stimulating electrodes have been implanted in the septal region in addition to stimulation in the sensory thalamus or the periventricular grey. In 19 patients subjected to septal stimulation, 12 experienced satisfactory relief of their spontaneous pain together with abolition of allodynia. There were no untoward side effects. The follow-up ranged from 1 to 10 years.

Our results suggest that the septal area may be a suitable alternative target for chronic brain stimulation.

Keywords: Stereotactic surgery; pain; brain stimulation; septal area.

Introduction

The introduction of chronic brain stimulation has been a major advance in the neurosurgical management of chronic pain. However, the physiological mechanisms underlying the pain relief are insufficiently understood. Therefore, it is not surprising that the clinical results obtained with stimulation in the 'traditional' targets are inconsistent. In an attempt to explore the possibility of deploying alternative stimulation targets in the brain, electrodes were implanted in the septal area in addition to an electrode placed in the sensory thalamus or in the periventricular grey (PVG).

Material and Methods

In short, the patients were operated on under light sedation using a Hitchcock stereotactic apparatus. The ventricular system was outlined by water-soluble positive contrast and a standard DBS electrode was placed in the PVG or in the sensory thalamus. An additional bipolar or tetrapolar electrode was placed in the septal area (Fig. 1). Its tip was aimed at 5 mm in front and 2.5 mm below the anterior commissure in the coronal and horizontal planes, at 5.5 mm from the midsagittal plane. Both electrodes were connected to percutaneous leads to permit trial stimulation prior to interualization.

Criteria for patient selection and methods of evaluating the results have previously been reported[13,14].

Nineteen patients presenting with a predominating neurogenic pain were subjected to septal stimulation. Six cases had pain due to neoplastic invasion of the brachial plexus, and 14 cases had pain associated with partial or total deafferentation. In all patients the pain was severe and incapacitating, and it was generally described as being of a deep burning character. In most cases there was also allodynia in the painful area. All had extensive analgesic medication but no narcotic drugs. The pain history ranged from 1 to 5 years.

Results

Stimulation of the septal region below the threshold for subjective sensations generally produced a reduction of pain which outlasted the period of stimulation. Stimulation with somewhat higher intensity produced a sensation of warmth and often also a feeling of well-being and relaxation. Six of the 19 patients reported significant pain relief, i.e. more than 75% pain reduction. In no patient was there complete suppression of pain. Six patients had moderate relief, i.e. 50–75% pain reduction, and seven failed to obtain useful alleviation. The follow-up period ranged from 0.5 to 10 years. There were no untoward side-effects.

In no case was there a clear-cut reversal of pain relief by naloxone. It was also observed that experimentally induced pain was unaffected by stimulation in the septal area.

Discussion

The aim of the present study has been to explore a new target site for intracerebral stimulation and to evaluate its efficacy in treating neurogenic pain in a long-term perspective[8,12,13]. Based on early neuropsy-

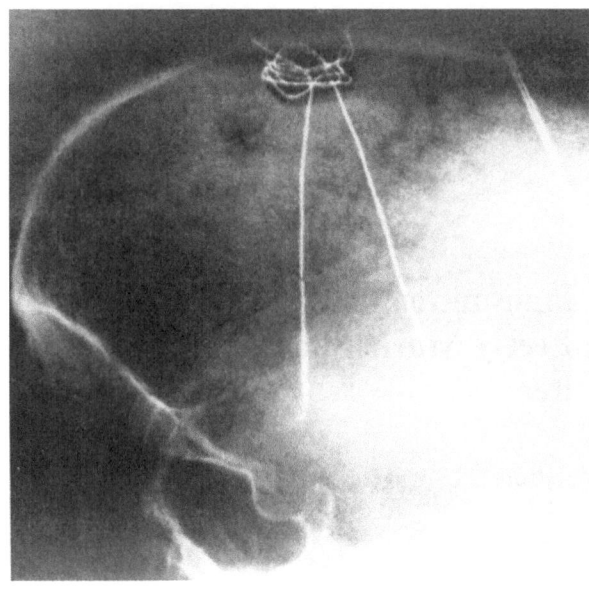

Fig. 1. Radiograph showing 'chronic' electrodes implanted in both the septal area and the periventricular grey region

chological experiments, previous attempts to correlate the positive rewarding effect of septal area stimulation with chronic pain in man have been made by Heath and Mickle[4], Gol[3] and Obrador et al.[5]. However, more recent research has shown that behavioural changes observed after stimulation of this region may be due to the activation of opioid mechanisms. Watson et al.[15] have demonstrated that the anatomical distribution of immunoreactive β-lipotropin in the rat includes the septal area, anterior hypothalamus, periventricular and periaqueductal grey. Moreover, Roldan et al.[11] found an increase in the plasma levels of immunoreactive β-endorphin after septal stimulation in the rat, and Cuello[1] and Pioro[7] have reviewed the enkephalinergic innervation of this same region in man.

Richardson[9] reported that acute stimulation of both the superior and inferior septal area produced significant relief of neurogenic pain in five patients. In a previous study, we gave preliminary data on chronic stimulation in the septal area in man[13]. In the present study, the early favourable results could be confirmed in that satisfactory pain relief could be produced on a long-term basis in 63% of our patients.

It is generally believed that deafferentation pain does not respond to stimulation which activates opioid mechanisms. This idea seems to be further substantiated by the finding in this study that naloxone did generally not counteract the pain relief obtained by septal stimulation. However, it can not be ignored that the stimulation certainly activates different neuronal circuits which may involve both opioid and non-opioid mech-

anisms. This may also be the background why for example PVG-stimulation may be effective for particular cases of neurogenic pain 2, 5, 10, 14. Although our results in this series of patients appear quite promising, the septal area has still to be further explored as a suitable target for chronic stimulation.

References

1. Cuello AC, Milstein C, Couture R, Wright B, Priestley JA, Jarvis, J (1984) Characterization and immunocytochemical application of monoclonal antibodies against enkephalins. J Histochem Cytochem 32: 947–957
2. Dieckmann G, Witzmann A (1982) Initial and long-term results of deep brain stimulation for chronic intractable pain. Appl Neurophysiol 45: 167–172
3. Gol A (1967) Relief of pain by electrical stimulation of the septal area. J Neurol Sci 5: 115–120
4. Heath RG, Mickle WA (1960) Evaluation of seven years' experience with depth electrode studies in human patients. In: Ramy, O'Doherty (eds) Electrical studies on the unanesthetized brain. Hoeber, New York, pp 214–247
5. Martín-Rodríguez JG, Obrador S (1979) Therapeutic electrical stimulation of the brain. Biochemical changes induced in the ventricular fluid with regard to opiate like substances. In: Hitchcock ER, Ballantine T, Meyerson B (eds) Modern concepts in psychiatric surgery. Elsevier, Amsterdam, pp 47–56
6. Obrador S, Delgado J, Martín Rodríguez JG (1974) The future of functional neurosurgery. In: Sano K, Ishii S (eds) Recent progress in neurological surgery. Excerpta Medica, Amsterdam, pp 265–269
7. Pioro EP, Mai JK, Cuello AC (1990) Distribution of Substance P- and Enkephalin-Immunoreactive Neurons and Fibers. In: Paxinos G (ed) The human nervous system. Academic Press, London, pp 1051–1094
8. Richardson DE (1987) Looking for better targets in the human endogenous opioid system. Pain [Suppl] 4: 330.
9. Richardson DE (1982) Analgesia produced by stimulation of various sites in the human β-endorphin system. Appl Neurophysiol 45: 116–122
10. Richardson DE, Akil H (1977) Pain reduction by electrical brain stimulation in man. Chronic self-administration in the periventricular gray matter. J Neurosurg 47: 184–194
11. Roldan P, Moreno J, Vincent E, Cerdá M, Ramos S, Barcia-Salorio JL (1984) Effect of electrical stimulation of periaqueductal gray and septal area on β-endorphin plasma levels in a model of deafferentation pain. Experimental study in rats. In: Gybels J, Hitchcock ER, Ostertag C, Rossi GF, Siegfried J, Szikla G (eds) Advances in stereotactic and functional neurosurgery, Vol 6. Acta Neurochir (Wien) [Suppl] 33: 501–503
12. Schvarcz JR (1987) Looking for better targets in the human endogenous opioid system. In: Pain [Suppl] 4: 330
13. Schvarcz JR (1985) Chronic stimulation of the septal area for the relief of intractable pain. Appl Neurophysiol 48: 191–194
14. Schvarcz JR (1980) Chronic self-stimulation of the medial posterior thalamus for the alleviation of deafferentation pain. In: Gillingham J, Gybels J, Hitchcock ER, Rossi GF, Szikla G. Advances in stereotactic and functional neurosurgery, Vol 4. Acta Neurochir (Wien) [Suppl] 30: 295–301
15. Watson SJ, Barchas JD, Li Ch (1977) Beta-lipotropin: localization of cells and axons in rat brain by immunocytochemistry. Proc Natl Acad Sci USA 74: 5155–5158

Correspondence: Prof. J. R. Schvarcz, M.D., Juncal 1845, Buenos Aires 1116, Argentina.

Acta Neurochir (1993) [Suppl] 58: 156–160
© Springer-Verlag 1993

An Animal Model for the Study of Brain Transmittor Release in Response to Spinal Cord Stimulation in the Awake, Freely Moving Rat: Preliminary Results from the Periaqueductal Grey Matter

B. Linderoth[1], C.-O. Stiller[2], W. T. O'Connor[2], G. Hammarström[1], U. Ungerstedt[2], and E. Brodin[2]

Departments of Neurosurgery[1] and Pharmacology[2], Karolinska Institute, Stockholm, Sweden

Summary

Electrical spinal cord stimulation (SCS) is an important method in the treatment of certain chronic pain syndromes which are difficult to manage with conventional techniques. The indications for this procedure have gradually narrowed to neuropathic pain states, especially those of peripheral origin, ischaemic pain due to peripheral vascular disease, and treatment-resistant angina pectoris. In spite of the clinical use of this method for more than 20 years, the mechanisms underlying the pain alleviating effect remain largely unknown.

For the effect on ischaemic pain, recent animal research indicates a mediation via autonomic pathways. Concerning the effect on neuropathic pain progress in knowledge has been scanty. Data from spinal microdialysis in decerebrated or anaesthetized animals indicate the possible importance of serotonin and substance P in the dorsal horn for pain inhibition by SCS. However, data from experiments on anaesthetized animals are, for several reasons, not likely to truely reflect the mechanisms active in conscious humans under treatment with SCS. To avoid the influence of anaesthesia and to approach the clinical situation, we have developed an animal model enabling simultaneous SCS and supraspinal microdialysis in awake, freely moving rats.

The animal model is described and some preliminary data indicating a release of gamma-amino butyric acid (GABA) induced by SCS in the periaqueductal grey matter (PAG), are presented.

Keywords: Animal model; GABA; microdialysis; periaqueductal grey matter; rat; spinal cord stimulation.

Introduction

Electrical stimulation applied to the posterior surface of the spinal cord (spinal cord stimulation, SCS or dorsal column stimulation, DCS) is an established treatment in certain chronic pain syndromes resistant to conventional therapeutic procedures. The method has proved very valuable in neuropathic pain conditions (e.g. Gybels and Kupers 1987; Gybels and Sweet 1989), especially those of peripheral origin (Meyerson 1990). Nociceptive pain states have generally been found to be unresponsive to this treatment modality, but the ischaemic pain in patients suffering from peripheral vascular disease constitutes an exception to this (Augustinsson *et al.* 1985; Galley *et al.* 1988; Meglio *et al.* 1989).

In spite of the clinical value of SCS, the mechanisms behind the efficacy of the method has been largely unknown to date. It seems probable that the mechanisms crucial for pain relief differ considerably between the application of SCS for neuropathic and for ischaemic pain (Linderoth *et al.* 1987a; Linderoth 1992). For the relief of ischaemic pain, induction of peripheral vasodilation by SCS appears to be mandatory. Furthermore, this vasodilatation seems to be the result of a stimulation-induced transitory inhibition of sympathetically maintained vasoconstriction (cf. Linderoth *et al.* 1991a, b). Regarding the mechanisms possibly involved in the alleviation of neuropathic pain, knowledge is fragmentary (cf. Meyerson 1983, 1990; Linderoth 1992).

The prolonged pain relief commonly observed after a 20–40 min SCS session has been interpreted to imply the activation of long-lasting neurochemical mechanisms in the CNS (cf. Meyerson 1983; Gybels and Sweet 1989). Several neurotransmitters in the dorsal horn and supraspinally are considered to be involved in the pain-alleviating effect of SCS (e.g. Bonica *et al.* 1990; Blumenkopf 1991; Linderoth 1992). At present there is little evidence that endogenous opioides are important for the effect of SCS (reviews cf. Meyerson 1983, 1990; Linderoth 1992), but recent data indicate a putative role for serotonin and substance P for the stimulation-induced pain inhibition (Broggi *et al.* 1985; Linderoth *et al.* 1987b, 1991c, 1992).

Inhibitory amino acids, especially gamma-amino butyric acid (GABA), have attracted some attention in relation to SCS. GABA appears to be involved in pre- and postsynaptic inhibition of primary afferent terminals within the dorsal horn, and the administration of GABA antagonists (bicucullin, strychnine) yields a behavioural reaction in experimental animals resembling sensory dysaesthesia or allodynia (cf. Matthews et al. 1988; Yaksh 1989; Blumenkopf 1991). Furthermore, Duggan and Foong (1985) reported that the $GABA_A$-antagonist bicucullin decreased the inhibitory effect on pain transmission of SCS. Such observations have resulted in a wide-spread belief that GABA in the dorsal horn is involved in the effects of SCS on pain (e.g. Duggan and Foong 1985; Blumenkopf 1991), as well as in some instances of acupuncture analgesia (e.g. Zhu et al. 1990).

In supraspinal regions GABA may also play a role for pain alleviation by SCS. GABA is found in abundance in the PAG, a structure considered to be involved in descending pain inhibition (cf. e.g. Basbaum and Fields 1984; Williams and Beitz 1990). Most of this GABA seems to be located in interneurons possibly modulating the activity in descending projections via the rostral ventral medulla, including the nucleus raphe magnus (Behbehani et al. 1990; Reichling and Basbaum 1990; Cho and Basbaum 1991). Furthermore, in the anterior pretectal nucleus a microinjection of GABA seems to decrease the inhibition on nociceptive transmission obtained by electrical spinal cord stimulation (Rees and Roberts 1989). Thus, there are several reasons to investigate whether GABA is released both in supraspinal regions associated with pain modulation and in the dorsal horn. Here, a study exploring the possible release of the substance in the PAG is presented.

A well known complicating factor in the interpretation of animal experiments on transmitter release is the possible influence of anaesthesia. There are many reports on the differences in neurotransmitter levels between animals under different types of anaesthesia and the awake preparation (cf. e.g. Osborne et al. 1990; Ståhle et al. 1990; Kelland et al. 1990). However, for some years it has been possible to monitor extracellular neurotransmitter levels by microdialysis in various brain regions in awake freely moving laboratory animals (cf. Zetterström et al. 1984).

In the present report we describe the combination of this technique with SCS via chronically implanted extradural electrodes. Some preliminary data on GABA release in the PAG induced by SCS are also presented.

Materials and Methods

Male Sprague-Dawley rats (weight 300–380 g; ALAB, Södertälje Sweden) with free access to food and water at all times were used.

Surgery. With the animal fixed in a sterotaxic unit (David Kopf Instr., Tujunga, Ca., USA) and under halothane anaesthesia, a microdialysis guide cannula was inserted into the PAG using the coordinates 7.6 mm caudal and 2.5 mm lateral to the bregma, with a depth of 4.6 mm from the dura. A trajectory at 22° angle from the sagittal plane was used to avoid aqueductal interference. The guide was fixed to the skull with screws and dental cement.

For SCS, a system consisting of an intraspinal silver cathode (diameter 2 mm) and a subcutaneous silver anode (diameter 5 mm) connected to a microcontact by 0.25 mm plaited stainless steel teflon-coated wires (Medtronic Inc., Minn, Mn USA) was used. The intraspinal electrode was placed extradurally onto the dorsal aspect of the cord in the lower thoracic region (T12-T13) and the position checked by x-ray postoperatively. The contact was sutured to the skin between the scapulae. An x-ray overview of the entire microdialysis and SCS system is shown in Fig. 1.

During SCS with simultaneous microdialysis, the animal was kept in a special hemispheric cage (CMA Microdialysis Stockholm Sweden), permitting the animal to circle in the same direction for any number of turns or chose any irregular movement pattern. A balance arm, at the end of which a liquid swivel is mounted, follows the movements of the animal. An electric swivel, adapted to the CMA/Microdialysis system and enabling simultaneous SCS, was designed for the present project.

Microdialysis system. The dialysis probes used (CMA/12) have a membrane diameter of 0.5 mm, a membrane length of 2.0 mm and a molecular cut-off at 20.000 Daltons. In each experiment, following 2–3 day of recovery from surgery, the sampling was started one hour after the insertion of the probe. The dialysis fluid consisted of modified Ringer solution (conc. in mM: NaCl 148.0; KCl 2.7; $MgCl_2$ 0.85; $CaCl_2$ 4.0). The perfusion rate was 7.0 µl/min and samples were collected every 15 minutes and kept cooled at + 4 °C throughout the experiment. Immediately after the experiments the samples were frozen to − 20 °C until the time of assay.

Spinal cord stimulation. Monopolar stimulation with a constant current system using 100 Hz; 0.2 msec pulses and 2/3 of the current required to induce a hind limb flexor response in the lightly anaesthetized animal (average 0.24 +/− 0.3 mA) was used. The stimulating current was generated by a Grass S88 stimulator (Grass Instr., Quincy, Mass., USA) connected to a Grass constant current unit (CCU1). SCS was applied twice for 30 min. during the collection of samples 5–6 and 13–14, respectively. In some experiments GABA-release was stimulated by increasing the potassium concentration in

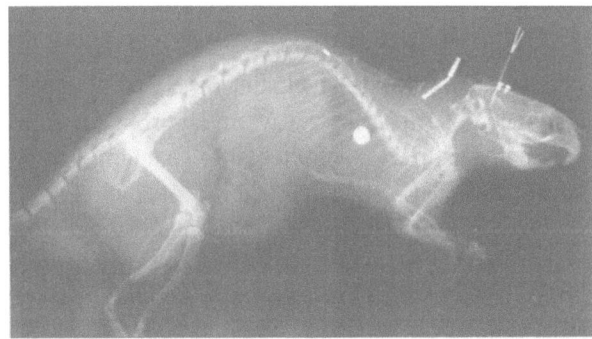

Fig. 1. Lateral x-ray of a rat with a chronically implanted microdialysis cannula in the PAG and a SCS system with the intraspinal cathode at level T12–T13

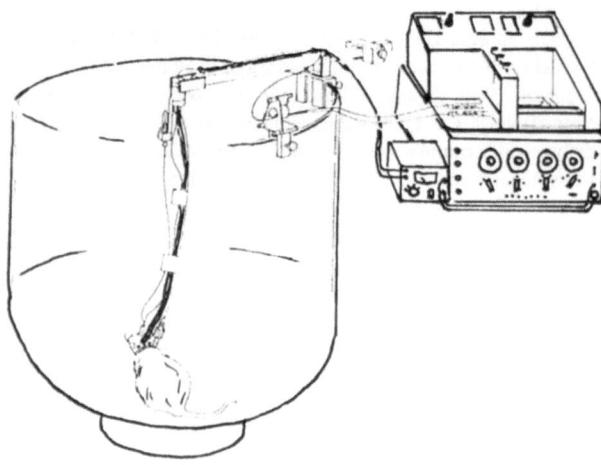

Fig. 2. Schematic drawing of experimental set-up with the rat in the CMA observation cage, the balance arm with the liquid and electrical swivels following the movements of the animal. The dialysis catheters and the electric wires follow a string connecting the balance arm to a collar. Infusion pump and Grass stimulator to the right (redrawn after Microdialysis users guide, 4th Ed, Carnegie Medicin Stockholm, Sweden)

the dialysis fluid to 100 mM during 15 min. The total dialysis time exceeded five hours. The entire set-up is schematically illustrated in Fig. 2.

Assay system. Several substances were analyzed but only some data on GABA will be given here. A full report will appear later (Stiller, Linderoth, O'Connor, Franck, Brodin, in preparation).

GABA was assayed as previously described by Kehr and Ungerstedt (1988). Briefly, the procedure is based on precolumn derivatization with an o-phthaldialdehyde/t-butylthiol reagent and separation by a reverse phase HPLC on a Nucleosil 3, C-18 column with electrochemical detection under isocratic conditions. The detection limit for GABA was approx. 0.5 nM at the injection volume of 10 µl as used here.

Histological control. The brain was dissected out after the experiments and immediately frozen to − 80 °C. It was later cut in a microtome for probe tract verification.

The studies reported here were examined and approved by the Local Ethical Committe for Animal Research.

Results

Technical Aspects

The stimulation system including the electrical swivel functioned in a satisfactory way. Initial problems with wire breakages were eliminated after changing to the above-mentioned plaited and teflon-coated wires.

Behavioural Response

The animals seemed to tolerate the stimulation well. At the onset of an SCS period they displayed a freezing reaction, possibly getting aware of the paraesthesiae, usually standing or sitting without moving for some

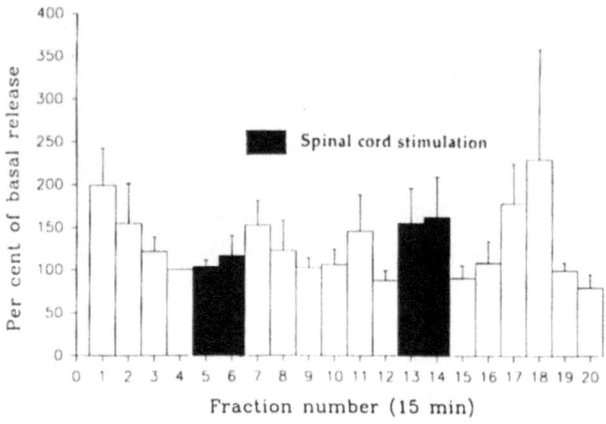

Fig. 3. GABA release in the PAG. Outcome of the GABA assay in nine animals. SCS was applied during fractions 5 + 6 and 13 + 14 respectively (black bars). All data are expressed in per cent of fraction four which is set to 100 (mean +/− SEM)

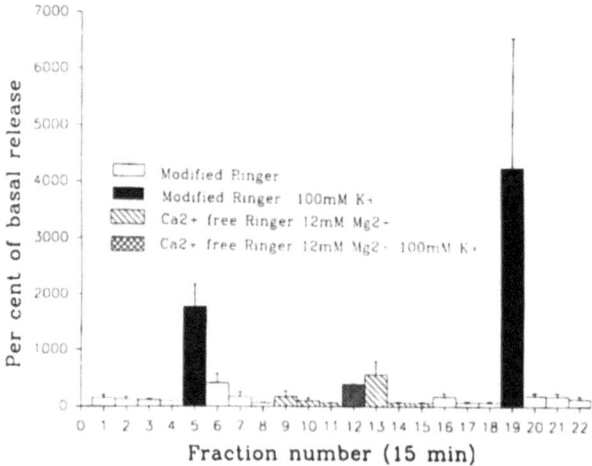

Fig. 4. GABA release in the PAG during the perfusion with different types of dialysis fluids as indicated in the figure. Data given is in per cent of fraction four. Elevation of the K^+ concentration induced an intense release (black bars), from 5.3 nM in fraction four to 88 nM in fraction five. In the absence of calcium and with 12 mM Mg in the perfusion medium, the GABA release induced by 100 mM potassium was only 20.8 nM (cross-hatched bar). After changing to the standard calcium-containing solution, 100 mM potassium again induced a marked GABA release to 128 nM (mean +/− SEM; n = 3)

minutes. After a while they behaved as prior to the SCS, investigating the environment or sometimes sleeping. During the perfusion with the 100 mM potassium solution the animals appeared aroused and on the alert.

GABA Assay

Data from nine animals have been analyzed to date. These preliminary results are illustrated in Fig. 3. The basal GABA level recorded before the first SCS period was 11.6 $+/-2.6$ nM (SEM; n = 9). During the first stimulation period only a slight tendency to increase of the extracellular GABA level could be observed. After the cessation of SCS the GABA concentration continued to increase to 19.6 $+/-7.9$ nM (fraction 7). During the second stimulation the GABA concentration increased from 8.2 $+/-1.4$ nM (fraction 12) to 13.4 $+/-3.7$ nM (fraction 14). Using the Wilcoxon nonparametrical test for related observations the GABA increase in fractions 13 and 14 compared to the preceeding basal value, was found to be significant ($P < 0.05$).

In Fig. 4 it is seen that increasing the K^+ concentration in the perfusion medium to 100 mM induced a marked elevation of the average GABA release. This potassium-induced release was markedly reduced in the absence of calcium and in the presence of 12 mM magnesium (acting as a calcium channel blocker) in the perfusion fluid.

Discussion

Thus, it is demonstrated that supraspinal microdialysis and electrical spinal cord stimulation may be combined in the awake unrestraind rat without major problems. This model may be more comparable to the situation in man during the clinical use of SCS than the experimental procedures on anaesthetized animals used earlier (e.g. Linderoth *et al.* 1987b, 1992).

As already mentioned GABA is presently considered as one of the most interesting neurotransmitter candidates for the mediation of the pain-alleviating effect of SCS (cf. Duggan and Foong 1985; Blumenkopf 1991). The preliminary results reported here suggest a release of GABA in the PAG induced by SCS. Furthermore the calcium-dependent release resulting from potassium stimulation indicates that the GABA originates from neuronal vesicular stores. However, the number of animals analyzed so far is too small to permit any further conclusions about the possible role of GABA in these circumstances.

It may be pointed out that in another on-going study (Linderoth, Gunasekera, Stiller, O'Connor and Brodin, in preparation) GABA release with SCS was also observed in the dorsal horn of halothane-anaesthetized rats. Such studies are difficult to perform in a mobile preparation because of the large movement artifacts that would result from the use of a stiff microdialysis system at the spinal level. Although spinal cord microdialysis using "soft" dialysis catheters has been described (e.g. Sorkin *et al.* 1988), there are, to the best of our knowledge, no reports of spinal microdialysis with soft catheters in an awake, freely moving animal as yet. However, some trials with stiff standard probes have been performed by Privat and collaborators (personal communication, 1992).

Thus, there are now some more observations supporting the view that GABA may play a role in pain inhibition induced by SCS. However, it must be kept in mind that GABA represents *only one* neurotransmitter in the bouquet of substances possibly released by the stimulation—and where only a few as yet may be known to us.

Further studies utilizing different experimental models are called for. The animal models of neuropathic pain recently developed by Bennett and Seltzer may prove to be valuable for the future investigation of the neurochemical mechanisms underlying the effects of SCS on pain.

Conclusions

Combined SCS and supraspinal microdialysis may be adequately performed in the awake, unrestrained rat. Preliminary observation, obtained with the model described here, suggest that GABA release is induced in the PAG by concomitant low thoracic SCS.

Acknowledgements

The studies reported here were supported by grants from the Karolinska Institute, Swedish Medical Research Council (project No 6836) and from the Svenska Sällskapet för Medicinsk Forskning. Technical support from CMA Microdialysis Stockholm, Sweden and from Medtronic Inc, Minneapolis Minn., USA is also gratefully acknowledged.

References

1. Augustinsson LE, Carlsson CA, Holm J, Jivegård L (1985) Epidural electrical stimulation in severe limb ischemia. Ann Surg 202: 104–110
2. Basbaum AI, Fields HL (1984) Endogenous pain control systems: brainstem spinal pathways and endorphin circuitry. Ann Rev Neurosci 7: 309–338
3. Behbehani MM, Jiang M, Chandler SD, Ennis M (1990) The effect of GABA and its antagonists on midbrain periaqueductal grey neurons in the rat. Pain 40: 195–204
4. Bonica JJ, Yaksh T, Liebeskind JC, Pechnick RN, Depaulis A (1990). Biochemistry and modulation of nociception and pain. In Bonica JJ (ed) The management of pain, 2nd Ed. Lea and Febiger, Philadelphia, pp 95–121
5. Blumenkopf B (1991) The general aspects of neuropharmacology of dorsal horn function. In: Nashold Jr BS, Ovelmen-Levitt J (eds) Deafferentation pain syndromes: Pathophysiology and treatment. Raven, New York, pp 151–162

6. Cho HJ, Basbaum AI (1991) GaBAergic circuitry in the rostral ventral medulla of the rat and its relationship to descending antinociceptive controls. J Comp Neurol 303: 316–328

7. Duggan AW, Foong FW (1985) Bicuculline and spinal inhibition produced by dorsal column stimulation in the cat. Pain 22: 249–259

8. Galley D, Elharrar C, Scheffer J, Rakotonarivo J, Medvedowski A, Barnay JC, Medvedowski L, Serena G (1988) Intéret de la neurostimulation épidurale dans les artériopathies des membres inférieurs. Arteres et Veines 7: 61–71

9. Gybels JM, Kupers R (1987) Central and peripheral electrical stimulation of the nervous system in the treatment of chronic pain. Acta Neurochir (Wien) [Suppl] 38: 64–75

10. Gybels JM, Sweet WH (1989) Neurosurgical treatment of persistent pain. In: Gildenberg PL (ed) Pain and headache. Karger, Basel, pp 442

11. Kehr J and Ungerstedt U (1988) Fast HPLC estimation of aminobutyric acid in microdialysis perfusates: Effects of nipecotic and 3-mercaptopropionic acids. J Neurochem 51: 1308–1310

12. Kelland MD, Chiodo LA, Freeman AS (1990) Anesthetic influences on the basal activity and pharmacological responsiveness of nigrostiatal dopamine neurons. Synapse 6: 207–209

13. Linderoth B (1992) Dorsal column stimulation and pain: Experimental studies of putative neurochemical and neurophysiological mechanisms. Published thesis. Karolinska Institute, Stockholm, 67 pp

14. Linderoth B, Meyerson BA, Skoglund CR (1987a) Spinal cord stimulation for treatment of peripheral vascular disease: Review and short-term effects. Ann Meeting Scand Ass Study of Pain, Kolding, Denmark (abstract)

15. Linderoth B, Gazelius B, Brodin E (1987b) Effect of noxious stimuli and dorsal column stimulation on release of substance P in cat dorsal horn. Pain [Suppl] 4: 193 (abstract)

16. Linderoth B, Fedorcsak I, Meyerson BA (1991a) Peripheral vasodilatation after spinal cord stimulation: Animal studies of putative effector mechanisms. Neurosurgery 28: 187–195

17. Linderoth B, Gunasekera L, Meyerson BA (1991b) Effects of sympathectomy on skin and muscle microcirculation during dorsal column stimulation: Animal studies. Neurosurgery 29: 874–879

18. Linderoth B, Gazelius B, Franck J, Brodin E (1991c) Dorsal column stimulation and neurotransmitters in the dorsal horn: Animal studies. In: Brock M, Banerji AK, Sambasivan M (eds) Modern neurosurgery, Vol 2. (Proc. 9th Int. Congr. of Neurosurgery, New Delhi 1989). World Federation of Neurosurgical Societies. Berlin, New Delhi, Trivandrum, pp 15–23

19. Linderoth B, Gazelius B, Franck J, Brodin E (1992) Dorsal column stimulation induces release of serotonin and substance P in the cat dorsal horn. Neurosurgery 31(2): 289–297

20. Matthews MA, McDonald GK, Hernandez TV (1988) GABA distribution in a painmodulating zone of trigeminal subnucleus interpolaris. Somatosens Res 5(3): 205–217

21. Meyerson BA (1983) Electrostimulation procedures: Effects, presumed rationale and possible mechanisms. In: Bonica JJ, Lindblom U, Iggo A (eds) Advances in pain research and therapy, Vol 5. Raven, New York, pp 495–534

22. Meyerson BA (1990) Electric stimulation of the spinal cord and brain. In: Bonica JJ, Loeser JD, Chapman RC, Fordyce WE (eds) The management of pain, 2nd Ed. Lea and Febiger, Philadelphia, pp 1862–1877

23. Osborne PG, O'Connor WT, Drew KL, Ungerstedt U (1990) An in vivo microdialysis characterization of extracellular dopamine and GABA in dorsolateral striatum of awake freely moving and halothane anesthetised rats. J Neurosci Methods 34: 99–105

24. Rees H, Roberts MHT (1989) Antinociceptive effects of dorsal column stimulation in the rat: Involvement of the anterior pretectal nucleus. J Physiol 417: 375–388

25. Reichling DR, Basbaum AI (1990) Contribution of brainstem GABAergic circuitry to descending antinociceptive controls: II. Electron microscopic immunocytochemical evidence of GABAergic control over the projection from the periaqueductal gray to the nucleus raphe magnus in the rat. J Comp Neurol 302: 378–393

26. Sorkin LS, Steinman JL, Hughes MG, Willis WD, McAdoo DJ (1988) Microdialysis recovery of serotonin released in spinal cord dorsal horn. J Neurosci Methods 23: 131–138

27. Ståhle L, Collin A-K, Ungerstedt U (1990) Effects of halothane anesthesia on extracellular levels of dopamine, dihydroxyphenylacetic acid, homovanillic acid and 5-hydroxyindoleacetic acid in rat striatum: a microdialysis study. Naunyn Schmidebergs Arch Pharmacol 342: 136–140

28. Williams FG, Beitz AJ (1990) Ultrastructural morphometric of GABA-immunoreactive terminals in the ventrocaudal periaqueductal grey: Analysis of the relationship of GABA terminals and the $GABA_A$ receptor to periaqueductal grey-raphe magnus projection neurons. J Neurocytol 19: 686–696

29. Yaksh TL (1989) Behavioral and autonomic correlates of the tactile evoked allodynia produced by spinal glycine inhibition: effects of modulatory receptor systems and excitatory amino acid antagonists. Pain 37: 111–123

30. Zetterström T, Sharp T, Ungerstedt U (1984) Effect of neuroleptic drugs on striatal dopamine release and metabolism in the awake rat studied by intracerebral dialysis. Eur J Pharmacol 106(1): 27–37

31. Zhu L, Li Ch, Ji Ch, Yu Q (1990) Involvement of GABA in acupuncture-induced segmental inhibition. Eur J Pain 11: 114–118

Correspondence: B. Linderoth, M.D., Ph.D., Department of Neurosurgery, Karolinska Hospital, S-10401 Stockholm, Sweden.

Acta Neurochir (1993) [Suppl] 58: 161–164

Preliminary Results of a Randomized Study on the Clinical Efficacy of Spinal Cord Stimulation for Refractory Severe Angina Pectoris

M. J. L. de Jongste[1] and **M. J. Staal**[2] (on behalf of the Working Group Neurocardiology)

[1]Department of Cardiology, Thoraxcenter, and [2]Department of Neurosurgery, University Hospital Groningen, The Netherlands

Summary

For the last 6 years Spinal Cord Stimulation (SCS) has been advocated for patients with therapeutic refractory angina pectoris[2–4]. We studied the efficacy of spinal cord stimulation on the relief of otherwise intractable angina pectoris in a 2 months' randomized study with 1 year follow-up by quality of life parameters, cardiac parameters and complications. Twenty four patients were randomized to either an actively treated group A (12 patients received the device within a 2 weeks' period) or a control group B (10 patients had an implantation after the study period). In both groups one patient dropped out before the implantation but after the randomization.

It is concluded that spinal cord stimulation improves both quality of life and cardiac parameters. The latter included a trend towards reduction in ischaemia after implantation of the device in both treadmill exercise and 24-hour ambulatory Holter recordings, with a concomitant better exercise capacity.

Keywords: Spinal cord stimulation; SCS; angina pectoris; randomized study; results.

Introduction

Since the introduction, by Shealy and Wall in 1970, of an implantable neurostimulator for electrical treatment of pain[1], an estimate of over 20.000 devices have been implanted worldwide for all varieties of pain. In 1987. Mannheimer *et al.* were the first to publish on the improvement of symptoms of patients with severe angina pectoris treated with Spinal Cord Stimulation (SCS)[2]. Similar results were independently realized by Murphy[3] and Sanderson[4].

For determination of the effects of SCS the patients have to become aware of the paraesthesias, which make blinding or cross-over design not feasible, while randomization is still conceivable.

Patients and Methods

This study was a randomized trial that attempted to determine whether patients with medically refractory angina pectoris can be effectively treated with spinal cord stimulation. Medically refractory encompasses those patients who are on pharmalogical optimal drug-treatment for at least 1 month, without any possibility for revascularization procedures such as coronary artery bypass surgery. Patients with angiographically documented coronary artery disease who have severe otherwise intractable angina pectoris with proven ischaemia (by either ST-T segment changes on the ECG, or in case of Bundle Branch Block by Thallium-201 scintigraphy or Positron Emission Tomography) were included in the study if they were between 18–76 years of age. Exclusion criteria mainly referred to a short life expectancy and or the inability to perform an exercise test. All patients signed a written informed consent certified by the Hospital Ethical Commission.

After selection and informed consent 24 patients (14 males, 10 females mean age 61) were enrolled in the study of whom 22 were operated on. Patients were eligible for the study if they fulfilled the above mentioned inclusion criteria.

In all patients a Medtronic Itrell* was percutaneously implanted under local anaesthesia and fluoroscopy. In 8 patients an Itrell I* with a monopolar lead and in 14 patients an Itrell II* with a quadripolar electrode was used. Stimulation was standardized to 3 hours three times per day and therapeutically.

After selection the patients were randomly divided into either group A (active treatment) or B (control) (see Fig. 1).

In group A (N = 12) the device was implanted and 'set' to active stimulation within 2 weeks after the randomization. In group B (N = 10) the stimulator was implanted after the study period.

Repeated treadmill exercise tests (TET) and 24-hour ECG recordings (24 h ECG) were studied as cardiac parameters in relation to baseline characteristics such as left ventricular ejection fraction and cardiac history. In addition, quality of life parameters were evaluated by a standardized and validated activities of daily life (ADL) scoring and a diary. In the diary the number of anginal attacks (AA) and glyceryl trinitrate intake (GTN) were counted.

During the 2 months' study period the same battery of tests was performed. After the study period the patients were their own controls. The follow-up was meant to evaluate the persistence of change in parameters.

Surgical Procedure of the Implantation Technique

Although the procedure is performed under local anesthesia we consider the attendance of an anesthetist as mandatory. Prophylactic antibiotics (cotrimoxazol t.i.d. 960 mg) are given in a 2 days' standard regime. With the patient in prone position, a small incision in the

---

* Trademark of Medtronic Inc. Minneapolis, Minn., U.S.A.

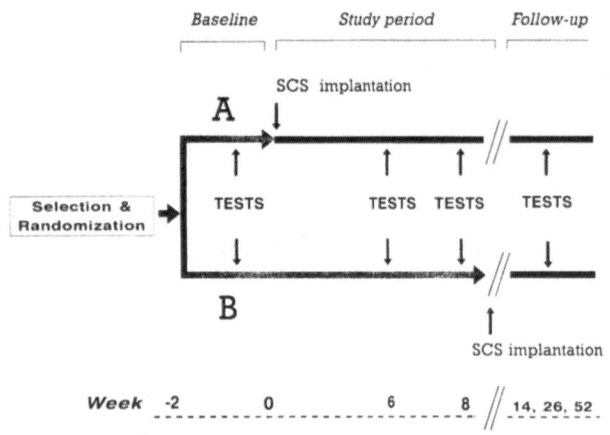

Fig. 1. Flowchart study

midline, at the T4-T5 level, is made. Then the epidural space is punctured with a Tuohy needle. Under fluoroscopic control an epidural electrode (either a unipolar Pisces Sigma* or a quadripolar Quad* electrode) is inserted through this needle in the dorsal epidural space, with the tip at the C7 level. With an external stimulation device paraesthesias are evoked.

By withdrawing the stylet from the electrode, the final position (mostly at T1 and slightly left from the midline) is determined by sensation of paraesthesias in the thoracic area, sometimes also including the left arm, according to the area of anginal pain as indicated individually by the patient.

After that, a left subcostal retrofascial pocket is created. An Itrell 1* (in case of unipolar stimulation), or an Itrell 2* pulse generator (in patients with a quadripolar electrode), is placed in this pocket and connected to the epidural electrode with an extension lead.

Results

In group B fewer patients were enrolled with on average per patient a longer period of coronary heart disease (10.9 versus 9.9 years), more myocardial infarctions (0.9 versus 0.7) and a worse left ventricular ejection fraction (46.5 versus 50.2). On the other hand the patients randomized to group A underwent more previous revascularization procedures (0.9 versus 0.8) and had more vessels involved (2.8 versus 2.5) (see Table 1).

All the 22 patients who had had surgery reported an improvement in quality of life measured by ADL scoring and diary (see Table 2) after implantation. The ADL

Table 1. *Baseline Characteristics*

Randomization No.	Group	Age	Sex	CAD since (yrs)	Previous MI	Previous CABS (+) PTCA (×)	LVEF in %
1	A	48	M	9		+	48
2	B	66	F	6	+	+	63
3	A	59	M	16	+	+ ×	28
4	B	64	M	10	+	+	35
5	B	64	M	8	+	+ ×	51
6	B	60	M	25	+		47
7	A	58	M	11	+	+	58
8	A	59	F	4		+ ×	53
9	B	63	M	15		+ ×	62
10	A	58	M	3	+		57
11	B	51	M	13	+	+	57
12	A	48	M	11	+	+	49
13	A	64	M	8	+	+	35
14	B	53	M	11	+	+	62
15	B	52	F	3	+	×	35
16	A	73	F	15		+ ×	67
17	B	73	M	8	+		47
18	A	49	M	5		+ ×	61
19	A	58	F	4	+	+	63
20	B	60	M	9	+	+	22
21	A	60	F	8		+ ×	60
22	A	66	F	17	+	×	34
23	B	66	M	12	+	+	31
24	A	75	M	17	+	+	40
13	A	59.6 ± 4.6	8M/5F	9.9 ± 5.1	0.7	0.9	50.2 ± 12.5
11	B	61.1 ± 5.6	9M/2F	10.9 ± 5.7	0.9	0.8	46.5 ± 14.8

Age age at entry study; *CAD* coronary artery disease; *CABS* coronary artery bypass grafting; *MI* myocarial infarction; LVEF in % left ventricle ejection fraction in percentage. d.o. drop out; ≠ diseased; Values are represented as means ± S.D.

Table 2. *Outcome of the Quality of Life Parameters*

	Group A Baseline	Study	Follow-up	Group B Baseline	Study	Follow-up
ADL	1.3 ± 0.3	2.0 ± 0.6	2.2 ± 0.5	1.3 ± 0.3	1.3 ± 0.3	2.5 ± 0.3
GTN/wk	13.3 ± 6.7	1.7 ± 1.7	1.8 ± 2.1	13.0 ± 16.3	12.0 ± 4.0^a	7.2 ± 3.3
AA/wk	16.8 ± 7.8	6.3 ± 5.1	3.5 ± 3.010	16.3 ± 7.9	13.6 ± 7.1	7.2 ± 7.3

[a] 1 patient spontaneously reduced his GTN-intake from 58 to 13 tablets per week.

ADL activities of daily life, scored by questionnaire; *GTN* shortacting glyceryl trinitrate intake; *AA* anginal attacks. Values are represented as means $\pm$ S.D.

Table 3. *Outcome of the Treadmill Exercise Tests*

	Group A Baseline	Study	Follow-up	Group B Baseline	Study	Follow-up
RPP × 100	125 ± 30	142 ± 44	$^{149} \pm 45$	144 ± 53	137 ± 34	158 ± 34
Σext (min)	9.7 ± 4.7	10.8 ± 4.1	9.9 ± 5.1	11.1 ± 5.2	10.8 ± 4.0	14.6 ± 5.8
AAt (min)	9.6 ± 4.5	10.9 ± 3.9^a	9.9 ± 2.6	11.0 ± 5.0	10.4 ± 4.0	13.8 ± 5.5
ΣST$\downarrow$ max	1.60	0.95	1.30	2.40	2.02	1.7

[a] Stopcriterion for 2 patients became fatigue. *RPP* rate pressure product; *Σext* total exercise time; *AAt* time to angina; *ΣST$\downarrow$* max total ST segment depression in mV for the entire group.

score improved during the study in group A and after the study in group B patients. There was a concomitant dramatic fall in both anginal attacks and GTN intake after the implantation.

According to the cardiac parameters an increase in rate pressure product was observed in the treated group empowered by a prolonged exercise capacity. In the control group there was a decrement in both rate pressure product and exercise capacity, possibly as a result of the natural history of the disease. In addition total ischaemic burden, expressed as ST-T segment depression at maximal exercise tended to decrease after implantation (see Table 3).

The 24-hour ambulatory Holter recordings demonstrated a clear but not yet significant trend in reduction of ischaemia. A total number of 135 Holters were performed (group A: 67 and group B: 68).

Ischaemia was demonstrated in 12 out of the 22 evaluated patients prior to SCS and in 6 patients with SCS. Before SCS in 23% of all 24-hour Holters ischaemia was demonstrated, after SCS in only 10%.

The effects of SCS on the measured parameters lasted for an average follow-up of 13.8 months (total patients' months: 332) to a maximum of 3 years.

Postoperative complications occurred in six patients mainly consisting of a dislocation of the lead requiring an operative reposition.

Two patients dropped out before the study (one in group A because of a myocardial infarction before

implantation and one due to a cerebrovascular accident before implantation). Two others died during follow-up, one as a consequence of worsening ischaemia and one because of heart failure.

Discussion

Patients with severe therapeutic refractory angina pectoris are somatically and socially disabled. For this group of intractable patients SCS might be an adjuvant therapy with hopeful perspectives. We evaluated in an open randomized study the efficacy of SCS in this subset of patients.

In a pilot study[5] we demonstrated that SCS improved the subjective parameters but did not influence the more objective cardiac parameters. These later findings of our pilot study were in conflict with the results reported by others[2-4]. This difference might be due to other selection criteria and or to different protocols. For instance Mannheimer et al.[2] exercised the patients with the stimulator "on". In the study discussed here all treadmill exercise tests after SCS implantation were performed during active stimulation.

Baseline characteristics for both groups were not significantly different. After implantation of the device the patients had a better exercise capacity represented by an increase in both exercise time and time to angina with a concomitant higher Rate Pressure Product (RPP) and less (but not significant) ST-T segment depression.

The reason that we could not uncover a significant reduction in ischaemia could be either a too small sample size, or that exercise testing is not a good parameter of ischaemia in this subset of patients. Eventually there may be no influence of SCS on ischaemia as suggested by Landsheere et al.[6]. The average, maximal and minimal heart rate appeared not to be influenced by SCS, but a decrease of 50% in total ischemic burden during 24-hour ambulatory ECG monitoring was observed.

Another explanation for the results might still be the difference in selection criteria. We did include patients with Bundle Branch Blocks and proven ischaemia on Thallium-201 scintigraphy. As a consequence we were not able to evaluate in these patients' ST-T segment changes. In addition patients with significant 3 vessel disease might have no ST-T segment changes due to cancellation of the ST-T segments, as is observed during acute myocardial infarction.

Quality of life parameters as expressed by the ADL score, the GTN intake and Anginal Attacks improved clearly. A placebo effect as an explanation for the lack of significance on exercise testing is unlikely since the improvement lasted for over one year.

From the results of our study we cannot unambiguously conclude that SCS is an effective treatment for otherwise intractable angina pectoris as far as ischaemia is concerned. However, although not all cardiac parameters improved substantially, a clear increase in all of the quality of life parameters was demonstrated. The results might even be biased by the operation itself. To analyze a possible bias from the operation we have changed the protocol. All implantations will be shortly after randomization, but the device in group A is set to the 'active' mode while the device in group B is not activated. Since the results we presented here were preliminary we did not perform statistical analysis.

Conclusions

Based on both subjective (Quality of Life) and objective cardiac improvement in group A (and after implantation in group B), SCS might be a valuable adjuvant therapy for patients with otherwise intractable angina pectoris. The question still remains valid if SCS has besides its electroanalgesic effect a more specific anti-ischaemic effect. An anti-ischemic influence of SCS on the heart will protect the patient from deleterious cardiac events.

References

1. Shealy CN, Mortimer JT, Hagfors NR (1970) Dorsal column electroanalgesia. J Neurosurg 32: 560–564
2. Mannheimer C, Augustinsson LE, Carlsson CA, Manhem K, Wilhelmsson C (1988) Epidural spinal electrical stimulation in severe angina pectoris. Br Heart J 59: 56–61
3. Murphy DF, Giles KE (1987) Dorsal column stimulation for pain relief from intractable angina pectoris. Pain 28: 365–368
4. Sanderson JE (1990) Electrical neurostimulators for pain relief in angina. Br Heart J 63: 141–143
5. Jongste de MJL, Lie KI, Linde van der MR, Meyler WJ, Mulder M, Staal MJ, Vries K, Zimmerman C, Zijlstra GJ (1989) A longterm randomized study on the efficacy of spinal cord stimulation as an adjuvant therapy for severe angina pectoris, Abstract. 1st I.C.E.S.S Congress 1989, Groningen, The Netherlands
6. Landsheere CH de, Mannheimer C, Habets A, Guillaume M, Bourgeois I, Augustinsson LA, Eliasson T, Lamotte D, Kulbertus H, Rigo P (1992) Effect of spinal cord stimulation on regional myocardial perfusion assessed by positron emission tomography. Am J Cardiol 69: 1143–1149
7. Saikawa T, Abe M, Nakagawa M, Omwia I, Takakwia T, Horita S, Ho S, Takaki R, Ho M (1987) The complete cancellation of abnormal Q waves due to an old outeroseptal interaction following subsequent acute posterior myocardial infarction. Jpn Heart J 28: 805–810

Correspondence: Mike J. L. de Jongste, M.D., Department of Cardiology, Thoraxcenter, University Hospital of Groningen, Oostersingel 59, P.O. Box 30.001, NL-9700 RB Groningen, The Netherlands.

Acta Neurochir (1993) [Suppl] 58: 165–167
© Springer-Verlag 1993

Microvascular Decompression for Trigeminal Neuralgia: Prognostic Factors

A. Puca, M. Meglio, B. Cioni, M. Visocchi, and **R. Vari**

Institute of Neurosurgery, Catholic University, Rome, Italy

Summary

A series of 66 patients with trigeminal neuralgia (TN) treated by microvascular decompression (MVD) is analyzed in order to define the importance of various prognostic factors. A typical pattern of pain, presence of facial hypaesthesia and previous ablative trigeminal procedures showed an unfavourable influence on the results of MVD. Duration of symptoms was longer in patients who failed to benefit from MVD. Pain distribution and the type of vascular compression did not influence the outcome.

Keywords: Microvascular decompression; trigeminal neuralgia; trigeminal nerve.

Introduction

Several effective surgical procedures are now available for the treatment of trigeminal neuralgia (TN).

Microvascular decompression (MVD) of the trigeminal nerve in the posterior fossa has been proposed as an etiological treatment of TN[4-7,10]. Nevertheless, controversies still exist about the rationale of MVD[1,11]. Because of the problem of pain recurrence requiring subsequent operations, there is a need for a reliable protocol for the treatment of TN.

In order to evaluate the relevant prognostic factors in MVD and to design our protocol of treatment, we reviewed a series of 66 patients with TN, all operated on by MVD by the same surgeon (M.M.).

Clinical Material and Methods

Between 1986 and 1991, 66 patients with TN were operated upon by MVD; there were 31 males and 35 females. Mean age was 60 years (range 39–78). All patients had been previously treated by drugs but were drug-resistant or intolerant. Pain affected the right side in 39 subjects and the left in 27. All three divisions were affected in 11 patients, two in 27 and one division in 28.

TN was defined as typical in 50 patients, i.e. it was characterized by episodes of paroxysmal pain, presence of trigger zones and of pain-free periods. An atypical pattern of pain was identified in 16 subjects. Facial hypaesthesia was present in 34 patients. The mean duration of history of trigeminal symptoms was 7.9 years (range 6 months–30 years).

Previous percutaneous procedures had been performed in 33 patients, generally radiofrequency thermocoagulation and microcompression of the Gasserian ganglion.

All patients underwent a CT scan or MRI before surgery.

MVD was performed in the supine position, with the head contralaterally rotated and flexed, through a small retromastoid craniectomy of 16 mm and a supracerebellar approach to the cerebello-pontine angle.

At surgery, a definite vascular compression of the trigeminal nerve was evident in 62 cases (93.4%); the superior cerebellar artery was the compressing vessel in 48 subjects. Venous compression was evident in 5 patients. MVD was achieved by interposing a Teflon felt prosthesis between the vessel and the nerve.

Outcome was assessed at discharge and at the end of the follow-up period (mean follow-up 22.60 months).

The outcome was correlated to: 1) pain distribution; 2) presence of hypaesthesia; 3) typical or atypical pattern of pain; 4) duration of symptoms; 5) previous percutaneous procedures; 6) operative findings.

For statistical evaluation, contingency table analysis was used by means of chi square test with Yates correction.

Results

Immediate and complete pain relief was reported by 50 subjects (75.7%), partial relief by 14 (21.2%), while 2 patients reported no improvement.

Pain distribution did not affect the operative results. In the group with facial hypaesthesia, pain relief was obtained in 67.7% of the patients. Among those without sensory deficit, 84.4% were relieved from their pain. 80% of the subjects with typical TN were pain-free, while only 62.5% of patients with atypical pain benefitted. Mean duration of symptoms was 7.7 years in the group with complete pain relief and 10.6 years in the remaining patients.

If we consider the group of 33 patients previously

treated by percutaneous procedures, one can note complete improvement in 66.6%; in the group without previous surgery, 84.8% experienced complete relief.

The outcome could not be correlated with the compressing vessel. However, among the four patients not having a definite compression, two were partially relieved and two did not benefit from surgery.

Twenty-five out of 66 subjects were considered as a distinct group because they suffered from typical neuralgia, did not show any facial sensory impairment and had not been previously operated upon. This group had complete relief in 88% of the cases and partial in the remaining 12%. On the other hand, patients not included in this group (41 cases) showed complete improvement in 68.3%, partial in 26.8% and no relief in 4.9%.

However, it should be emphazized that all differences reported, were not statistically significant.

Postoperative complications include persistent headache (6 cases), persistent hypacusia (2 cases) and transitory fourth nerve palsy (1 patient). There was no operative mortality.

At the last evaluation permanent complete pain relief was achieved in 65.4% of patients and partial relief (in some cases pain was controlled by low doses of drugs) in 9.6%. 25.0% of the subjects never improved or suffered recurrence of their pain. Two of them underwent a second MVD. In the first case, the decompression from the previous operation remained intact, but the patient again did not benefit from surgery. In the second case, the trigeminal nerve was still compressed by an arterial loop and decompression was achieved by interposition of a new Teflon felt prosthesis. Post-operatively, the patient was pain free.

The differences reported between the various groups in the early postoperative period were still present, but less marked, at the end of the follow-up period.

Discussion

Various prognostic factors in MVD for TN have been evaluated by several authors. The influence of sex[4], duration of symptoms[2-4], prior surgical ablative procedures on the trigeminal nerve[3,4,7,14], presence of facial hypaesthesia[14], pattern and distribution of pain[14] and operative findings[2,5,7,14] have been reported. Some authors, however, were not able to confirm the significance of pain duration[7], previous procedures[5] and pain distribution[4].

Our results suggest a less favourable outcome in patients with atypical pain, facial hypaesthesia and

previous surgical procedures. History of disease was longer in the group who failed to benefit. The different surgical procedures now available are highly effective in producing immediate pain relief in TN; however, the recurrence rate is comparatively high following any of these operations, either percutaneous or MVD[2,4-6,9,12,13]. A repeated posterior fossa exploration for pain recurrence after MVD rarely reveals a persistent neurovascular compression[4,7]. Ablative procedures on the nerve are complicated by a relatively high incidence of trigeminal side effects, which are difficult to treat[5,8,9,13]. The risk of postoperative complications with the fifth nerve after MVD is increased in patients with prior ablative trigeminal procedures[4].

MVD is our treatment of choice in patients with drug resistant TN. However, previous ablative procedures could decrease the chance of relieving pain.

Pain recurrence requiring surgical treatment is best controlled by percutaneous procedures, such as radiofrequency thermocoagulation or microcompression of the Gasserian ganglion. The latter is particulary useful if the pain affects the first division.

References

1. Adams CBT (1989) Microvascular compression: An alternative view and hypothesis. J Neurosurg 70: 1–12
2. Apfelbaum RI (1984) Surgery for tic douloureux. Clin Neurosurg 31: 351–368
3. Barba D, Alksne JF (1984) Success of microvascular decompression with and without prior surgical therapy for trigeminal neuralgia. J Neurosurg 60: 104–107
4. Bederson JB, Wilson CB (1989) Evaluation of microvascular decompression and partial sensory rhizotomy in 252 cases of trigeminal neuralgia. J Neurosurg 71: 359–367
5. Burchiel KJ, Clarke H, Haglund M, Loeser JD (1988) Longterm efficacy of microvascular decompression in trigeminal neuralgia. J Neurosurg 69: 35–38
6. Jannetta PJ (1990) Microsurgical decompression of the trigeminal nerve root entry zone. In: Rovit RL, Murali R, Jannetta PJ (eds) Trigeminal neuralgia. Williams and Wilkins, Baltimore, pp 201–222
7. Klun B (1992) Microvascular decompression and partial sensory rhizotomy in the treatment of trigeminal neuralgia: Personal experience with 220 patients. Neurosurgery 30: 49–52
8. Latchaw JP Jr, Hardy RW Jr, Forsythe SB, Cook AF (1983) Trigeminal neuralgia treated by radiofrequency coagulation. J Neurosurg 59: 479–484
9. Meglio M, Cioni B, Moles A, Visocchi M (1990) Microvascular decompression versus percutaneous procedures for typical trigeminal neuralgia: Personal experience. Stereotact Funct Neurosurg 54 and 55: 76–79
10. Moller AR (1991) The cranial nerve vascular compression syndrome. I. A review of treatment. Acta Neurochir (Wien) 113: 18–23
11. Morley TP (1985) Case against microvascular decompression in the treatment of trigeminal neuralgia. Arch Neurol 42: 801–802
12. Sanders M, Henny CP (1992) Results of selective percutaneous

controlled radiofrequency lesion for treatment of trigeminal neuralgia in 240 patients. The Clinical Journal of Pain 8: 23–27

13. Sweet WH (1986) The treatment of trigeminal neuralgia (tic douloureux). N Engl J Med 315: 174–177

14. Szapiro J, Jr, Sindou M, Szapiro J (1985) Prognostic factors in microvascular decompression for trigeminal neuralgia. Neurosurgery 17: 920–929

Correspondence: Dr. Alfredo Puca, Instituto di Neurochirurgia, Università Cattolica, Largo A. Gemelli 8, I-00168 Roma, Italia.

Acta Neurochir (1993) [Suppl] 58: 168–170

Microsurgical Vascular Decompression (MVD) in Trigeminal and Glosso-Vago-Pharyngeal Neuralgias. A Twenty Year Experience

M. Sindou and **P. Mertens**

Department of Neurosurgery, University of Lyon, Hôpital Neurologique, Lyon, France

Summary

Report of the results of treatment in 1380 cases with trigeminal neuralgia and 14 cases with glosso-vago-pharyngeal neuralgia.

Trigeminal neuralgia was treated by percutaneous thermorhizotomy in 960 cases and by open micro-approach to the cerebellopontine angle in 420 cases. In cases treated by microsurgical vascular decompression, cure rate was 91%, partial relief 5%, failure 4%. Recurrence occured in 6%.

Glossopharyngial neuralgia was treated by percutaneous thermocoagulation of the Andersch ganglion in 3 cases and by a direct approach to the jugular foramen in 11 cases, in 9 of them with micro-vascular decompression. With microsurgical vascular decompression, all cases had total pain relief without recurrence.

Keywords: Trigeminal neuralgia; glosso-vago-pharyngeal neuralgia; microvascular decompression.

Microsurgical Vascular Decompression (MVD) for Trigeminal Neuralgias

Since 1973, 1380 trigeminal neuralgias were operated on by the senior author. 960 underwent a percutaneous thermorhizotomy[7], whilst in the 420 others an open microsurgical approach to the cerebello-pontine angle was preferred. In 24 of these, a tumour or a vascular malformation was found and successfully removed with suppression of pain in all cases. It should be noted that in one-third of these 24 patients the neuralgia was typical, with no sensory deficits. In another 17 patients with a huge atherosclerotic basilar artery, a selective juxta-pontine rhizotomy of the pars major was necessary since it was not possible to mobilize the compressive vessel. Among the remaining 379 patients, a compressing vessel was recognized in 97% of cases and dislodged (S.C.A 76%; A.I.C.A 17%; embedded vein 7%)[8].

Cure was total in 91%, partial in 5% and none in 4%.

Recurrence occurred in 6%. Two patients died because of a cerebellar infarction due to severe vasospasm secondary to excessive arterial manipulations. We later became more cautious when dissecting the arteries and started to use irrigation with papaverine whereby this complication could be avoided.

Two patients had permanent and 8 transient hearing loss after surgery. Therefore, we undertook intra-operative BAEP monitoring to correlate the electrophysiological changes with the surgical manoeuvres potentially responsible for hearing disturbances. This method demonstrated that auditory function was at risk during cerebellar retraction, especially if the cerebello-pontine angle was approached laterally with stretching of the VIIIth nerve[2]. These findings led us to modify the technique, i.e. to approach the trigeminal nerve via a superior retro-mastoid opening (Fig. 1) and a supra-cerebellar route below the tentorium with preservation of the superior petrosal veins (Fig. 2). Since this change was introduced, no more hearing loss, even transient, was observed, and BAEP monitoring has no longer been necessary.

In order to know whether the MVC procedure is effective because of a true decompression or is the result of a traumatizing and/or a neocompressive mechanism[1], the release of the nerve from its vascular compression has been performed with the utmost caution, avoiding all mechanical contact with the nerve. In addition, the offending vessel was dislodged by pulling it with small tapes of Teflon (2 mm in width, 4 cm in length) passed around the artery(-ies) (Fig. 2)[3], and maintained with a rectangular piece of Dacron (7 × 10 mm) sustained by the superior petrosal vein(s), without any contact with the nerve (Fig. 2)[5].

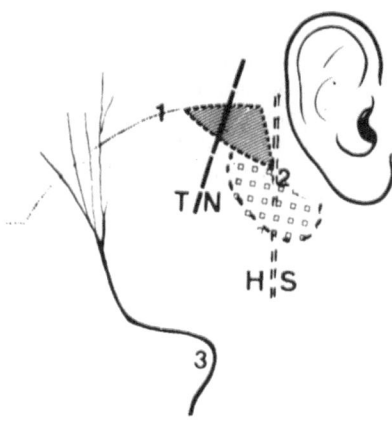

Fig. 1. Retromastoid approach (on right side) in the lateral position, Superior nuchal line (1), mastoid process (2). Skin incision (dotted line) for the trigeminal (TN) and the facial (HS) (or glosso-pharyngeal) nerve approaches. Triangular craniectomy (hatched area) for the trigeminal nerve and ovoid craniectomy (dotted area) for the facial or glosso-pharyngeal nerves. Skin incisions avoid injury to the greater occipital nerve (3)

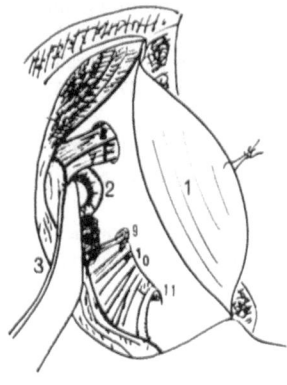

Fig. 3. Right infero-lateral approach (3), posterior to the mastoid tip (1), along the nerves of jugular foramen (9, 10, 11). The retractor tip is located in between the flocculus (above) and the choroid plexus of the foramen of Luschka (below)

In the latter part of the series in which the microsurgical decompression was performed using an "atraumatic non-compressive" technique, better results and fewer recurrences were observed than in the former groups. This indicates that MVD does act through a true decompressive mechanism.

Microsurgical Vascular Decompression for Glossopharyngeal Neuralgias

Since 1973, 14 cases of glosso-vago-pharyngeal neuralgia have been surgically treated[6]. Three had a percutaneous thermocoagulation of the Andersch ganglion. In 11, a direct approach to the jugular foramen—brain stem region was preferred. A microsurgical vascular decompression was carried out 9 times (alone in 8, associated with a sensory rhizotomy in 1). Total pain relief was achieved in all the 11 cases without any side-effects, apart from the 3 patients with rhizotomy who suffered from a moderate hypaesthesia. To date there has been no recurrence.

To avoid stretching the VIIIth nerve, while approaching the IXth, an inferior retromastoid opening (Fig. 1) and an infero-lateral cerebellar route has to be preferred (Fig. 3). With this technique there is no lateral retraction of the cerebellar hemisphere[2].

On the basis of the results reported above we are convinced that microsurgical decompression is the treatment of choice for neuralgias of the Vth and IXth nerves[4].

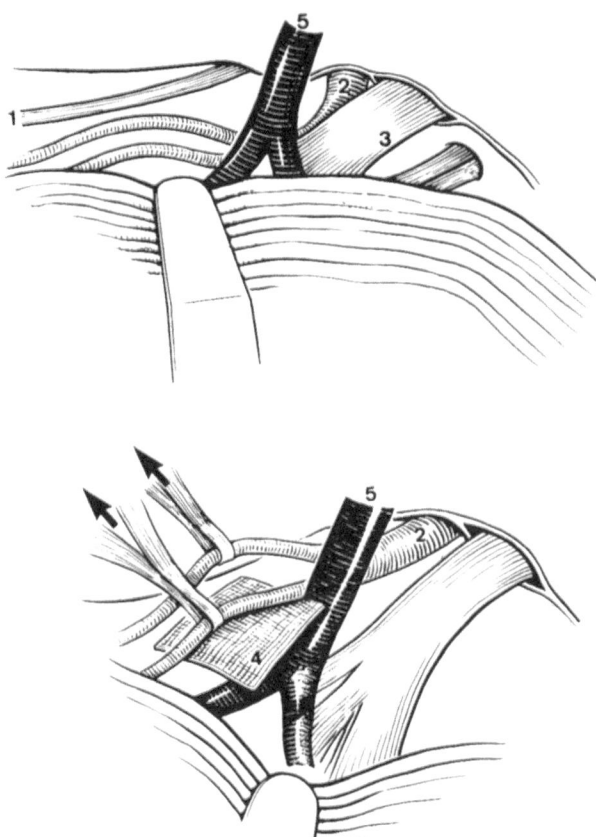

Fig. 2. Top: after retraction of the supero-lateral aspect of the right cerebellar hemisphere (with preservation of the superior petrosal vein) (5), the dorso-lateral part of the peripeduncular cistern along the trochlear nerve (1) is opened, so as to release the superior cerebellar artery (2) which compresses the trigeminal nerve (3). Bottom: Decompression of the trigeminal nerve by transposition of the superior cerebellar artery (2) without interposition of material. The two branches of the artery are maintained apart by two tapes of Teflon (arrows) flattened along the inferior wall of the tentorium cerebelli and a piece of Dacron (4) supported by the superior petrosal vein (5)

References

1. Adams CBT (1989) Microvascular compression: An alternative view and hypothesis. J Neurosurg 70: 1–12

2. Ciriano D, Sindou M, Fischer C (1991) Apport du monitorage per-operatoire des potentiels évoqués auditifs précoces dans la décompression vasculaire microchirurgicale pour névralgie du trijumeau ou spasme hémifacial. Neurochirurgie 37: 323–329

3. Fukushima T (1988) Personal communication

4. Jannetta PJ (1981) Vascular decompression in trigeminal neuralgia. In: Samii M, Jannetta PJ (eds) The cranial nerves. Springer, Wien New York, pp 331–340

5. Sindou M, Amrani F, Mertens P (1990) Décompression vasculaire microchirurgicale pour névralgie du trijumeau. Comparaison de deux modalités techniques et déductions physiopathologiques. Etude sur 120 cas. Neurochirurgie 36: 16–26

6. Sindou M, Henry JF, Blanchard P (1991) Névralgie essentielle du glosso-pharyngien. Etude d'une série de 14 cas et revue de la littérature. Neurochirurgie 37: 18–25

7. Sindou M, Keravel Y, Abdennebi B, Szapiro J Jr (1987) Traitement neurochirurgical de la névralgie trigéminale. Abord direct ou méthode percutanée? Neurochirurgie 33: 89–11

8. Szapiro J Jr, Sindou M, Szapiro J (1985) Pronostic factors in micro-vascular decompression for trigeminal neuralgia. Neurosurgery 17: 920–929

Correspondence: Prof. M. Sindou, M.D., D.Sc., Hôpital Neurologique et Neuro-Chirurgical Pierre Wertheimer, 59 boulevard Pinel, F-69003 Lyn, France.

Acta Neurochir (1993) [Suppl] 58: 171–173

Trigeminal Neuralgia: New Surgical Strategies

G. Broggi, A. Franzini, C. Giorgi, D. Servello, and **S. Brock**

Department of Neurosurgery, Istituto Nazionale Neurologico "C Besta", Milano, Italy

Summary

Presentation of the results of treatment in trigeminal neuralgia, using percutaneous radiofrequency coagulation in 712 cases, percutaneous microcompression in 206 cases, and microvascular decompression in 22 cases.

Based on the results the following management strategy is proposed:

pts. 65 years or younger = percutaneous baloon compression or, if neuroradiological evidence of neurovascular compression) is given, microvascular decompression.

Pts. elder than 65 years = thermorhizotomy. It may be repeated in case of recurrence.

If the initial operation was percutaneous compression, the second one should be microvascular decompression or, depending on age or other clinical circumstances of the patient, thermorhizotomy.

Keywords: Trigeminal neuralgia; thermorhizotomy; percutaneous microcompression; microvascular decompression; results; management strategy.

Introduction

Trigeminal neuralgia is a condition known to have afflicted humanity since ancient times, but, to this day, the pathophysiological basis of this disease remains poorly understood. Several techniques have been developed over the last 50 years to treat the symptoms of this disease. Three techniques are currently used at the Neurosurgical Department of the Istituto Nazionale Neurologico "C Besta" of Milan. We strongly believe that both CT and MR examinations are mandatory prior to surgery in order to exclude the presence of tumours in the cerebellopontine angle or at the edge of the tentorium. Moreover, a demyelinating disease, that can provoke symptomatic trigeminal neuralgia, can often be demonstrated with these techniques.

Since 1974, Sweet's technique[7] for percutaneous radiofrequency rhizotomy (thermorhizotomy, TRZ) has been employed, and to date, 1870 idiopathic cases have been operated on using this technique. In this report our experience with the first 1000 consecutive cases[2] will be discussed.

More recently, two other operations have been introduced. The first is Mullan's technique of percutaneous microcompression (PMC) of the Gasserian ganglion, performed via the foramen ovale[5]. The second is microvascular decompression (MVD) in the pontine angle according to the technique described by Jannetta[4].

All idiopathic cases operated on had previously been treated medically for long periods of time. Carbamazepine and diphenylhydantoin were the most commonly used drugs; clonazepam and baclofen were used more rarely. The patients were considered for surgery only when medical treatment had failed and side effects had become a problem.

Materials, Methods and Results

Percutaneous Radiofrequency Coagulation (Thermorhizotomy)

The technique we employ in 712 cases (Table 1) is that described by Sweet[7] with some modifications. On

Table 1. *Surgical Treatment of Idiopathic Trigeminal Neuralgia.* Cases operated on at Istituto Neurologico "C. Besta" from 1986 to 1992

Operation	No. of operations	Comment
TRZ	712	1870 since 1974
PMC	206	since 1986
MVD	22 + 1 for IX nerve	
	+ 5 for VII nerve	since 1990

TRZ Thermorhizotomy; *PMC* percutaneous microcompression; *MVD* microvascular decompression.

the basis of experimental data a coagulation temperature of 65–70 °C was chosen[3]. Careful intra-operative clinical and neurophysiological controls are necessary to obtain good results and to avoid undesired sequelae[2], and these include the evaluation of the sensory response when the patient is conscious and, after barbiturate anaesthesia, the determination of the threshold for a motor response which must be at least 5 times greater than that for the paraesthesia[2]. Once these conditions are obtained, one can be reasonably sure that the electrode probe is in the roots of the Gasserian ganglion and that lesions of the trigeminal motor root, or of the oculomotor nerves (III, IV and VI) which run nearby, will be avoided.

In our series of TRZ operations[2] the follow-up varied from 2 to 14 years and in 95% the pain was relieved immediately. In the 5% of cases where pain did not disappear immediately or reappeared within 24 hours, re-operations were successful. The rate of relapse was 18%, dysaesthesia occurred in 6.7% and masseter weakness was found in 10.5%. The lesion can be selectively placed in the branches of the trigeminal nerve correlated with the pain during surgery. In these patients (about 2/3 of total) in whom there was a postoperative anaesthesia the relapse rate was 7.5%, while the corresponding figure was 41% when hypalgesia was present. Thus, the amount of sensory damage is prognostic for the long-term result (see Table 2).

Keratitis appeared in only 0.6% of cases, but a corneal reflex deficit occurred (without keratitis) in 19.7% of cases. The incidence of corneal sensory deficit was higher (31.4%) when the pain involved the first branch of the trigeminal nerve; in cases without involvement of the first ophthalmic division, the incidence of corneal deficit was 14.1%.

Percutaneous Microcompression (PMC)

This procedure has been used in 206 cases. It is not as selective as thermorhizotomy, but produced much less sensory damage. The technique of introducing the needle is the same. Once the dura of Meckel's cavity has been perforated, a 3G Fogarty catheter is inserted and this is positioned in order to exert pressure on the root proximal to the Gasserian ganglion. The balloon is inflated by injecting 0.35–0.5 ml of fluid (Iopamiro) and the pressure is maintained for 4–5 minutes, after which the balloon is deflated and withdrawn.

Since 1986 we have operated on 206 patients using this technique, with immediate and beneficial result in 79%. After two years, pain reappeared in 34% of the patients. In about 10% of the patients a distressing dysaesthesia similar to that after thermorhizotomy remained, and in about 50% there was some sensory deficit. Patients in whom the pain reappeared underwent thermal rhizotomy or repeated microcompression. A small number underwent the Jannetta operation. The advantage of percutaneous microcompression is that it is easy to perform and it usually spares sensation. When the first brach of the trigeminal nerve is involved the operation is considerably more successful than when the third is affected. Both TRZ and PMC are useful also in patients with demyelinating disease because they are simple to perform and generally relieve the pain, even though they may have to be repeated several times[1] (see Table 3).

Microvascular Decompression (MVD)

In our opinion, microvascular decompression following Jannetta's technique[3], should preferably be used as a first choice when neurophysiological and radiological evidence suggest the presence of pathology along the course of the nerve root or in patients under 65 years. Since 1990, we employed the Jannetta and Sindou surgical technique[5] in 22 patients. In 20 of these the nerve was found to be in conflict with an artery, in 1

Table 2. *Hypalgesia as a Prognostic Factor for Relapse in 1000 Cases Operated on by TRZ*

	No. of cases	No. of relapses	Percentage of relapses
Hypalgesia present at 24 h post-op.	175	71	41%
Analgesia present at 24 h post-op.	825	110	7.5%

Table 3. *Trigeminal Neuralgia in Association with Demyelinating Disease*

121 cases	54 pre-MRI
	76 post-MRI
Relapse rate	100% pre-MRI
	68% post-MRI
31 with bilateral pain 10 received TRZ 21 received TRZ + PMC	
Strategy	TRZ or PMC as first operation PMC in bilateral cases Consider glycerol?
No indication for MVD	

Abbreviations see Table 1.

Table 4. *Findings in MVD Operations*

Total MVD operations:		22
Findings:		
arterial conflict		20
venous conflict		3
no conflict		—
arachnoiditis	after TRZ	2
	after glycerol	1
"microangiomatosis"	after TRZ	2

Abbreviations see Table 1.

the nerve was in conflict only with a vein and in 2 both a vein and an artery were involved (Table 4). In all cases but one the pain disappeared immediately and in the short follow-up time (3 months–2 years) there has been no relapse. The main problem has been leakage of cerebrospinal fluid, which accumulates extradurally or even may occur as rhinorrhea. The appearance of this complication in our series is probably explained by the lack of postoperative drainage of spinal CSF.

References

1. Broggi G, Franzini A (1982) Radiofrequency trigeminal rhizotomy in treatment of symptomatic non-neoplastic facial pain. J Neurosurg 57: 483–486
2. Broggi G, Franzini A, Lasio G, Giorgi C, Servello D (1990) Long-term results of percutaneous retrogasserian thermorhizotomy for "essential" trigeminal neuralgia: Considerations in 1000 consecutive patients. Neurosurgery 26: 783–787
3. Frigyesi T, Siegfried J, Broggi G (1975) The selective vulnerability of evoked potentials in the trigeminal sensory root to graded thermocoagulation. Exp Neurol 49: 11–21
4. Jannetta PJ (1967) Arterial compression of the trigeminal nerve at the pons in patients with trigeminal neuralgia. J Neurosurg 26: 159–162
5. Mullan SF, Lichtor T (1983) Percutaneous microcompression of the trigeminal ganglion for trigeminal neuralgia. J Neurosurg 59: 1007–1012
6. Sindou M, Amrani F, Mertens P (1990) Decompression vasculaire microchirurgicale pour la névralgie du trijumeau. Neurochirurgie 36: 16–26
7. Sweet WH, Wepsic JG (1974) Controlled thermocoagulation of trigeminal ganglion and rootlets for differential destruction of pain fibers. J Neurosurg 40: 143–156

Correspondence: Giovanni Broggi, M.D., Istituto Neurologico "C Besta", Via Celoria 11, I-20133 Milano, Italy.

Acta Neurochir (1993) [Suppl] 58: 174–177
© Springer-Verlag 1993

Retrogasserian Glycerol Rhizotomy and its Selectivity in the Treatment of Trigeminal Neuralgia

A. T. Bergenheim, M. I. Hariz, L. V. Laitinen[1]

Departments of Neurosurgery, University Hospital, Umeå, and [1] Sophiahemmet Hospital, Stockholm, Sweden

Summary

In the treatment of trigeminal neuralgia, the possibility of obtaining a selective effect on different trigeminal branches by glycerol rhizotomy was studied. Transcutaneous electrical stimulation was used to quantify sensory impairment. An attempt was made to obtain a localized neurotoxic effect of the glycerol on the different trigeminal branches by keeping the patient's head in different positions during and for one hour after glycerol injection. The amount of glycerol injected varied according to the estimated size of the trigeminal cistern and/or to which branch that was involved. The study demonstrated a highly selective effect on the ophthalmic branch, less selective on the maxillary, and a low selective effect on the mandibular branch. However, the clinical results were equal regardless of the affected trigeminal division.

Keywords: Trigeminal neuralgia; glycerol; rhizotomy.

Introduction

In the treatment of trigeminal neuralgia retrogasserian glycerol rhizotomy is a widely used method. In an attempt to selectively affect only the painful trigeminal division several techniques have been developed[1,8,12]. We have used a modification of methods which originally were described by Håkansson[8] and later developed by Arias[1]. These methods relied on variations in the tilting of the patient's head, depending on the trigeminal branch involved, during and for one hour following the glycerol injection. The injected amount of glycerol varied according to which branch was affected and to the size of the trigeminal cistern.

It is well known that retrogasserian glycerol rhizotomy often provokes a facial sensory disturbance as an undue effect[1,2,4,15]. This glycerol-induced sensory impairment was assessed, quantitatively, using transcutaneous electrical stimulation, in order to evaluate the distribution and the degree of the neurotoxic glycerol effect. We have earlier reported on this issue and the present paper is an expanded report from a substantially larger group of patients[3].

Patients and Methods

One hundred six consecutive patients with trigeminal neuralgia treated by retrogasserian glycerol rhizotomy were included in the study. The mean age was 67.6 (38–85) years and the male/female ratio was 50/56. Ten patients suffered from multiple sclerosis. Six patients had pain in the first branch, 73 patients in the second branch, and 27 patients in the third division.

The free-hand technique using cisternography as described by Håkansson[8] was used in 88 patients. Eighteen patients were operated with stereotactic technique using the Laitinen Trigeminal Stereoguide without cisternography[9]. The amount of glycerol injected varied depending on the estimated size of the trigeminal cistern and/or on which branch was affected. If the cistern appeared large at positive contrast cisternography up to 0.35 ml was injected, and if it appeared small 0.20 ml was injected. If the pain was restricted to the third branch, usually 0.25 ml or less was injected[1,8]. During and for one hour after the glycerol injection, the head of the patient was kept maximally flexed, about 40°, if the pain was within the first branch, flexed about 25° if the second branch was involved, and almost upright if the pain was restricted to the third branch as demonstrated in Fig. 1[1]. The specific weight of the glycerol was 1.260 g/ml with a water content less than 0.5%.

Facial sensory thresholds for perception and pain were measured by transcutaneous electrical stimulation using a portable constant current stimulator (ISSAL 1412)[10]. It has a bipolar electrode delivering monophasic square-wave pulses with a frequency of 100 Hz and a duration of 0.2 ms. The perception threshold was defined as the lowest current at which the patient felt the very first sensation. After further increase in current the patient usually experienced a slight unpleasant, pricking, sensation and this current value was defined as the pain threshold. The thresholds were expressed in mA.

The thresholds for perception and pain were measured at six standard sites on each side of the face the day before and the day after surgery. Three months after surgery the clinical results were evaluated. When analysing the results the patients were divided into three groups depending on which branch was *mainly* affected by the pain. Statistical analysis was done using Students paired t-test and chi-square test.

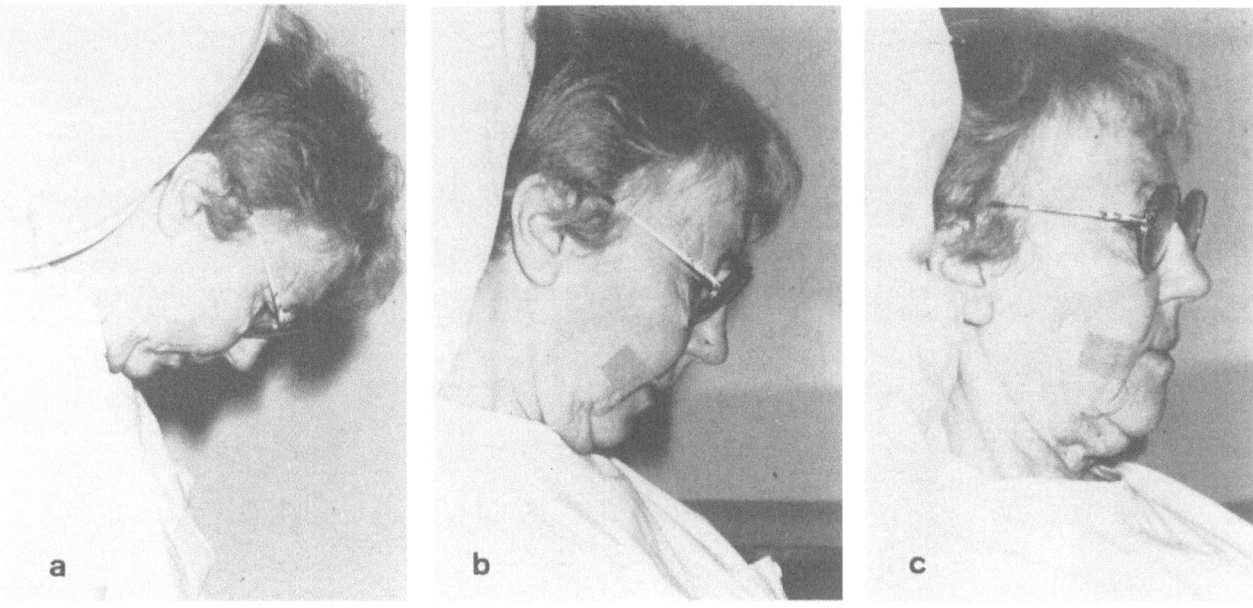

Fig. 1. The different positions of the patient's head during and for one hour following the retrogasserian glycerol injection. (a) Ophthalmic pain; (b) maxillary pain; (c) mandibular pain

Results

The degree of postoperative facial sensory impairment at the painful site compared to the contralateral site is shown in Fig. 2. There was a significant postoperative increase in thresholds for both perception and pain on the painful side. The distribution of the sensory impairment according to which trigeminal branch was *mainly* affected is demonstrated in Fig. 3. The three groups of patients showed different patterns

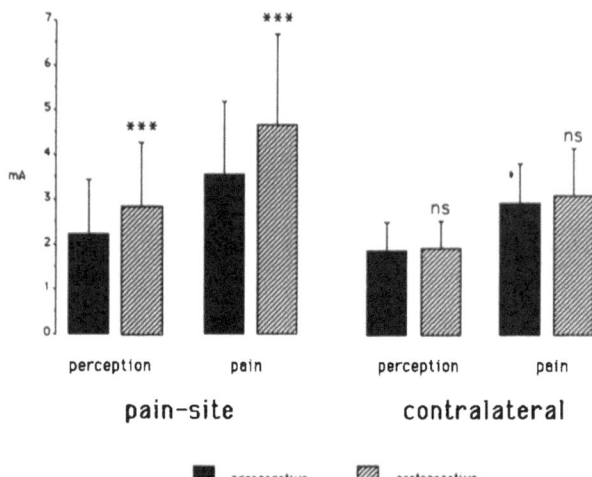

Fig. 2. Mean and S.D. sensory thresholds for perception and pain, pre-operative and one day post-operative, in 106 glycerol rhizotomies for trigeminal neuralgia. The thresholds are measured at the pain-site and contralateral*** p < 0.001; *ns* not significant

in their sensory disturbance: In patients with pain in the first branch the sensory impairment was well localized within the first trigeminal branch which contrasted with patients whose pain was in the second branch. These patients had a sensory impairment which was more widespread affecting not only the second but also the lower first branch and the third branch. For patients with pain in their third branch, the sensory impairment was most pronounced within the second branch with a less marked affection of the third and lower first branch.

Three months after surgery 87.6% of the patients were still painfree and 73% were without any medication. There were no statistical differences in the results, neither according to which trigeminal branch was affected, nor between patients operated on with or without cisternography.

Discussion

It has been shown that after retrogasserian glycerol rhizotomy a significant facial sensory disturbance usually appears[1,2,4,15]. The sensory affection could be more or less pronounced and it usually improved with time[1,2]. We have used transcutaneous electrical stimulation, allowing *quantitative* assessment of cutaneous sensibility[10], for measurement of the degree and distribution of the sensory disturbance in order to assess the ability to selectively affect any specific

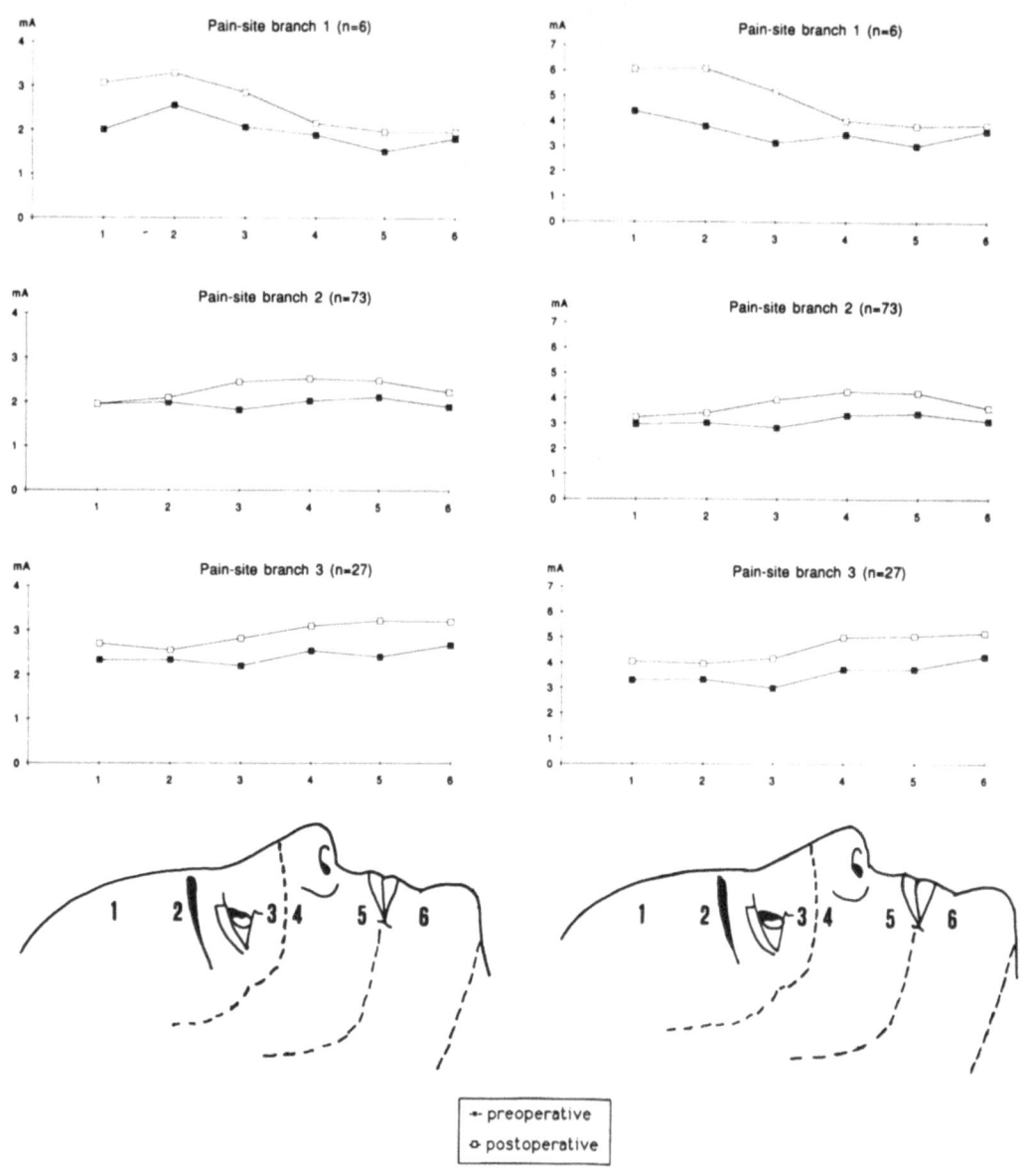

Fig. 3. Mean sensory thresholds for perception and pain, before and after glycerol rhizotomy for trigeminal neuralgia, at six standard areas in the painful side of the face. The patients are divided into three groups according to which trigeminal branch was mainly affected. Numbers 1–6 correspond to the sites of measurements

trigeminal branch with glycerol injection. The thresholds for perception and pain were believed to reflect the functional properties of thick (Aβ) and thin (A∂) myelinated nerve fibers, respectively[5]. Our hypothesis was that the sensory loss induced by glycerol would be most pronounced in the distribution area of the trigeminal branch where the glycerol was applied.

A highly selective effect on the ophthalmic trigeminal branch after glycerol rhizotomy for patients with pain in this area is demonstrated in Fig. 3. For patients with pain in the maxillary or mandibular branch our described technique showed a poor selective neurotoxic effect. When trying to reach selectively the maxillary branch we obtained a more wide-spread sensory disturbance with affection of the lower ophthalmic branch and to a minor degree also of the mandibular branch. When attempting to selectively reach the mandibular branch, the sensory disturbance became widespread with affection of all branches with the exception of the upper ophthalmic branch. However, the affection was most pronounced in the lower maxillary branch. Thus, the mandibular branch was the most difficult branch to reach selectively.

In the treatment of trigeminal neuralgia, especially in the elderly, the choice of treatment often stands between glycerol injection or radiofrequency (RF) lesioning of the gasserian ganglion. One of the major disadvantages with RF lesions is the high incidence of severe sensory disturbances[6]. In treating patients with pain in their ophthalmic division RF lesions have been reported to give an incidence of 18 to 28 per cent of corneal anaesthesia[13,14] and the risk for this complication was stressed also by other authors[7,11]. No corneal anaesthesia occurred in our 6 patients with first branch pain. However, 2 patients showed a moderately decreased corneal sensitivity. Most side-effects reported after RF lesions are not analysed in relation to the affected branch. However, due to the sometimes troublesome side-effects of sensory alterations, including corneal anaesthesia, it seems that some authors prefered RF lesions for pain in the third branch while reserving glycerol injection or gasserian compression for pain in the first and sometimes second branch[6,7]. When comparing the results in pain relief it seems that RF lesions may have a slightly higher success rate than glycerol injection[6,13].

Conclusions

Our study clearly demonstrates the possibility for a highly selective effect in treating pain in the ophthalmic branch which points to the advantage of glycerol injection for this group of patients. However, for patients with pain in the mandibular branch, where the glycerol effect is less selective, RF lesioning may be considered thanks to its higher selectivity and success rate. For pain in the maxillary branch where RF lesions provides not only a higher rate of success but also a higher rate of sensory disturbances compared to glycerol injection, the choice of treatment modality is less obvious.

References

1. Arias MJ (1986) Percutaneous retrogasserian glycerol rhizotomy for trigeminal neuralgia. J Neurosurg 65: 32–36
2. Bergenheim AT, Hariz MI, Laitinen LV, Olivecrona M, Rabow L (1991) Relation between sensory disturbance and outcome after retrogasserian glycerol rhizotomy. Acta Neurochir (Wien) 111: 114–118
3. Bergenheim AT, Hariz MI, Laitinen LV (1991) Selectivity of retrogasserian glycerol rhizotomy in the treatment of trigeminal neuralgia. Stereotac Funct Neurosurg 56: 159–165
4. Burchiel KJ (1988) Percutaneous retrogasserian glycerol rhizolysis in the management of trigeminal neuralgia. J Neurosurg 69: 361–366
5. Collins WF Jr, Nulsen FE, Randt CT (1960) Relation of peripheral nerve fiber size and sensation in man. Arch Neurol 3: 381–385
6. Fraioli B, Esposito V, Guidetti B, Cruccu G, Manfredi M (1989) Treatment of trigeminal neuralgia by thermocoagulation, glycerolization, and percutaneous compression of the gasserian ganglion and/or retrogasserian rootlets: Long-term results and therapeutic protocol. Neurosurgery 24: 239–245
7. Frank F, Fabrizi (1989) Percutaneous surgical treatment of trigeminal neuralgia. Acta Neurochir (Wien) 97: 128–130
8. Håkansson S (1981) Trigeminal neuralgia treated by the injection of glycerol into the trigeminal cistern. Neurosurgery 9: 638–646
9. Laitinen LV (1984) Trigeminus stereoguide: An instrument for stereotactic approach through the foramen ovale and foramen jugulare. Surg Neurol 22: 519–523
10. Laitinen LV, Eriksson AT (1995) Electrical stimulation in the measurement of cutaneous sensibility. Pain 22: 139–150
11. Rovit RL (1990) Percutaneous radiofrequency thermal coagulation of the gasserian ganglion. In: Rovit RL, Murali R, Jannetta PJ (eds) Trigeminal neuralgia. Williams and Wilkins, Baltimore, pp 109–136
12. Sweet WH, Poletti CE, Macon JB (1981) Treatment of trigeminal neuralgia and other facial pains by retrogasserian injection of glycerol. Neurosurgery 9: 647–653
13. Sweet WH (1990) Treatment of trigeminal neuralgia by percutaneous rhizotomy. In: Youmans JR (ed) Neurological surgery, 3rd Ed. Saunders, Philadelphia, pp 3888–3921
14. Tew JM, van Loveren H (1988) Percutaneous rhizotomy in the treatment of intractable facial pain (trigeminal, glossopharyngeal, and vagal nerves). In: Schmidek HH, Sweet WH (eds) Operative neurosurgical techniques, Vol 2, 2nd Ed. Grune and Stratton, Philadelphia, pp 1111–1123
15. Young RF (1988) Glycerol rhizolysis for the treatment of trigeminal neuralgia. J Neurosurg 69: 39–45

Correspondence: A. Tommy Bergenheim, M.D., Department of Neurosurgery, University Hospital, S-901 85 Umeå, Sweden.

Epilepsy

Acta Neurochir (1993) [Suppl] 58: 181–185
© Springer-Verlag 1993

Intracerebral Low Frequency Electrical Stimulation: a New Tool for the Definition of the "Epileptogenic Area"?

C. Munari[1,2], P. Kahane[1,2], L. Tassi[1], S. Francione[1], D. Hoffmann[1], G. Lo Russo[1,3], and A. L. Benabid[1,2]

[1] Neurosciences Department, CHRU of Grenoble, France, [2] INSERM U 318, CHRU of Grenoble, France, and [3] Istituto di Neurochirurgia I, Università di Torino, Italy

Summary

Low Frequency (1 Hz) Electrical Stimulation (LFES) has been systematically utilized, during stereo-EEG investigations, in 24 consecutive young adult patients considered for surgical treatment of severe drug-resistant partial epilepsy.

Ninety seizures (1–14/patient) identical to the spontaneous ones previously recorded were thus obtained in 19 patients (79%). LFES is less effective for induction of seizures than high frequency (50 Hz) stimulation (5.9% vs 22.9%), and it also provokes less "false positive" responses (1% vs 17%). The main "sensitive" structures to LFES are the hippocampus, the amygdala, and the hippocampal gyrus. However, seizures were also induced by stimulating the temporal lobe white matter, the temporal pole, and the temporal neocortex, as well as the orbito-frontal cortex (in the only patient with fronto-temporal epilepsy). The more frequently observed electrical pattern is a gradual increase of spikes and spikes and waves frequency, with or without occurrence of low voltage fast activity. The high percentage of early "subjective" manifestations similar to the spontaneous ones, the lack of major electrical artifact, and the good visualization of the spatial evolution of the induced-discharge, strongly suggest that LFES is of great help for defining the "epileptogenic area".

Keywords: Partial epilepsy; epilepsy surgery; stereo-EEG; intracerebral electrical stimulation.

Introduction

There is presently a general agreement that very good results may be obtained by surgical treatment of medically intractable partial epilepsies. In contrast, presurgical diagnostic procedures vary among epilepsy surgery centers. Thus, up to now, some teams still decide to operate on patients on the only basis of interictal and frequently ictal EEG recordings and neuro imaging techniques[4]. Cortical excision is then generally performed according to per-operative electro-cortico-graphy data[8]. On the other hand, several "invasive" procedures are currently conducted by means of intracranial[5,13] or intracerebral[1] electrodes. The major aim of such "invasive" procedures is to record spontaneous seizures in order to identify cortical areas from where the ictal discharges originate; different tools may also be useful, among which electrical stimulation (ES) is of particular interest:

— High Frequency ES (HFES, 50 Hz) of selected brain structures are well-known and largely employed to induce seizures[2], with apparently good results[3,12];
— Low Frequency ES (LFES, 1 Hz) are classically used to delineate functional regions[10], but we have recently suggested that this kind of stimulation is useful for the provocation of seizures in patients with temporal lobe epilepsy[6,11].

This report presents the results of a prospective study concerning the effects of LFES during stereo-EEG procedures and we discuss its value for the definition of the "Epileptogenic Area" (EA)[1].

Materials and Methods

The methodology used since January 1990 at the CHRU and INSERM U 318 in Grenoble is directly derived from that proposed by J Talairach and J Bancaud[9]. Figure 1 schematically summarized the possible different steps in the presurgical investigation adapted to the individual, taking into account the clinical, electrical, and anatomical characteristics[7]. Stereo-EEG strategy is defined in accordance with the previously collected data and implies individualized stereotactic placement, mostly unilateral, of 7 to 12 intracerebral semi-rigid electrodes (diameter: 0.9 mm —5 to 15 contacts of 1.5 mm length, 2 mm apart). "Chronic" recordings are made during a period of about 15 days, allowing thorough evaluation of electroclinical features.

Our study concerns 24 consecutive patients with severe drug-resistant partial epilepsy who are candidates for surgical treatment, and in whom LFES has been applied on almost all paired contiguous leads of each electrode. LFES was performed after the recording of at least one spontaneous seizure in 22/24 patients. HFES was used

Severe
drug-resistant
partial epilepsies

- anamnestic data
- clinical & neuro-psychological findings (+ / - Wada test)
- ictal & inter-icta scalp-EEG
- CT scan & MRI (+ / - SPECT & PET)

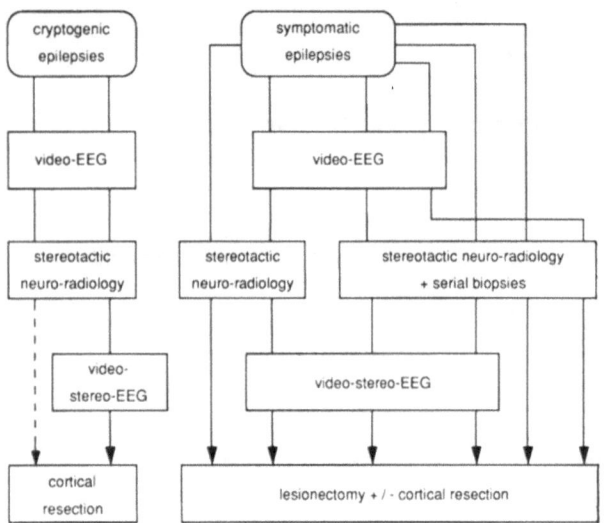

Fig. 1. Possible sequences of presurgical investigations

Table 1. *Parameters Used During Low and High Frequency Electrical Stimulations (LFES and HFES)*

	LFES	HFES
Frequency (pulse / sec)	1	50
Pulse width (msec)	3	1
Intensity (mA)	0.25 - 4	0.25 - 4
Duration (sec)	40	5 - 7

Results

Number of Seizures Recorded (24 Patients)

Three hundred and twenty six seizures have been recorded:

— 289 in group A: 90 spontaneous (0–15/patient), 109 elicited by HFES (2–14/patient), and 90 induced by LFES (1–14/patient);
— 37 in group B: 27 spontaneous (1–12/patient), 10 pro-voked by HFES (0–5/patient).

Between the two groups, there was a significant difference concerning the number of HFES-elicited seizures ($p = 0.037$, using unpaired Student's t-test), but no difference in the number of spontaneous fits recorded.

in 23/24 patients. In most patients, several ES were performed on the same structures (and sometimes on the same contiguous paired leads) with different intensities. The main electrical parameters used for these two types of ES are given in Table 1.

Two groups have been identified and their general characteristics are summarized in Table 2. Group A comprises the 19 patients in whom seizures have been elicited by LFES, and group B the 5 patients in whom LFES was ineffective. Twenty-one patients have been operated on (group A: 17, group B: 4) with a mean follow up of 15,4 months (3–24); all of them were "class I" according to Engel's Classification (1987).

Table 2. *General Characteristics of Patients in whom LFES have Induced at Least 1 Seizure (Gr A: Group A), and of Those for whom LFES have been Ineffective (Gr B: Group B)*

	Number of patients	Sex (F / M)	Mean age in years (SD)	Mean duration of epilepsy in years (SD)	Seizures frequency per month (SD)	Treatment reduction (number)	Pure temporal lobe epilepsy (number)	Symptomatic epilepsy (number)
Gr A	19	12 / 7	27.3 (8.9)	14.3 (8.1)	19,7 (21,8)	16	15	10
Gr B	5	2 / 3	23 (9.5)	15.6 (8.8)	8,2 (2,5)	5	1	1
p		ns **	ns *	ns *	0.037 *	ns **	ns **	ns **

F female; *M* male; *SD* standard deviation; * using unpaired Student's t-test; ** using Chi² test.

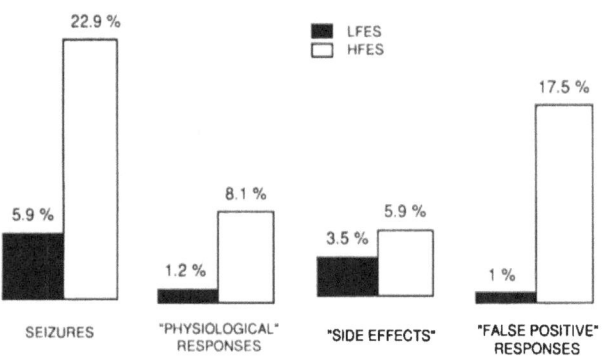

Fig. 2. Comparison of electrically-induced clinical responses by LFES and HFES in 23 out of the 24 patients studied (see text for explanations concerning the different terms used)

Electrically-Induced Clinical Responses (23 Patients)

In order to compare the clinical responses provoked by the two modalities of SE, we have only selected those of paired leads stimulated both by HFES and LFES, since LFES have been used on almost all contacts of each electrode, and HFES on only selected brain structures chosen on the basis of spontaneous seizures previously recorded. Clinical responses were classified as follows:

a) seizures: both subjective and objective clinical manifestations, either previously described by the patient himself and/or his parents, and/or previously observed during spontaneous seizure recordings;

b) "physiological" responses: unknown symptoms, clearly related to the anatomical site of stimulation (e.g. auditory illusions by ES of the superior temporal gyrus);

c) "side effects": minor well-localized cephalic symptoms, related to the vicinity of the dura;

d) "false positive" responses: clinical manifestations neither recognized by the patients, nor observed before, and not clearly related to the anatomical site of SE.

As shown in Fig. 2, efficacy of HFES is greater than that of LFES (22.9% vs 5.9%), but "false positive" responses are more frequent with HFES than with LFES (17.5% vs 1%). These differences are significant ($p < 0.0001$) using paired Student's t-test.

Anatomical Structures where LFES have Induced Seizures (19 Patients)

The main "sensitive" structures for this kind of ES seem to be the hippocampus (45 seizures/12 patients), the amygdaloid nucleus (13 seizures/7 patients), and the hippocampal gyrus (10 seizures/5 patients). Nevertheless, LFES has also provoked a great number of seizures in temporal lobe white matter (8 seizures/6 patients), temporal pole (6 seizures/4 patients), and temporal

Table 3. *Distribution of Anatomical Structures where LFES have Induced Seizures (19 Patients)*

Anatomical site	15 patients with a pure temporal lobe epilepsy															T-PS	T-P	F-T	O-P-T	Mean intensity	Mean duration
AN	.	3	.	.	2	2	.	2	.	.	.	.	1	.	.	.	.	2	1	2.5 mA	26.6 sec
aHc		.	4	4	8	.	4	.	.	.	3	1								1.6 ma	19.7 sec
mHc	3		.	.	.	.	.	.	.	.	.	.	.	1	1	2				2.3 mA	22.7 sec
pHc	.	.	.	.	1	.	.	.	2	3	.	4	.	2	2	.	.			2.5 mA	27.8 sec
HcG	.	.	.	.	.	.	.	.	1	1	.	1	.	6	.	.	.	1		1.9 mA	29.1 sec
T NeoCx	.		.	1	.	.	.	.	1	.	.	.	.	2	.	.	.	.		2.75 mA	37.2 sec
T Pole	.	.	.	.	1	.	2	.			.	1	2	.	.	.	.			2.6 mA	29.2 sec
T Wm	.	.	1	.	2	.	.	1	.	.	1	2	1	.	.	.	.			2.6 mA	20.8 sec
Orb Cx		.	.	.	.	.	.	.	.	.	.	.	.	.	.	.	4			3 mA	45.2 sec

Each column represents 1 patient, and each row an anatomical structure where at least 1 seizure has been provoked by LFES. Mean intensity and average duration (stimulation was stopped when the first clinical sign occured) of effective LFES have been calculated, for each structure, on the basis of total number of elicited fits. Filled squares mean that the structure has not been stimulated, sign (−) that LFEs has not been effective. *AN* Amygdaloid nucleus; *aHc, mHc,* and *pHc* anterior, mid, and posterior hippocampus; *Hcg G* hippocampal gyrus; *TNeoCx* temporal neocortex; *T Pole* temporal pole; *T Wm* temporal lobe white matter; *Orb Cx* orbito-frontal cortex; *T-PS* temporo-peri-Sylvian epilepsy; *T-P* temporo-parietal epilepsy; *F-T* fronto-temporal epilepsy; *O-PT* occipito-parietotemporal epilepsy

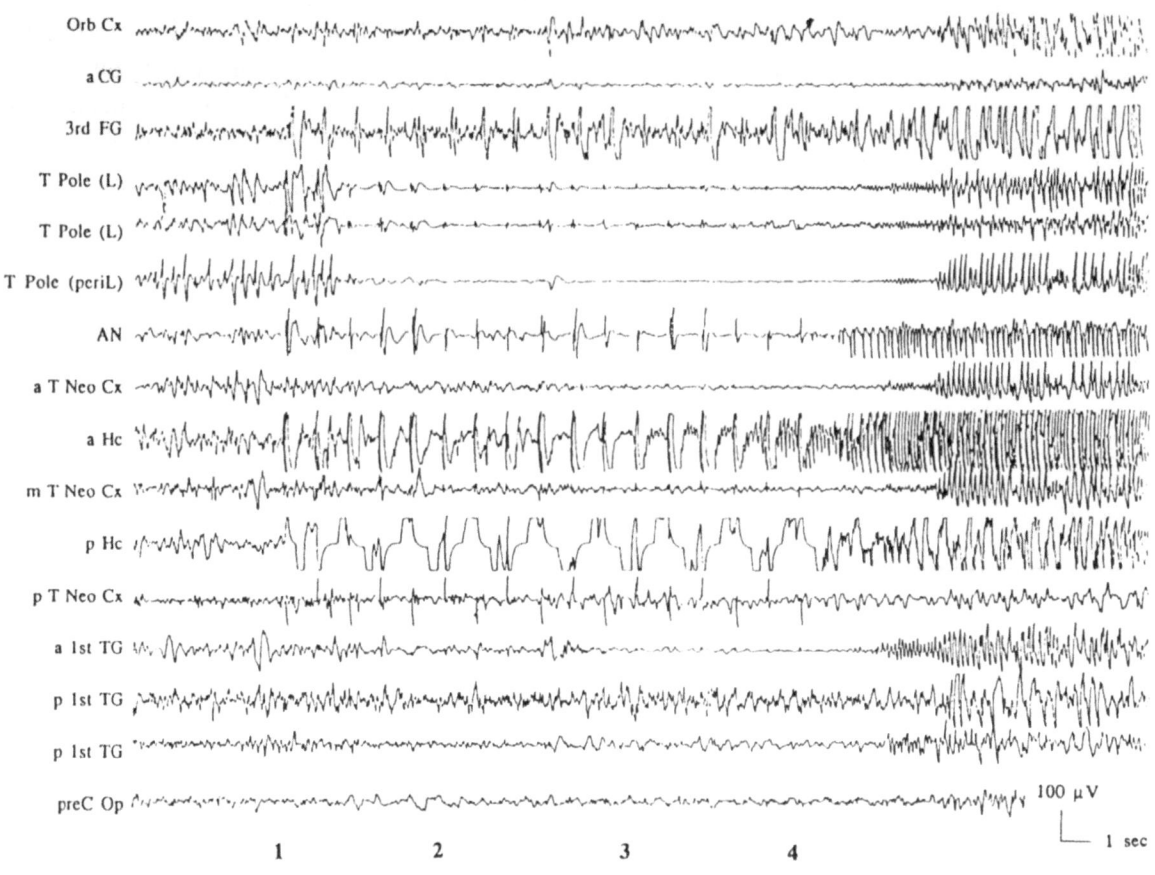

Fig. 3. Seizure induced by low frequency electrical stimulation (3 mA, 18 sec) in the posterior part of the hippocampus. The patient was a 26 years-old male with left temporal lobe epilepsy associated with a left temporo-polar low grade glioma; he is seizures-free since the anterior temporal lobectomy performed 5 months ago. *Orb Cx*: orbito-frontal cortex; *a CG*: anterior cingulate gyrus; *3rd FG*: third frontal gyrus; *T Pole*: temporal pole; *L*: lesion; *AN*: amygdaloid nucleus; *a, m,* and *p T Neo Cx*: anterior, mid and posterior temporal neocortex; *a* and *p Hc*: anterior and posterior hippocampus; *a* and *p 1st TG*: anterior and posterior temporal gyrus; *preC Op*: pre-central opercular region. All these structures have been implanted in the left hemisphere. *1* At stimulation onset, spikes and waves immediately appear at T Pole (L + periL) and AN, and polyspikes and waves on aHc; 2 sec later, a low voltage fast activity occurs on both lesional and mainly peri-lesional area where stimulus artifact disappears; at the same time, a flattening on the 1st TG can be seen. *2.* Questioned, the patient says "that's all right", 4 sec later, a polyspikes discharge appears on aHc, with a gradual increase in voltage; at that time, a clear low voltage fast activity appears on the 1st TG, and perhaps on preC Op. *3* The patient grasps the "push button" but does not answer, makes faces and mumbles 1 sec later, and then has oroalimentary movements, contraction of the left corner of the lips, aggressive behaviour and partial loss of contact for almost 50 sec. *4* End of stimulation, that does not seem to affect spatial evolution of ictal discharge which spreads over both mesial and lateral temporal regions in 50 sec, while frontal and opercular regions are much less involved; there is a post-ictal aphasia and marked electrical depression in the temporal lobe structures for almost 2 min

neocortex (4 seizures/3 patients). The only one patient in whom LFES was effective in an extra-temporal region (orbito-frontal cortex) had fronto-temporal lobe epilepsy. All these data are detailed, for each patient, in Table 3.

Electrical Modifications During LFES-Elicited Seizures (19 Patients)

Following each pulse of LFES, spikes and waves (SW) and polyspikes and waves (PSW) often appear with gradually increasing frequency during the stimulation, leading to a SW and PSW discharge which progressively spreads at the end of stimulation.

In some cases, a well-localized low voltage fast activity occurs during LFES, sometimes associated with a well-localized spike (S) discharge, and followed by a polyspike (PS) discharge which gradually spreads at the end of stimulation (Fig. 3).

A well-localized S, SW and PSW discharge at the onset of LFES may also be observed, followed by a high frequency PS discharge.

Discussion

Up to now, recording of spontaneous seizures during stereo-EEG procedures remains the best tool for defining the EA. Nevertheless, ES is widely employed, and HFES has been applied in order to provoke seizures, whereas LFES is mainly used to "map" functional regions.

In fact, our data clearly demonstrate that LFES is useful for inducing seizures in many patients. This "new" modality of ES seems to have numerous advantages, among which we would like to emphasize the following:

— the high percentage of "subjective" manifestations obtained[6];
— the lack of stimulus artifact during the stimulation itself allowing recognition of the ictal changes, and a good visualization of the spatial evolution of induced discharge;
— the more precise evaluation of anatomo-electro-clinical correlations.

Although LFES is less "effective" than HFES, it appears to be more reliable since "false positive" responses are exceptional. It appears to be more "effective" in patients with a high seizure frequency (see Table 2) and a high sensitivity to HFES, and this may indicate a lower "epileptogenic threshold" in such patients. However, it seems to be particularly "effective" in mesial temporal lobe structures (see Table 3), and we have just began a systematic study in order to explore variable excitability thresholds of different anatomical structures[14].

It is perhaps too early to propose LFES as the tool for identifying EA, avoiding the often time-consuming observation of spontaneous seizures. However, it should be noted that all the patients subjected to surgery, and in whom LFES was effective, corresponded to "class I" (IA: 12; IB: 4; ID: 1) according to Engel's classification. LEFS must be further evaluated and compared to HFES. It is also necessary to assess whether LFES-induced seizures are always identical to those appearing spontaneously.

References

1. Bancaud J, Talairach J, Bonis A, Schaub C, Szikla G, Morel P, Bordas-Ferrer M (1965) La stéréo-électro-encéphalographie dans l'épilepsie. Informations neuro-physio-pathologiques apportées par l'investigation fonctionnelle stéréotaxique. Masson, Paris
2. Bancaud J, Talairach J, Geier S, Scarabin JM (1973) EEG et SEEG dans les tumeurs cérébrales et l'épilepsie. Edifor, Paris
3. Bernier GP, Richer F, Giard N, Bouvier G, Mercier M, Turmel A, Saint-Hilaire JM (1990) Electrical stimulation of the human brain in epilepsy. Epilepsia 31: 513–520
4. Engel J Jr (1987) Surgical treatment of the epilepsies. Raven, New York
5. Goldring S, Gregoric EM (1984) Surgical management of epilepsy using epidural recordings to localize the seizure focus. J Neurosurg 60: 457–466
6. Kahane P, Tassi L, Francione S, Hoffmann D, Lo Russo G, Munari C Manifestations électro-cliniques induites par la stimulation électrique intra-cérébrale par "chocs" dans les épilepsies temporales. Neurophysiol Clin (in press)
7. Munari C, Lo Russo G, Hoffmann D, Tassi L, Kahane P, Lebas JF, Pasquier B, Joannard A, Garrel S, Perret J, Benabid AL (1991) Stratégie diagnostique et thérapeutique dans l'approche des épilepsies partielles graves pharmaco-résistantes. In: Crises épileptiques et épilepsies du lobe temporal (Tome I). Documentation médicale Labaz, pp 139–151
8. Penfield W, Jasper H (1954) Epilepsy and the functional anatomy of the human brain. Little, Brown and Company, Boston
9. Talairach J, Bancaud J, Szikla G, Bonis A, Geier S (1974) Approche nouvelle de la neurochirurgie de l'épilepsie. Méthodologie stéréotaxique et résultats thérapeutiques. Neurochirurgie 20 [Suppl 1]: 1–274
10. Talairach J, Szikla G, Tournoux P, Prossalentis A, Bordas-Ferrer M, Covello L, Jacob M, Mempel E (1967) Atlas d'anatomie stéréotaxique du télencéphale. Etudes anatomo-radiologiques. Masson, Paris
11. Tassi L, Kahane P, Munari C, Lo Russo G, Hoffmann D, Garrel S, Feuerstein C, Benabid AL, Perret J (1991) Role of Single Shocks Electrical Stimulation (SSES) for the anatomical definition of the "Epileptogenic Area" (EA) in patients with severe drug-resistant temporal lobe epilepsy, candidates to the surgical treatment. Epilepsia 32 [Suppl 1]: 93
12. Wieser HG, Bancaud J, Talairach J, Bonis A, Szikla G (1979) Comparative value of spontaneous and chemically and electrically induced seizures in establishing the lateralization of temporal lobe seizures. Epilepsia 20: 47–59
13. Wyler AR, Ojemann GA, Lettich E, Ward AA (1984) Subdural strip electrodes for localizing epileptogenic foci. J Neurosurg 60: 1195–1200
14. Wyler AR, Ward AA Jr (1981) Neurons in human epileptic cortex. Response to direct cortical stimulation. J Neurosurg 55: 904–908

Correspondence: Claudio Munari, M.D., Neuro-physio-pathologie du Sommeil et e l'Epilepsie, INSERM U 318, CHRU de Grenoble, BP 217 X, F-38043 Grenoble Cédex, France.

Acta Neurochir (1993) [Suppl] 58: 186–189

Multimodal Imaging Integration and Stereotactic Intracerebral Electrode Insertion in the Investigation of Drug Resistant Epilepsy

G. P. Kratimenos and **D. G. T. Thomas**

The National Hospital for Neurology and Neurosurgery, Queen Square and Maida Vale, London, U.K.

Summary

Insertion of intracerebral electrodes for EEG recording is some-times necessary during the pre-operative evaluation of patients with drug resistant epilepsy to define the site of seizure onsets. The precise and accurate placement of the electrodes requires a stereotactic technique of insertion based on correlated information derived from computerised imaging and stereotactic angiography. Described methods of multimodal stereotactic image integration present limi-tations in terms of satisfactory relocation and ability to spread data acquisition over a period of time.

An alternative method of stereotactic acquisition of multimodal image information using the Gill-Thomas stereotactic repeat localiser is presented. Digital Angiographic (DSA), Computerised (CT) and Magnetic Resonance Imaging (MRI) data were correlated and used for target selection. The positional accuracy of the electrodes was confirmed repeatedly during the recording period with standard radiographic and MRI means and found to be satisfactory. There were no permanent complications in any of the patients included in the study. Stereoangiography correlative to computerised neuro-imaging offered a high degree of safety during the operation. Non-invasive relocation was an important feature of the combined system which was particularly helpful and duly appreciated by the patients. The temporal freedom provided during the investigative and operative period offers the advantage of an unhurried multi-image integration and targeting combined with less discomfort for the patient. The positional accuracy of the electrodes was easily verified during the post-operative period and this information added to the electroence-phalographic localising value of the technique.

Keywords: Electroencephalography; electrode implantation; epi-lepsy; stereotactic surgery; neuro-imaging; image integration.

Introduction

Various methods of multimodal stereotactic image integration have been described and among the tech-niques that have been used for the precise determination of the intracerebral targets are the "stereo-encephalo-graphic method"[1,9-11], the use of the corpus callosum for the transposition of stereotactic information obtained by Digital Subtraction Angiography (DSA) to non stereotaxic Magnetic Resonance (MR) and Positron Emission Tomographic (PET) imaging[6,7] and finally the computerized multimodal stereotactic image analysis system[4].

Computerised analysis and image integration are inherent parts of the methods currently in use but a common limitation is the need for the serial perfor-mance of imaging studies after the application of the stereotactic frame. The operation usually takes place immediately following the target definition or if the frame is to be removed between studies, invasive devices like localising pins penetrating the outer skull table must remain in place. Both methods represent a com-promise either in the sense of temporal freedom during the integration and correlation of the images or by increasing patients' discomfort and protracting inva-siveness.

An alternative system of non-invasive repeat stereo-tactic localisation which will effectively lead to multi-modality image integration would therefore be highly desirable.

Material and Methods

Image Correlation and Technique

The Gill-Thomas repeat localiser system enables the non-invasive attachment of an image localising device onto the patient's head in precisely the same position and on repeated occasions. The system is applied using a specially designed dental tray carrying an impres-sion of the upper dentition and an occipital receptacle tray for an impression of the occiput. Added security is provided with the use of straps over the head preventing lateral and downwards movement. The dental trays, which come in three sizes, are filled with a silicon based ultraviolet light curing paste. Once the impressions are made, the repeat localiser can be quickly and repeatedly positioned accu-rately to the head by setting the same scale values on the posterior and the lateral slides. The CRW/BRW base ring attaches to the base frame thus allowing fixation of the CT, MRI and angiographic

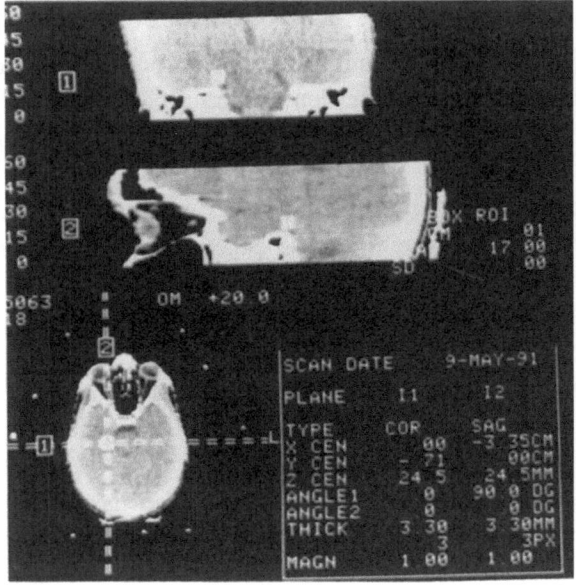

A

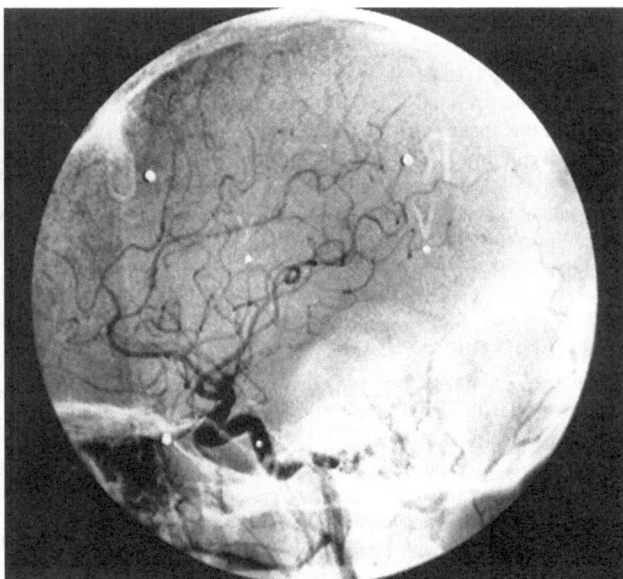

B

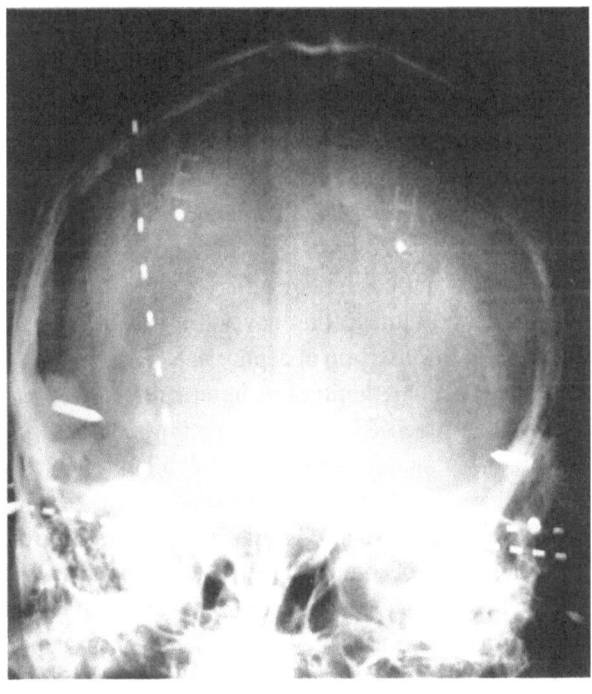

C

Fig. 1. Right posterior mesial temporal targets are selected on axial, coronal and sagittal CT stereotactic images (A) and the coordinates of the targets are correlated with stereotactic angiograms (B). The post-operative confirmation of the positional accuracy of the electrodes is easily verified with standard radiography following the application of the G-T repeat locator and the angiographic localiser (C)

localising components of the stereotactic system or of the stereotaxic arc system itself. Anterio-posterior and lateral radiographs with the SGVAL angiographic localiser in place may easily confirm the accuracy of the positioning.

Following application of the G-T repeat localiser with the appropriate localising devise digital angiographic (DSA) images are obtained. Re-application of the combined system with the standard fiducial component for CT or MRI will give the spatial coordinates for the deepest contacts of each electrode. Targets are selected on axial CT slices of 1.5 mm thickness with 1.5 mm separation to allow for high definition reconstruction and also on sagittal, coronal and

axial contiguous slices of 5 mm thickness on a 0.5 Tesla MRI unit. Five to seven targets are usually selected depending on the epileptic history of the patient and the reasons for the depth electrodes' implantation. Four of the targets are usually located in the mesial temporal structures bilaterally, anteriorly and posteriorly in the amygdalo-hippocampal complex and one to three targets are selected from the frontal pre-motor areas. One of the frontal electrodes with the distal contact sampling the medial orbito-frontal cortex is used as the reference electrode.

The integration and correlation of the computerised stereotactic data, the final decision about the electroencephalographic validity of the selected targets and the confirmation of the accuracy and safety of the proposed trajectories may take place in an unhurried fashion over the next few days while the patient either remains in the ward or may even be discharged home.

Prior to the operation the G-T repeat localiser and the CRW base frame is reapplied under general anaesthesia and the positional accuracy is confirmed with anterio-posterior and lateral radiographs with the SGVAL angiographic localiser in place. The stereotactic arc system of the CRW-3 frame is placed in the lateral rather than the anterior-posterior position to allow for a lateral implantation of the electrodes sampling the mesial temporal structures.

The electrodes used are flexible multi-contact platinum Spencer probes (Ad. Tech) with six-contact electrodes inserted in the temporal regions and eight or ten contact electrodes usually reserved for the frontal areas. Frontal electrodes are implanted through a single 1/4 inch pre coronal drill hole, while for the temporal electrodes and for the medio-basal frontal electrode an 1/8 inch drill hole is used. The accuracy of the final position of the electrodes is confirmed in the operating theatre with lateral and anterio-posterior radiographs immediately following the insertion and also at any time during the post-operative period with Radiographs and Magnetic Resonance Tomographs (MRI).

Patients and Results

A total number of 41 electrodes were implanted in seven patients during the development of the technique. The indications for chronic intracerebral electroencephalographic recording were problems of localisation in four patients, discrimination of temporal from extra-temporal seizure onsets in two patients and determination of the required extent of cortical resection in one patient. The recording period extended from nine to eighteen days with an average of 296 hours of continuous intracranial electroencephalographic recording.

There were no permanent operative or post-operative complications and electroencephalo graphic video telemetry began as soon as eight hours following the operation. One patient, however, had a transient change in non dominant psychometry and was found on MRI to have an area of signal change around the tip of the right temporal electrode. This change recovered on subsequent testing. The patients' antiepileptic medication was reduced post-operatively and the recording continued until enough habitual seizures were recorded. Using the technique described by Gotman, spike detection, counting and computer analysis and phase studies of ictal electroencephalograms led to a standard anterior temporal lobectomy in three patients, a frontal partial lobectomy in one patient and a selective cortical resection in one patient. In two patients the electroencephalographic data excluded the possibility of a successful cortical resection and no operation was performed.

Discussion

Talairach, Bancaud and collaborators popularised the technique of insertion of intracerebral electrodes for the precise localisation of epileptic onsets[1,9,10] using the method of "stereoencephalography". Stereotactic data were obtained from a series of contrast neuroradiological studies (carotid angiograms, pneumo-encephalograms and gas and lipiodol ventriculograms) and targeting definition was indirect, based on intracerebral landmarks and using the Talairach stereotaxic atlases[9-11] while their stereotactic frame allowed only orthogonal approaches for multiple electrode insertion[1,9].

With the introduction of computerised neuroimaging, target determination was achieved with the use of multiplanar reformatted CT images constructing an individual computerised tomographic atlas of the brain of each patient[5]. Olivier and his colleagues[6,7] developed stereotactic procedures for use in epilepsy surgery by incorporating the newer computerised imaging modalities of CT scanning, MRI and digital subtraction angiography (DSA) using the method of "anatomical cross-correlation with non-stereotactic MRI images"[6]. They used the anatomical relations of the Corpus Callosum to transpose stereotactic angiographic data onto MRI and CT imaging performed in non-stereotactic conditions. Computerised systems for stereotactic image analysis and integration have also been developed and used for the insertion of depth electrodes[4] although they present equaly limitations in terms of cost effectiveness and the need for serial performance of the stereotactic imaging studies. A modified Leksell or the Talairach's stereotactic frame is usually used[4-7]. The modifications of the Leksell frame mainly aim towards the achievement of compatibility of the system with the modern neuroimaging techniques and to allow for either the orthogonal or the arc-radius approach for the depth electrodes' insertion, although it is doubtful if the design permits the targeting of the posterior mesial temporal structures from an inferior temporal entry point.

The CRW-3 design is based on the arc-centred Cartesian movement principle thus permitting infinite number of entry points for each target or use of the same entry point for multiple targets and making sampling possible from important regions that are relatively inaccessible to the orthogonal approach, such as the supplementary motor area, the orbitofrontal cortex and the posterior hippocampal complex. A further advantage is that the arc can be rotated clear off the immediate operative field and re-introduced at any time without the loss of target localisation thus allowing unrestricted access for haemostasis and securing of the depth electrodes to the skin. Mesial temporal foci may be approached from a frontal, lateral or an occipital entry site. This latter approach is said to allow detection of a posterior temporal focus, since the electrode lies along the longitudinal axis of the temporal lobe. Although a lateral approach for temporal lobe implantation risks puncturing the Sylvian vessels[5], this may be avoided, if three dimensional stereoangiography is used[8] or by direct visualisation of the surface through a burr-hole.

Non-invasive accurate relocation becomes an essential feature for the integration and correlation of the stereotactic data with the precision and the consideration of so many parameters that the intracerebral electrophysiological studies require. The stereotactic accuracy of the G-T repeat localiser is well proven[2,3] in clinical studies and the design features of the system have been exploited in other fields of functional neurosurgery and radiotherapy. In routine clinical use the frame relocates within 1 to 3 minutes offering significant labour and cost effectiveness. In the authors' series the frame has proved to be a satisfactorily reproducible method of stereotactic relocation providing good head fixation for depth electrodes' insertion purposes. The temporal freedom that the G-T repeat localising base provides, permits an unhurried study of the stereotactic data, their correlation with the requirements of the clinical problem, their integration into a common stereotactic apparatus and the advantage of specialist contributions of neurologists and electrophysiologists during the target selection to maximise the chances for successful outcome. Finally, the ability to confirm the accuracy of the implanted electrodes at any time during the postoperative period of recording provides a new level of precision in stereotactic surgery for epilepsy and the potential for direct correlation of the electrophysiological data with the functional and the neuroanatomical "real time" information.

Conclusion

The Gill-Thomas repeat localiser offers some important advantages for intracerebral electrode insertion in the investigation of medically resistant epilepsy. Accu-

rate targeting following repeat relocation, temporal freedom during the integration of information derived from multiple imaging studies, less discomfort for patients and unhurried decision-making for the surgeon, orthogonal or non-orthogonal operative approach, easy and non-invasive postoperative confirmation of the accuracy of targeting and excellent localising capabilities are some of their most important features. These features can potentially make the invasive electrophysiological investigation of the epileptic patient a safer procedure thus offering this superior diagnostic and localising tool to a greater number of epileptic patients entering their pre-surgical evaluation.

References

1. Bancaud J, Talairach J, Bonis A, Schaub C, Szikla G, et al (1967) La stereoelectroencephalographie dans l'epilepsie. Masson, Paris.
2. Gill SS, Thomas DGT, Warrington AP, Brada M (1991) Relocatable frame for stereotactic external beam radiotherapy. Int J Radiat Oncol Biol Physics 20: 599–603
3. Graham JD, Warrington AP, Gill SS, Brada M (1991) A noninvasive, relocatable stereotactic frame for fractionated radiotherapy and multiple imaging. Radiother Oncol 21: 60–62
4. Levesque MF, Zhang J, Wilson CL, Behnke EJ, Harper RM, Lufkin RB, et al (1990) Stereotactic investigation of limbic epilepsy using a multinodal image analysis system. J Neurosurg 73: 792–797
5. Lunsford LD, Latchaw RE, Vries JK (1983) Stereotactic implantation of deep brain electrodes using computed tomography. Neurosurgery 13: 280–286
6. Olivier A, de Lotbiniere A (1987) Stereotactic techniques in epilepsy. In: Tasker RR (ed) Neurosurgery: State of the art reviews, Vol 2, No 1. pp 257–285.
7. Olivier A, Peters TM, Clark JE, et al (1987) Integration de l'angiographie numerique, de la resonance magnetique, de la tomodensitometrie et de la tomographie par emission de positrons en stereotaxie. Rev EEG Neurophysiol Clin 17: 25–43
8. Szikla G (1979) Stereotactic neuroradiology and functional neurosurgery: Localisation of cortical structures by three dimensional angiography. In: Ransmussen T, Marino R (ed) Functional neurosurgery. New York, Raven Press, pp 197–217
9. Talairach J, Bancaud J (1974) Stereotactic exploration and therapy in epilepsy. In: Vinken PJ, Bruyn GW (eds) The epilepsies. Amsterdam, North-Holland, pp 758–782
10. Talairach J, Bancaud J (1973) Stereotaxic approach to Epilepsy: Methodology of anatomico-functional stereotaxic investigations. In: Progress in Neurological Surgery. Karger, Basel, pp 297–354
11. Talairach J, Szikla G, Tournoux P, et al (1967) Atlas d'anatomie stereotaxique du telencepfale. Masson, Paris

Correspondence: Prof. David G. T. Thomas, M.A., F.R.C.P., F.R.C.S., The National Hospital for Neurology and Neurosurgery, Queen Square, London WC 1N 3BG, U.K.

Acta Neurochir (1993) [Suppl] 58: 190–192

Sub-Occipital Approach for Implantation of Recording Multi-Electrodes over the Medial Surface of the Temporal Lobe

L. González-Feria and **V. García-Marín**

Department of Neurosurgery, University Hospital of Canary Islands, Medicine Faculty, La Laguna, Tenerife, Spain

Summary

Recording the electrical activity from the medial aspects of the temporal lobe, including uncus and hippocampal convolution, has an important role in the preoperative evaluation of patients with drug-resistant temporal lobe epilepsy.

We report our experience with subdural cylindrical multi-electrodes placed along the medial, basal and lateral aspects of the temporal lobe through a suboccipital approach. Eight patients have been examined with this technique for 3 to 8 days. The quality of the recordings was excellent. No displacement of the electrodes has been noticed. One patient developed meningitis which was successfully treated. The electrodes have been kept in place until surgery and have served as useful landmarks for the resection enabling also neurophysiological monitoring throughout the operation.

Keywords: Epilepsy; temporal lobe; electrode placement.

Introduction

Despite new diagnostic tools, the precise localization of an epileptic focal discharge remains an important step in the surgical treatment of epilepsy. A large proportion of cases referred for neurosurgical evaluation suffer from socalled temporal lobe epilepsy. The medial aspect of the temporal lobe, the uncus, amygdala and hippocampus are crucial in epilepsy because of their high excitability and their importance for production or spreading of epileptic discharge. Since Foerster[1] and Penfield[2] performed recordings from the surface of the brain, neurosurgeons have continued to make use of intracranial electrodes placed in different locations: epidural, subdural and depth electrodes. Cylindrical multiple contact (rigid or soft), strip or grid electrodes have been utilized. Placing electrodes at, or near, the medial aspects of the temporal lobes may be done with stereotactic technique[3,4]. Strip electrodes have also been used for this purpose[5].

An interesting approach is the one used by Wieser and Yaşargil[6] who introduced the electrodes towards the medial aspect of the temporal lobe through the foramen ovale. The same authors have also performed selective removal of the medial aspect of the epileptic temporal lobe, sparing the normal neocortex of the convexity. With this approach the outcome regarding epilepsy was favourable and memory deficit was minimal[7].

Since 1986 we have used multi-contact cylindrical, soft subdural electrodes implanted at the medial aspect of the temporal lobes via a suboccipital approach. We have found this technique to be safe and reliable, and it has improved our surgical results.

Surgical Technique

The patients are placed in the sitting or prone position under general anaesthesia. A 5 cm vertical incision is made, one inch from the midline, and centered just above the transverse sinus. A slightly elongated craniectomy, 1 × 2 cm, is performed. The dura is opened in inverted V fashion, close to the sinus, and the basis of the occipital lobe is elevated with a small spatula. A Nelaton catheter, about 2.5 mm diameter, with a metal mark on its end, is gently introduced between the tentorium and the basal occipital lobe under X-ray control. The catheter easily follows a route close to the incisura tentorii around the brain stem, then goes along the lateral aspect of the cavernous sinus and usually turns around the temporal pole under the lesser wing of the sphenoid. The catheter is useful for the subsequent introduction of an electrode with 8–10 contacts placed 1 cm apart. These electrodes have been made by one of us (V.G.M.)[8] but they have recently been made commercially available. The electrode is inserted first percutaneously through a trocar some 3–4 cm away from the skin incision and passed along the Nelaton catheter which is then removed. One or two more electrodes are placed along the basis or onto the convexity of the temporal lobe.

This method of electrode introduction can be done bilaterally and in one case we have used a similar approach for placing an interhemispheric electrode around the corpus callosum.

Results

A total of 8 consecutive patients have been studied with electrodes inserted intracranially through this route. In all of them the electrode position has been checked by X-rays and CT scan (Fig. 1). All of them were in the intended position along the medial aspect of the temporal lobe and around the temporal pole. The electrodes were kept in place 3–8 days until the amygdalo-hippocampectomy was performed.

The quality of the recordings has been excellent and confirmed the suspected diagnosis of unilateral temporal lobe epilepsy, seven of the cases having a medial type of seizures and one a basal temporo-occipital focus. The electrodes have been kept in place during the operation allowing for a more precise anatomical control of the extension of the surgical resection using the electrode contacts as landmarks. Moreover, continuous EEG recording from the remaining part of the hippocampus could be done. In each case the resection

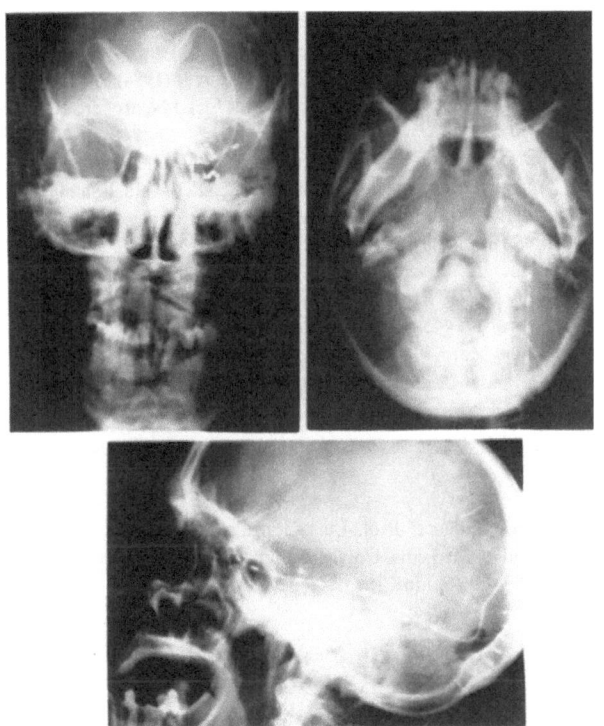
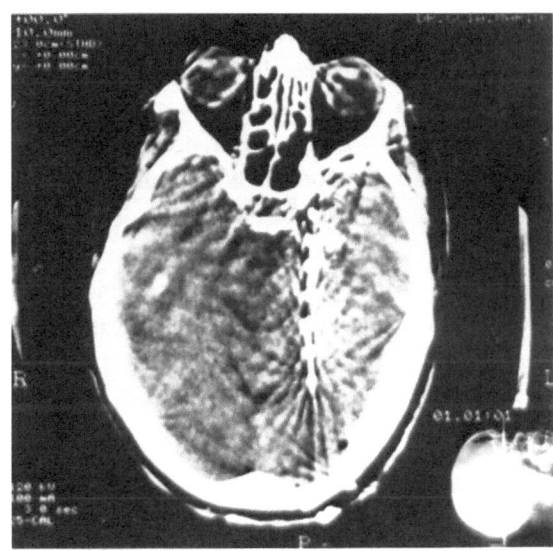

Fig. 1. Postoperative control of the position of the electrodes on X-ray (A) and CT-scan (B)

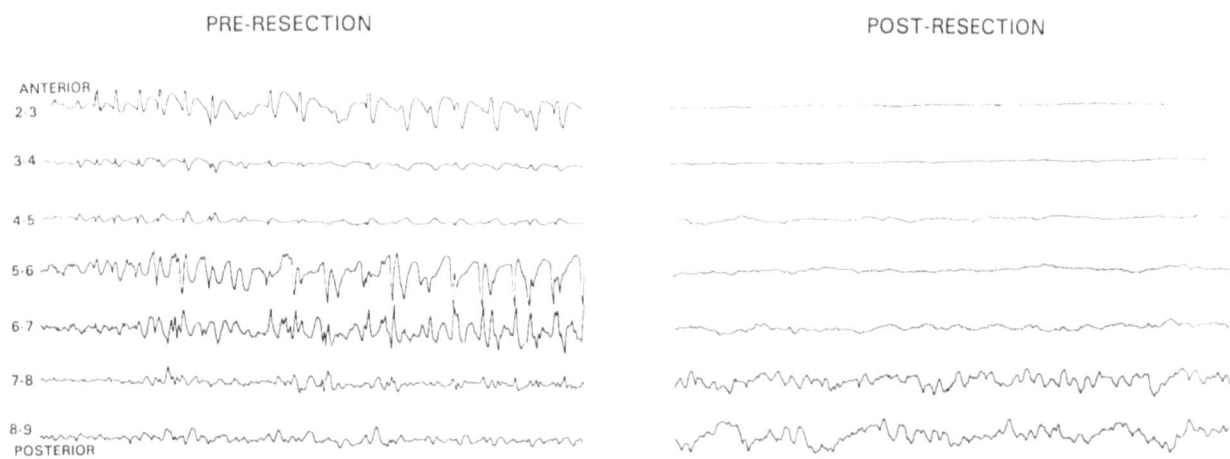

Fig. 2. Electrical activity monitored during the resection of the hippocampus. A focus can be observed on electrodes 5–6 and 6–7 which are located just at the level of the uncus. After resection extending to the level of the 7th electrode pole the abnormal activity disappeared and remained normal at the level of the 8th and 9th pole

was extended posteriorly until the electrical abnormality disappeared, or until the level of the corpora quadrigemina was reached (Fig. 2).

The follow-up is only 6–26 months and does not permit a definite evaluation of the results. However, of the eight patients operated on with this technique, seven are seizure-free and one, who had several seizures per day, has now about one seizure every six months.

Comments

The method of introducing intracranial recording electrodes for the exploration of temporal lobe epilepsy has been found to be simple, safe and reliable. In all cases studied so far it appeared that the anatomy of the tentorium and the cavernous sinus made it possible to place the electrodes along the medial surface of the temporal lobe all the way to the temporal pole. In some cases the electrode has deviated somewhat due to a dural vein or at the level of the anterior part of the cavernous sinus. The posterior, occipital approach for electrode introduction has been found useful also for the placement of electrodes onto the occipital lobe or into the interhemispheric fissure. In cases of temporal lobe epilepsy the electrode has been found useful as a guide to limit the extension of the surgical resection to comprise only that part which exhibits abnormal electrical activity. In our hands the suboccipital approach for introducing electrodes has been found to be most useful and has improved our technique when performing selective amygdalo-hippocampectomy.

References

1. Foerster O, Altenburger H (1935) Elektrobiologische Vorgänge an der Menschlichen Hirnrinde. Deutsche Ztschr 135: 277–288
2. Penfield W, Jasper H (1954) Epilepsy and functional anatomy of the human brain. Little, Brown, Boston
3. Bancaud J, Talairach J, Bonis A, Schaub C, Szikla G, Morel P, Bordas-Ferrer M (1965) La Stéréoencéphalographie dans l'épilepsie. Masson, Paris
4. Spencer SS (1981) Depth electroencephalography in selection of refractory epilepsy for surgery. Ann Neurol 9: 207–214
5. Wyler A, Ojemann G, Lettich E, Ward A Jr (1984) Subdural strip electrodes for localizing epileptogenic foci. J Neurosurg 60: 1195–1200
6. Wieser H, Elger C, Stodieck S (1985) The foramen ovale electrode: A new recording method for the preoperative evaluation of patients suffering from mediobasal temporal lobe epilepsy. Electroencephal Clin Neurophysiol 661: 314–322
7. Wieser H, Yasargil M (1982) Selective amygdalohippocampectomy as a surgical treatment of medio-basal limbic epilepsy. Surg Neurol 17: 445–457
8. Garcia-Marin V, Rodriguez Palmero M, González-Feria L, Ginoves Sierra M, Martel Barth-Hansen D, Ravina Cabrera J (1989) Design of an electrode for monitoring. Act Neurochir (Wien) [Suppl] 46: 25–27

Correspondence: Prof. Luis González-Feria, Dept. de Cirugía, Facultad de Medicina, Universidad de La Laguna, Enrique Wolfson, 17-4, E-38006 Santa Cruz de Tenerife, Spain.

Acta Neurochir (1993) [Suppl] 58: 193–194

Multiple Contact Foramen Ovale Electrode in the Presurgical Evaluation of Epileptic Patients for Selective Amygdala-Hippocampectomy

U. Steude, **S. Stodieck**, and **P. Schmiedek**

Neurosurgical and Neurological Clinic of the University of Munich, Klinikum Großhadern, Munich, Federal Republic of Germany

Summary

For stereo-EEG evaluation of patients suffering from drug resistent mesiobasal temporal lobe epilepsy a method is described to place multi-contact electrodes percutaneously through the foramen ovale near the brain stem and medial surface of the temporal lobe.

The experience with 41 cases is reported. In 28 of them a unilateral well localized focus could be identified. All of them were treated by selective amygdala-hippocampectomy according to Wieser and Yaşargil. During follow-up for at least one year 22 of patients were seizurefree. An additional 5 had a reduction of seizure frequency. Complications were subarachnoid haemorrhage in one case, transient hypaesthesia in another, and transient herpes simplex of the lips in 7 cases.

Keywords: Epilepsy; temporal lobe; stereo-EEG; electrode placement; foramen ovale; amygdala-hippocampectomy; results.

Introduction

Patients suffering from drugresistent mesiobasal temporal lobe epilepsy can be successfully operated on by anterior temporal lobe resection or selective amygdala-hippocampectomy (Wieser and Yaşargil 1982; Ward 1983). Especially the latter microsurgical approach, however, requires a precise identification of the epileptogenic focus. Surface EEG-records, as well as records with nasopharyngeal and sphenoidal electrodes have in our hands yielded insufficient information on the exact localization and extent of the ictal focus. Therefore, stereo-EEG evaluation was introduced. This diagnostic approach is not without risks, because two stereotactic operations under general anaesthesia are necessary, the first with carotid angiography and ventriculography and the second with implantation of several depth electrodes (Wieser 1983). However, stereo-EEG yields the most valuable information by identifying and delineating the epileptic focus. In the individual patient the increased risk of inserting additional depth or foramen ovale electrodes (FO) is justified by the greater precision in the identification of the epileptic focus, enabling the surgeon to perform a "tailored" and minimal resection.

The usage of a monopolar foramen ovale electrode was first described by Wieser (1983). To improve the precision in the identification of epileptic foci in the area of the amygdala-hippocampus we developed a multicontact electrode with six contacts inserted via the foramen ovale.

Operative Technique

During a shortlasting general anaesthetic with barbiturates the foramen ovale is percutaneously punctured by a Touhy needle which is advanced to the level of the clivus where CSF flow is obtained. The recording multicontact electrode is introduced and pushed forward under X-ray control until the first contact is located near the brainstem. The proximal contact is located in the foramen ovale. After withdrawal of the cannula, the electrode is fixed to the skin by adhesive tape. An electrode is also implanted in the other side. A lateral X-ray confirms the subtemporal location of the multicontact electrodes and an a.p. view their positions above the incisura trigemini (see Fig. 1). The electrodes can also be demonstrated by CT-scan showing the tip of the electrode along side the brainstem.

Results and Conclusion

In the period 1986–91 we have implanted bilateral foramen ovale electrodes in 39 patients. In an additional two patients an electrode could be implanted on one side only and in two others it was not possible to puncture via the foramen ovale on both sides. The only

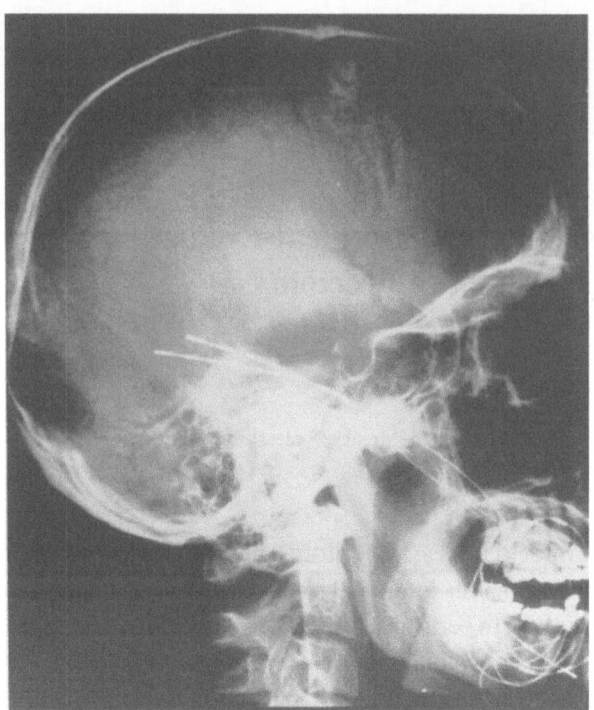

Fig. 1. Lateral x-ray view of the skull with the implanted electrodes

Of the 39 patients 28 were found to have a unilateral, well localized ictal focus. In the remaining 11 patients no epileptogenic focus could be demonstrated. In the 28 patients with a demonstrated focus we performed selective amygdala-hippocampectomy according to the method described by Wieser and Yaşargil (1982). Twenty two of these patients were seizure-free with a follow-up of more than one year, five patients had reduction of their seizure frequency and only one patient had no benefit from the procedure.

In our experience the usage of foramen ovale electrodes provide valuable information about the location of an ictal focus which may be suitable for anterior temporal lobe resection or a selective amygdala-hippocampectomy. The method of implanting foramen ovale electrodes is simple and quite familiar to all neurosurgeons who are used to performing percutaneous thermocoagulation for trigeminal neuralgia. Moreover, it is a safe and reliable approach to the exploration of patients considered for epilepsy surgery.

serious complication was a subarachnoid haemorrhage which occurred in one patient after withdrawing the electrode following 14 days of recording. Seven patients suffered from transient herpes simplex on the lips. There were no other infections.

In one patient there was a transient hypaesthesia in the third trigeminal branch lasting for three months.

References

1. Ward Jr AA (1983) Surgical management of epilepsy. In: Browne TR, Feldman RG (eds) Epilepsy. Little, Brown, Boston, MA, pp 281–296
2. Wieser HG, Yaşargil G (1982) Selective amygdalo-hippocampectomy as a surgical treatment of mesiobasal limbic epilepsy. Surg Neurol 17, 445–457
3. Wieser HG (1983) Electroclinical features of the psychomotor seizure. Gustav Fischer/Butterworth, Stuttgart, New York

Correspondence: P. D. Dr. Ulrich Steude, Neurosurgical Clinic of the Ludwig Maximilians-University of Munich, Marchioninistrasse 15, D-81377 München, Federal Republic of Germany.

Acta Neurochir (1993) [Suppl] 58: 195–197

Radiosurgery of Epilepsy

J. L. Barcia-Salorio[1,2], **J. A. Barcia**[1,2], **P. Roldán**[1,2], **G. Hernández**[3], **L. López-Gómez**[4]

[1] Servicio de Neurocirugía, Hospital Clínico Universitario, [2] Departamento de Cirugía, Universidad de Valencia, [3] Servicio de Terapéutica Física, Hospital Clínico Universitario, and Servicio de Electroencefalografía, Hospital Clinico Universitario, Valencia, Spain

Summary

Since 1982 a series of 11 epileptic patients have been treated with stereotactic radiosurgery. Patients were intracranially recorded with cortical and deep electrodes until the location of the epileptogenic focus was determined. A deep electrode was stereotactically placed at this point to confirm the accuracy of the location. All patients received radiosurgery with a gamma source and a dose of 10 to 20 Gy, except two of them in which a betatron was used. The results were: Total disappearance of the crises and withdrawal of medication: 4 cases (36%). More than 80% reduction of crises: 3 cases (27%). More than 50% reduction of crises: 2 cases (18%). Less than 50% reduction of crises: 2 cases (18%).

No complications were observed, even in those cases in which the focus was located near critical areas of the brain. The efficacy of the low doses used accounts for non-destructive mechanisms, probably mediated by a neuronal plasticity phenomenon, as experimental studies suggest. The lack of complications can make this therapeutic approach an alternative to be considered when critical areas are involved.

Keywords: Epilepsy; radiosurgery.

Introduction

There are some clinical and experimental facts suggesting that irradiation of epileptic foci may be favourable for their clinical evolution.

These include the improvement or cure of epileptic seizures in patients with AVMs or low-grade tumours after conventional or stereotactic irradiation, as reported by several authors[5–7,10] and the disappearance of convulsive crises after focal irradiation, in an experimental cobalt epileptic model in the cat[2]. Based on these observations, the authors performed stereotactic radiosurgery in 11 cases of idiopathic focal epilepsy from 1982 to 1991[1].

Material and Methods

This series consists of 11 (4 male, 7 female) patients presenting with epileptic seizures, their ages ranging from 16 to 42 years old. The preoperative symptomatic period varied from 23 to 24 years. The clinical features of each patient are summarized in Table 1.

Every patient was carefully studied to rule out any primary treatable cause of the seizures. The preoperative protocol included clinical and psychological studies, standard EEG, CT scanning, angiography in some cases and, in the most recent cases, MR imaging. Patients were considered candidates for this program only when more than one year of treatment with a single oral anticonvulsive drug and more than one year of treatment with a multidrug regimen had elapsed, provided that the serum drug concentrations had been regularly monitored.

According to the results of the initial studies, patients were subjected to burr-hole electrocorticography. The technique consisted in introducing several uni- or multielectrode probes* through a single (or bilateral) burr hole into the subarachnoid space, and guiding them with the aid of radioscopical imaging to the problem areas previously selected by the non-invasive studies.

The ECoG ictal and interictal recordings were carefully studied to determine the area of the cortex which was the origin of the seizures. This sometimes required the introduction of a new series of cortical electrodes to the problematic areas. The search for the most probable area where the focus could be located was guided by a computer-assisted method based on the mathematical model of dipole generators as has been previously reported[3]. When evidence existed that a certain cortical area was responsible for the epileptic activity, deep multicontact electrodes were stereotactically introduced to the area in order to accurately localise and estimate the size of the focus.

Those patients with an estimated epileptogenic area of 15 mm of diameter or less were irradiated with photons from a single conventional ^{60}Co gamma source using collimators of 10 mm. with an estimated dose of 10 Gy using the cross-fire technique, as described elsewhere[1].

Only one patient received bilateral irradiation because of a right temporal focus and a contralateral phenomenon. Two patients with cortical foci of 4 cm or more were irradiated with electrons of 10–15 MeV from a 45 MeV betatron, receiving 10 Gy from only a one beam entrance direction, and another one received 2 sessions of 10 Gy from the cobalt source. Patients continued receiving their antiepileptic medication for one year after irradiation, which was progressively tapered off when possible.

---

* 4-contact Medtronic Quad electrodes, supplied by Medtronic Hispania S.A.).

Table 1

Case	Clinical features	Focus	Dose	Follow-up	Preop. seizure freq.	Postop. seizure freq.	%Red.	Med	Psy-L.
r.m.g.	c.p.s.	rt. mes. temp.	20 Gy	30 mo.	4/mo.	1/mo.	75%	=	+
v.l.m.	c.p.s.	lt. mes. temp.	20 Gy	22 mo.	10/mo.	10/mo.	0%	=	=
a.c.a.	c.p.s.	lt. mes. temp.	20 Gy	19 mo.	4/mo.	1/mo.	75%	=	+
d.d.n.	c.p.s., s.g.	lt. mes. temp.	10 Gy	9 yrs.	0.5/mo.	0	100%	0	=
c.c.a.	c.p.s., s.g.	lt. mes. temp.	10 Gy	8 yrs.	75/mo.	0	100%	0	+
m.f.b.	s.s., s.g.	rt. temp.	10 Gy	9 yrs.	110/mo.	2/mo.	98.18%	<	=
i.p.r.	s.s., s.g.	lt. occ.	10 Gy	8 yrs.	30/mo.	0	100%	0	+
c.b.c.	s.s., s.g.	lt. occ.	20 Gy	8 yrs.	80/mo.	80/mo.	0%	=	=
e.c.m.	s.s., s.g.	lt. occ.	20 Gy	5 yrs.	30/mo.	4/mo.	86.67%	<	=
c.t.m.	s.s.	rt. par.	20 Gy	5 yrs.	8/mo.	0	100%	0	=
a.m.m.	p.g.s.	rt. mes. temp.	20 Gy	19 mo.	100/mo.	12/mo.	89.09%	=	=

c.p.s. complex partial seizures; *c.p.s.*, *s.g.* complex partial seizures, secondarily generalized; *s.s.*, *s.g.* sensitive seizures, secondarily generalized; *s.s.* sensitive seizures; *p.g.s.* primarily generalized seizures; *% Red.* percentage of reduction in seizure frequency; *Med* change in medication postoperatively (= same as before the operation, < reduction of medication, 0 no medication); *Psy-L.* psychosocial and professional integration (+ ameliorated, = similar to prior the operation).

Results

The results were as follows: Total disappearance of the crises and withdrawal of medication: 4 cases (36%). More than 80% reduction of crises: 3 cases (27%). More than 50% reduction of crises: 2 cases (18%). Less than 50% reduction of crises: 2 cases (18%) (these are summarized in Table 1).

Discussion

The total disappearance of crises in 4 cases of criptogenic focal epilepsy previously resistant to medication without recurrence after several years of follow-up suggests a permanent structural or physiologic change in the neurons at the epileptogenic focus. Elomaa[4] proposed an action of radiation upon somatostatin and the loss of its effects on cell membrane. However, our results with focal irradiation in cobalt experimental epilepsy in the cat, a synaptic model, suggested changes in synaptic plasticity[2].

Most authors think of deafferentation as the basic mechanism involved in epileptogenesis, the most striking feature being the loss of dendritic spines. On the other hand, Malis *et al.*[8] observed a great dendritic proliferation after producing a deuteron laminar lesion at the IV cortical layer in the rabbit. Several hypotheses may therefore be presented. Irradiation may induce synaptic neoformation, or the epileptic neurons are hypersensitive to irradiation or, rather, as Monnier and Krupp[9] pointed out, 10 Gy irradiation diminishes cortical activity.

In one of the cases (case v.l.m.), the target was selected by the structural changes (gliosis) seen on MRI, rather than by the results of electrophysiological studies. This patient did not improve after irradiation. On the other hand, the doses used are too low to have an effect the lesion which provokes the seizures. These two facts suggest that for the focal irradiation to be effective the target point must be placed not on the lesion, but on the epileptogenic brain tissue which surrounds it. The volume of cerebral tissue irradiated by radiosurgery is small. Thus, in our opinion crucial for radiosurgery of epilepsy is a very accurate localisation of the epileptogenic area, and this must be sufficiently small. This makes the localisation procedure long and painstaking, and may be the reason for the poor results in some of the cases. In the future, non-invasive localisation procedures such as magneto-encephalography may be the ideal complement for this non-operative treatment method.

The lack of secondary effects achieved by this low-dose irradiation may indicate its use in foci located very close to functionally important areas of the brain.

References

1. Barcia-Salorio JL, Roldán P, Hernández G, López-Gómez L (1985) Radiosurgical treatment of epilepsy. Appl Neurophysiol 48: 400
2. Barcia-Salorio JL, Vanaclocha V, Cerdá M, Ciudad J, López-Gómez L (1987) Response of experimental epileptic focus to focal ionizing radiation. Appl Neurophysiol 50: 359
3. Barcia-Salorio JL, Barcia JA, Ciudad J, Such V (1987) Automatic calculation of epileptogenic focus location within the brain. Appl Neurophysiol 50: 600

4. Elomaa E (1980) Focal irradiation of the brain: An alternative to temporal lobe resection in intractable focal epilepsy. Med Hypoth 6: 501

5. Fabrikant JI, Lyman JT, Hosobuchi Y (1984) Stereotactic heavy-ion bragg-peak radiosurgery for intra-cranial vascular disorders: Method for treatment of deep arteriovenous malformations. Br J Radiol 57: 479

6. Heikkinen ER, Konnov B, Melnikov L, et al (1989). Relief of epilepsy by radiosurgery of cerebral arteriovenous malformations. Stereotact Funct Neurosurg 53: 157

7. Kjellberg RN, Davis KR, Lyons S, et al (1983) Bragg-peak proton beam theraphy for arteriovenous malformation of the brain. Clin Neurosurg 31: 248

8. Malis LI, Rose JE, Kruger L, Baker CP (1962) Production of laminar lesions in the cerebral cortex by deuteron irradiation. In: Haley TJ, Snider RS (eds) Response of the nervous system to ionizing radiation. Academic Press, New York, pp 359–368

9. Monnier M, Krupp P (1962) Action of Gamma irradiation on electrical brain activity In: Haley, Snider RS (eds) Response of the nervous system to ionizing radiation. Academic Press, New York, pp 607–620

10. Rossi, GF (1985) Epileptogenic cerebral low-grade tumors: Effect of interstitial stereotactic irradiation on seizures. Appl Neurophysiol 48: 127

Correspondence: Dr. Juan A. Barcia, Servicio de Neurocirugia, Hospital Clínico Universitario, Av. Blasco Ibañez, 17, E-46010 Valencia, Spain.

Acta Neurochir (1993) [Suppl] 58: 198–200

Multiple Subpial Cortical Transections for the Control of Intractable Epilepsy in Exquisite Cortex

M. Dogali[1], O. Devinsky[2], D. Luciano[2], K. Perrine[2], and A. Beric[2]

Departments of Neurosurgery[1] and Neurology[2], NYU Medical Center-Hospital for Joint Diseases, New York, NY, U.S.A.

Summary

In 5 cases suffering from intractable seizures and ictal onset in exquisite (primary somatosensory or language related) cortex, surgical therapy has been done consisting wholly or in part of multiple subpial transections.

In two cases with involvement of the primary somatosensory cortex, good seizure control without detectable neurological deficit was achieved. In the other three cases with involvement of the language cortex, deficits were minimal and cleared with time. Patients became seizure-free.

Keywords: Epilepsy; exquisite cortex; cortical transections; operative therapy.

Introduction

Since Victor Horsely, surgical management of epilepsy has been concerned with identification of a seizure focus in a "resectable" area of brain. For Horsely structural pathology and clinical findings localized the seizure focus. Foci in motor cortex, however, could only be resected at the cost of weakness[1]. With the introduction of effective antiepileptic drugs (AEDs), the concept of resectable cortex has been restricted to those areas with poorly defined, redundant, or non-exquisite function. The surgical management of focal epilepsy, particularly of the anterior temporal lobe and nondominant frontal regions, was pioneered by Penfield, Rasmussen, Talairach, and Falconer[2–5]. Invasive intracranial monitoring has allowed patients with more complex seizure disorders to be considered candidates for surgery[6,7]. Ictal intracranial seizure recording has led to a clearer understanding of the diverse nature and localization of epileptic activity, particularly in extratemporal and nonstructurally-related seizure cases[8].

While focal resection, particularly in cases related to structural lesions, has the highest surgical success rate, many patients with intractable epilepsy have ictal onsets in exquisite cortex. Often these patients are rejected for surgical therapy, or in cases with ictal activity in the primary somatosensory cortex, undergo resection which leaves a significant neurological deficit—a necessary but sad compromise. Morrell *et al.* postulated that horizontal intracortical fibers are involved in generating epileptogenic activity[9]. Further, if such fibers were interrupted, the critical mass needed for clinical seizures would not be recruited. As Morrell pointed out, observations by Sperry and his colleagues regarding the lack of behavioral disturbance with mica chips inserted vertically into the cortex further supported previous findings by Mountcastle, Asanuma, Hubell and Weisel, and Morell[10–14] that the functional components of cerebral cortical tissue are organized and dependent upon vertically oriented cortical columns. These observations and laboratory experiments led to the multiple subpial transection (MST) procedure.

We are reporting our experience with 5 patients with intractable seizures and ictal onset in exquisite cortex whose surgical therapy has consisted either wholly or in part of multiple subpial transections.

Technique

A craniotomy is routinely performed under general anesthesia. A 32 or 64-contact subdural grid, often with additional 4- to 6-contact strips, is placed over the abnormal cortical area, which has been ascertained by the clinical features and previous interictal and ictal scalp EEG recordings. The wound is closed over the grid, with the bone flap being left out and frozen sterilely until replacement at the second stage of the procedure. The grid wires are tunneled for 6 to 7 cm to prevent CSF leakage. The patient is monitored in an Intensive Care area for a period of 7 to 10 days. Medications are withdrawn and at least 3 typical seizures are recorded using continuous 64-channel video EEG monitoring (Telefactor). Correlation between the grid electrodes and the clinical seizure pattern is then made. Once the typical ictal event is observed and documented to arise from exquisite cortex, the area is functionally mapped and the neuropsychological data obtained via stimulation trials.

Stimulation is conducted with a Grass instrument S12 cortical stimulator with built-in stimulus isolation and constant current circuitry. A biphasic square wave at a frequency of 50 Hz and pulse duration of 0.3 milliseconds is delivered. Stimulator current ranges from 1 mA to 15 mA, and the duration of stimulation for any given trial does not exceed seven seconds. A 30 second intertrial interval without stimulation is provided. Afterdischarges are monitored with 21-channel hardcopy EEG utilizing a bipolar montage from contacts on the subdural grid. Functional mapping includes stimulation during automatic speech tasks such as counting or reciting the alphabet. For more elaborate mapping of language cortex, the patient is interactive with a computer paradigm presenting stimuli to name, read, and recall. The patient's motor functions are observed for the elicitation of movement during stimulation. Sensory phenomena are obtained using similar stimulation parameters. The patient's experiences are documented and related to the specific grid contact. The stimulations proceed from the most anterior to the most posterior contact along each row of the grid. For afterdischarge setting, stimulation begins at 1 or 2 mA and is increased in 1 to 2 mA increments until a functional response is elicited, an afterdischarge is observed, or until approximately 10 to 15 mA is reached. For functional tasks, a series of at least 3 trials is carried out at an amperage 1 mA below the afterdischarge threshold at each site.

The ictal and neuropsychological data are then transferred to a schematic cortical map and the epileptologist, neuropsychologist, and neurosurgeon confer preoperatively. In those cases that involve both cortical and limbic onset, appropriate resections are planned along with the subpial transections. The patient is returned to the operating room and fixed in a headholder under general anaesthesia, and the results of preoperative studies, namely the active seizure and functional site(s), are transferred to the cortical surface via a series of tickets indicating the area(s) for subpial transection. Using the operating microscope and an operating chair with arm support, a microarachnoid dissection of the sulci is performed, with care taken to spare the microscopic cortical vessels.

It has been our approach to carry out the dissection of the gyral areas prior to beginning the subpial transections. Once the dissection has been accomplished, the Morrell–Whisler instruments are utilized in the fashion which they described, with the instruments being passed from the sulcus subpially at 90 degrees and a depth of 4 to 5 mm. The transections are made at 5 mm intervals.

Clinical Summaries

Case 1: MP, a 13 year old right handed male, suffered from severe incapacitating post-traumatic seizures. A CT scan revealed increase in volume of his right frontal horn. The patient suffered from 2 seizure patterns — 1) focal sensory seizures beginning in his left foot and calf, ascending to his chest and then generalizing, and 2) classic supplemental motor area seizures with headturning to the right and fencing posture. Invasive monitoring revealed a classic supplemental motor focus, which was resected. Subsequent to that resection the grid was replaced. The patient was again monitored and continued to have significant clinical seizures originating from the primary sensory cortex. This 4 cm × 4 cm epileptogenic area was subpially transected. He tolerated this well and has no detectable neurological deficit. He occasionally suffers from simple partial seizures characterized by intermittent numbness of the foot. He has returned to school. Follow up is six months.

Case 2: SB, a 44 year old left-handed woman, suffered partial complex seizures beginning with an inability to speak or impaired speech, often progressing to a generalized seizure. This patient had a left frontal anaplastic astrocytoma resected 8 years prior without evidence of recurrent or residual tumor. She was toxic on two AEDs, unable to carry out activities of daily life, and had at baseline a mild expressive aphasia. Ictal and interictal activity were monitored invasively in this patient. There was extensive involvement of Broca's area, primary somatosensory cortex, and Wernicke's area. A prominent aphasia reflecting a Todd's paralysis occurred after a seizure during monitoring. An area 9 cm × 4 cm encompassing the entire region was subpially transected. The severe postictal aphasia was absent upon clearing from anesthesia. The patient is now maintained on one AED without toxicity, has had greater than 50% reduction in complex partial seizures, suffers no new measurable neurological deficit, and has had no change from her preoperative neurological status. Follow up is five months.

Case 3: NG, a 29 year old right handed male, suffered from rapidly generalizing partial complex seizures. He was found to have a benign neoplasm in his frontal region and ictal activity in his primary motor cortex. The lesion was totally resected and a 3 cm × 5 cm area of primary motor cortex and adjacent frontal region was transected. A mild disinhibition occurred postoperatively, but cleared over several months. He has no neurological deficit and is seizure-free. Follow up is four months.

Case 4: JV, a 47 year old right handed male, suffered complex partial seizures for 45 years. He had several episodes of convulsive status epilepticus. His ictal onset revealed independent left (dominant) mesial temporal and Wernicke's area ictal onsets. A 6 cm left temporal lobectomy was carried out, and then the inferior parietal lobe, measuring 4.5 cm × 5 cm, was subpially transected. The patient has been seizure-free. A postoperative aphasia cleared within several weeks, and a mild dyslexia is clearing.

Case 5: PG, a 28 year old right handed male, suffered partial seizures from age 17. Grid placement revealed diffuse left temporal interictal activity with diffuse ictal onset over the temporal lobe, including Wernicke's area. An anterior temporal lobectomy was carried out and subpial transections of Wernicke's area were performed. For approximately one to two weeks postoperatively this patient had a global aphasia, which slowly cleared over six weeks. He now shows a mild dysnomia, and is seizure-free. Follow up is five months.

Discussion and Conclusion

Patients with ictal onset in exquisite cortex, either primary somatosensory or language related cortex, have been either previously rejected for surgical therapy or suffered significant neurological loss with resections of sensitive cortex. Subpial transections theoretically appear to deny lateral recruitment of neurons and therefore abort complex partial seizure and secondary generalization while preserving the critical vertically oriented functions of the cortex. In two of our transection cases involving the primary somatosensory cortex there have been no detectable postoperative neurological deficits and good seizure control. In the three cases involving language cortex the deficits appear to be minimal and are clearing with time. These procedures have offered good seizure control. The dissection may be tedious, and an extensive area for multiple transections may require several hours. There is, however, surprisingly little hemorrhage, and we have experienced no significant complications related to this procedure. We remain are concerned about the possibilities of future pathological processes due to induced gliosis or cascular disruption of the cortex.

While our experience is short and limited, subpial transection has resulted in excellent short-term control of intractable seizures arising from exquisite areas of cortex without the induction of major neurological deficits. Its long-term effects on the cerebral cortex must be evaluated, including the possibility of significant late complications. Nonetheless it may be an effective adjunct to the armamentarium for the control of intractable seizures in desperate cases.

References

1. Horsely V (1886) Brain-surgery. Br Med J 2: 670–675
2. Penfield W, Rasmussen T (1950) The cerebral cortex of man Macmillan, New York
3. Penfield W, Jasper HH (1954) Epilepsy and the functional anatomy of the human brain. Little, Brown, Boston
4. Talairach J, Bancuad J (1974) Stereotaxic exploration and therapy in epilepsy. In: Vinken PJ, Bruyn GW (eds) Hb clinical neurology, Vol 15. North-Holland, Amsterdam, pp 758–780
5. Falconer MA, Hill D, Meyer R, Mitchell W, Pond DS (1955) Treatment of temporal lobe epilepsy by temporal lobectomy, a survey of findings and results. Lancet 1: 827–835
6. Lüders H, Lesser RP, Dinner DS, Morris HH, Wylie E, Godoy J (1988) Localization of cortical function: new information from extraoperative monitoring of patients with epilepsy. Epilepsia 29 [Suppl 2]: S56-S65
7. Lesser RP, Lüders H, Klein G, Dinner DS, Morris HH, Hahn JF, Wylie E (1979) Extraoperative cortical functional localization in patients with epilepsy. J Clin Neurophysiol 4: 27–53
8. Spencer SS, Spencer DD Williamson PD, Mattson R (1990) Combined depth and subdural electrode investigation in uncontrolled epilepsy. Neurology 40: 74–79
9. Morrell F, Whisler WW, Bleck TP (1989) Multiple subpial transection: A new approach to the surgical treatment of epilepsy. J Neurosurg 70: 231–239
10. Sperry RW, Miner W (1955) Pattern perception following insertion of mica plates in visual cortex. J Comp Physiol Psychol 48: 463–469
11. Mountcastle VB (1957) Modality and topographic properties of single neurons of cat's somatic sensory cortex. J Neurophysiol 20: 408–434
12. Asanuma H (1975) Recent developments in the study of the columnar arrangement of neurons within the motor cortex. Physiol Rev 55: 143–156
13. Hubel DH, Wiesel TN (1962) Receptive fields, binocular interaction and functional architecture in the cat's visual cortex. J Physiol 160: 106–154
14. Morell F, Whisler WW (1982) Multiple subpial transection for epilepsy eliminates seizures without destroying the function of the transected zone. Epilepsia 23: 440

Correspondence: Michael Dogali, M.D., Department of Neurosurgery, NYU Medical Center-Hospital for faint Diseases, 550 1st Avenue, New York, NY 10016, U.S.A.

Acta Neurochir (1993) [Suppl] 58: 201–204
© Springer-Verlag 1993

Role of the Centromedian Thalamic Nucleus in the Genesis, Propagation and Arrest of Epileptic Activity. An Electrophysiological Study in Man*

F. Velasco, M. Velasco, I. Márquez, and G. Velasco

Units of Neurology and Neurosurgery, General and Children's Hospitals, SSA, and Division of Neurophysiology, National Medical Center IMSS, Mexico City, Mexico

Summary

Twentyeight patients underwent electrical stimulation of the centromedian (CM) thalamic nucleus as a neuro-augmentative procedure to control intractable seizures of various types. To assess the correct placement of electrodes, electrical stimulation at 3 and 6 Hz, 1.0 msec and 600–2000 uA pulses in trains of 30–60 sec were used while the EEG was recorded by scalp electrodes. Spontaneous seizure activities of various types were simultaneously recorded. In cases of focal epilepsy depth electrodes were used in order to study the sequence of activation of different structures. Three Hz electrical stimulation induced spike-and-wave complexes accompanied by a typical absence. Eight Hz induced recruiting-like responses with ipsilateral predominance and no clinical manifestations. Interictal spike activity was recorded in the CM in cases of generalized tonic-clonic convulsions (GTCC) and ictal spike pattern in the CM preceded discharges in other areas. In atypical absences cortical and CM discharges occurred simultaneously.

It is concluded that the CM participates in the onset of GTCC and typical absences, in the propagation of secondary GTCC but not in myoclonic convulsions.

Keywords: Centro median thalamic nucleus; electric stimulation of the brain; pathophysiology of epilepsy; recruiting responses.

Introduction

A preliminary study by Velasco *et al.* (1987) showed that chronic electrical stimulation of the Centro Median Thalamic Nucleus (CM) alleviated clinical seizures and improved the psychological performance in five patients with intractable generalized tonic-clonic seizures. Subsequently, (Velasco *et al.* 1993b and c) this treatment was given to an additional group of 23 patients with various intractable seizure patterns, classified in four groups according the predominant seizure type and to the response to CM stimulation. Thus, group A with generalized tonic-clonic seizures and typical absences and group B with focal motor and secondary generalized seizures responded; group C with partial complex and secondary generalized seizures and group D with tonic seizures and atypical absences did not respond to CM stimulation.

These studies provided us with the unique opportunity to investigate the role of CM in the generation, propagation and suppression of different types of seizures. Recording and stimulation were made in order to determine: 1) the initiation and propagation of spontaneous epileptiform activities monitored by simultaneous CM and scalp EEG recordings and 2) the epileptiform and anti-epileptiform clinical and EEG responses elicited by low and high frequency stimulation of CM.

Methods

Multicontact platinum teflon-coated electrodes were stereotactically placed bilaterally in the CM (Velasco *et al.* 1972). Experiments were performed from 10 to 20 days after implantation and prior to the chronic, therapeutic CM stimulation. In each patient, two hours' daily recording sessions in the morning and at least one sleep recording during night were performed. In addition, eight four hours' daily stimulation sessions were performed using 0.1 msec square pulses in trains of 0.3 to 1 min with three frequencies 3, 6 and 60 Hz and amplitudes ranging from 6 to 30 V (400 to 2000 uA). Concomitant subjective and objective responses of the patients and changes in reaction time were systematically investigated.

Results

Spontaneous Ictal EEG Activities in the CM

CM epileptiform EEG activities correlating with widespread scalp EEG discharges were present during generalized seizures in all recordings. They included

* Partially supported by the Contract 0013-M9105.

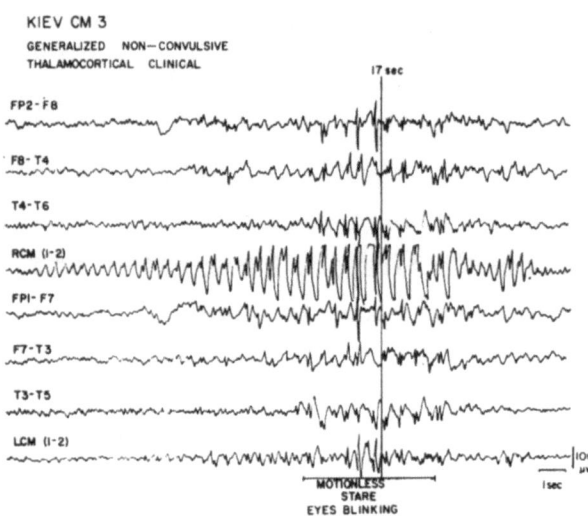

KIEV CM 3
GENERALIZED   NON—CONVULSIVE
THALAMOCORTICAL   CLINICAL

Fig. 1. Spontaneous absences. Conventional EEG recordings from bilateral scalp frontal and temporal and thalamic CM regions (RCM 1–2, LCM 1–2) during a typical absence. The horizontal line corresponds to the duration of the attack. Vertical line indicates a 17 secs cut-off of the records. Note that paroxysmal 3 Hz spike-wave EEG activity in the right CM precedes that of the left CM and scalp. (Modified from Velasco *et al.* 1989)

cases of tonic-clonic seizures and typical absences in adults, and tonic, atonic, tonic-clonic, myoclonic and combined tonic-atonic-myoclonic seizures and atypical absences in children.

CM epileptiform EEG activities preceded the appearence in the scalp recordings of tonic-clonic seizures and absences in adults (Fig. 1) and myoclonic attacks in children. Conversely, CM and scalp activities occurred simultaneously at the onset of other generalized seizures in children particularly in those presenting with Lennox-Gastaut syndrome.

These children also showed abnormal hyperdense mesencephalic areas surrounding the red nucleus at MR examinations. In contrast, no primary involvement of CM was observed in focal motor seizures starting in the cortical convexity or in partial complex seizures originating from the amygdala and hippocampus (Velasco *et al.* 1993a).

Scalp Responses Induced by CM Stimulation

Low Frequency CM Stimulation

Unilateral low frequency (6 Hz) and low intensity (8 V, 600 uA) CM stimulation elicited three types of incremental responses:

Type A, recruiting-like responses elicited by stimulation of the ventral CM regions (magnocellular CM and mesencephalic lemniscus). There were monophasic, negative potentials with a widespread bilateral distribution dominant in the ipsilateral frontal region. *Type B*, augmenting-like responses elicited by stimulation of the central CM regions (magno- and parvocellular CM). Biphasic, positive-negative potentials appeared with widespread bilateral distribution particularly in the

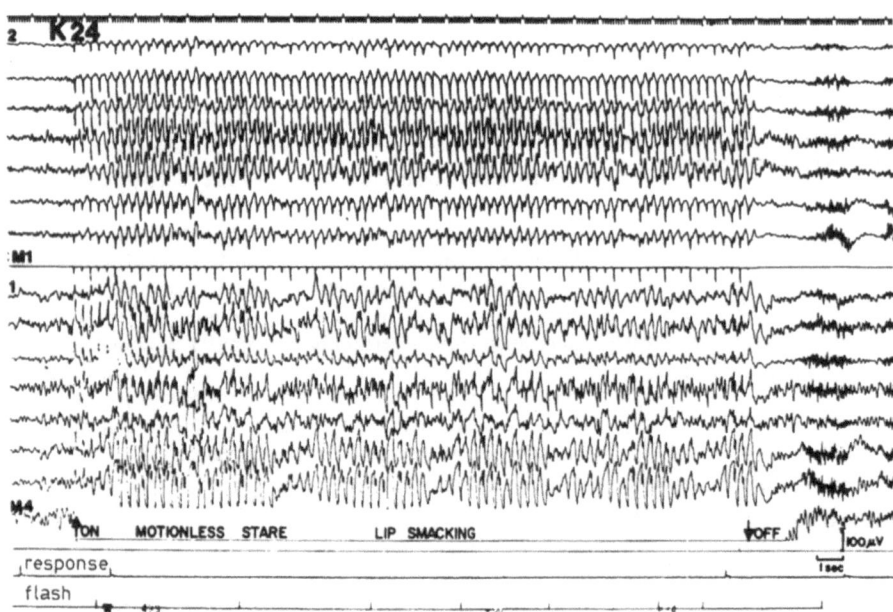

Fig. 2. Low frequency CM stimulation. EEG and clinical features of an absence-like attack elicited by simultaneous stimulation of the right central and left ventral portions of the CM at 3 Hz and 30 V. ON and OFF indicate the duration of the CM stimulation, and flash-response indicates the lack of response to a single visual reaction time task. (Modified from Velasco *et al.* 1993c)

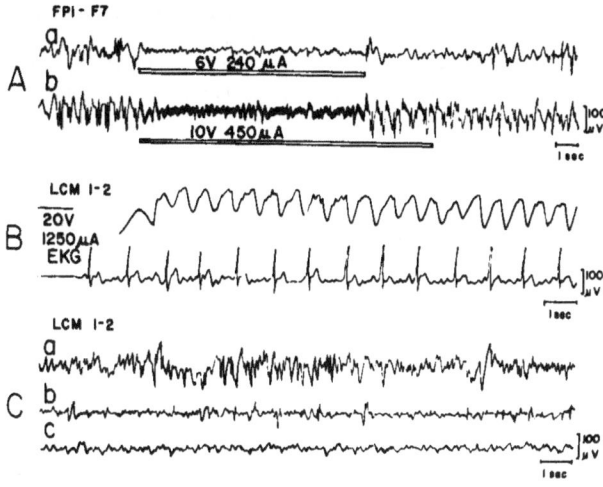

Fig. 3. High frequency CM stimulation. (A) *a* and *b*, desynchronization and attenuation of the ongoing paroxysmal activities during acute CM stimulation. (B) monomorphic delta after discharge, and (C), progressive blocking of the ongoing CM paroxysmal activities after daily repetition of CM stimulation. In (A), horizontal lines indicate the duration of stimulus. In (B), the EKG signal was simultaneously recorded to rule out the possibility of a superimposed pulse artefact and in (C), records were obtained before (*a*) and 22 hrs after the last stimulation trial at day 15 (*b*) and 25 (*c*) of daily electrical stimultion of the CM. (Modified from Velasco *et al.* 1993c)

ipsilateral central region. *Type C*, primary-like responses elicited by stimulation of the dorsal CM regions (parvocellular CM close to VPL). The responses were monophasic positive confined to the ipsilateal parietal region. Bilateral, low frequency stimulation of CM at supra-threshold intensities (30 V, 2000 uA) elicited EEG and clinical signs of an absence-like attack with 3 Hz spike-wave complex discharges, motionless stare, lip smacking and lack of response to flash (Fig. 2).

High Frequency CM Stimulation

Unilateral, high frequency CM stimulation with variable intensity (6–20 V; 120–1250 uA) elicited EEG desynchronization and partial blockage of spontaneous ongoing paroxysmal EEG activities (Fig. 3A). At low intensities this EEG arousal response was not accompanied by clinical symptoms whereas high intensity stimulation induced ipsilateral strabismus, diplopia and unpleasant sensations or contralateral paraesthesiae. Daily repetition of electrical stimulation of CM induced other responses at the stimulated site: one was the presence of a monomorphic delta afterdischarge lasting from 30 to 100 sec (Fig. 3B) and another the progressive suppression of the thalamic interictal EEG spike activity (Fig. 3C).

Discussion

The present results suggest that the CM has a crucial role in the initiation and propagation of epileptiform activities in generalized tonic-clonic seizures in children, since the epileptiform EEG activities in CM preceded the seizure onset in the cortex. The CM also seems to participate in the propagation of most other forms of generalized convulsions arising from the upper brain stem, since paroxysmal CM and cortical activities appeared simultaneously. In contrast, the CM does not appear to be involved in focal motor seizures originating in the cortex as well as in partial complex seizures originating from the amygdala and hippocampus or in infantile spasms originating from the lower brain stem.

Stimulation experiments suggest that, as in animals, the human CM is part of a thalamic diffuse projection (Dempsey and Morison 1942) or an ascending reticular thalamic system (Moruzzi and Magoun 1949, Jasper 1949), because electrical stimulation elicited cortical, widespread incremental potentials, spike-wave complexes (low frequency), and EEG desynchronization (high frequency stimulations). Therefore, the electrical activation of CM may produce a double response on epileptogenesis: A hypersynchronizing response of cortical neurons, (elicited by 3 Hz, high intensity stimulation) manifested as a typical absence-like attack, and a desynchronizing response of the cortical neurons, (elicited by 60 Hz, medium intensity stimulation), manifested by total or partial blocking of the ongoing background or paroxysmal EEG activities.

This last type of response is perhaps responsible for the seizure control by the use of chronic electrical stimulation of the CM (Velasco *et al.* 1987, 1993b).

References

1. Dempsey EW, Morison RS (1942) The mechanism of the thalamo-cortical augmentation and repetition. Am J Physiol 158: 297–308
2. Jasper HH (1949) Diffuse projection systems. The integrative action of the thalamic reticular system. Electroencephal Clin Neurophysiol 1: 405–419
3. Moruzzi G, Magoun HW (1949) Brain stem reticular formation and activation of the EEG. Electroencephal Clin Neurophysiol 1: 455–473
4. Velasco F, Velasco M Machado JP (1975) Statistical outline of the subthalamic target for the arrest of tremor. Appl Neurophysiol 38: 38–45
5. Velasco F, Velasco M, Ogarrio C, Fanghanel G (1987) Electrical stimulation of the centro median thalamic nucleus in the treatment of convulsive seizures: A preliminary report. Epilepsia 28: 421–430
6. Velasco M, Velasco F, Velasco AL, Luján M, Vázquez del Mercado J (1989) Epileptiform EEG activities of the centromedian thalamic nuclei in patients with intractable partial motor, complex partial and generalized seizures. Epilepsia 30: 295–306

7. Velasco M, Velasco F, Alcalá H, Dávila G, Díaz de León AE (1991) Epileptiform EEG activity of the centromedian thalamic nuclei in children with intractable generalized seizures of the Lennox-Gastaut syndrome. Epilepsia 32: 310–321

8. Velasco AL, Boleaga B, Santos N, Velasco F, Velasco M (1993a) Electroencephalographic and magnetic resonance correlations in children with intractable seizures of the Lennox-Gastaut syndrome and epilepsia partialis continua. Epilepsia 34(2): 262–270

9. Velasco F, Velasco M, Velasco AL, Jiménez F (1993b) Effect of chronic electrical stimulation of the centromedian thalamic nuclei on various intractable seizure patterns. I. Clinical seizures and paroxysmal EEG activities. Epilepsia (in press)

10. Velasco M, Velasco F, Márquez I, Velasco AL (1993c) Electroencephalographic responses elicited by acute electrical stimulation of the centro median thalamic nuclei in man. Epilepsia (submitted for publication)

Correspondence: Francisco Velasco, M.D., National Medical Center UNIBIC-IMSS, P.O. Box 73-032, Mexico City, Mexico.

Index of Keywords

Advances and Technical Standards in Neurosurgery

This series, sponsored by the European Association of Neurosurgical Societies, has already become a classic. In general, one volume is published per year.

The Advances section presents fields of neuro-surgery and related areas in which important recent progress has been made.

The Technical Standards section features detailed descript of standard procedures to assist young neurosurgeons in their post-graduate training. The contributions are written by experienced clinicians and are reviewed by all members of the Editorial Board.

Volume 20

1993. 96 figures. XIII, 308 pages.
Cloth DM 240,–, öS 1680,–, approx. US $ 160.00
ISBN 3-211-82383-2

Advances:

R.D. Lobato: Post-traumatic Brain Swelling.

K.-F. Lindegaard, W. Sorteberg, and H. Nornes: Transcranial Doppler in Neurosurgery.

A.E. Harding: Clinical and Molecular Neurogenetics in Neurosurgery.

Technical Standards:

B. Williams: Surgery for Hindbrain Related Syringomyelia.

J.-F. Hirsch and E. Hoppe-Hirsch: Medulloblastoma.

F. Resche, J.P. Moisan, J. Mantoura, A.de Kersaint-Gilly, M.J. Andre, I. Perrin-Resche, D. Menegalli-Boggelli, Y. Lajat, and S. Richard: Haemangioblastoma, Haemangioblastomatosis, and von Hippel-Lindau Disease.

Volume 19

1992. 44 partly colored figures. XIV, 224 pages.
Cloth DM 186,–, öS 1300,–, approx. US $ 129.00
ISBN 3-211-82287-9

Advances:

J.W. Berkelbach van der Sprenkel, N.M.J.Knufman, P.C. van Rijen, P.R. Luyten, J. A. den Hollander, and C.A.F. Tulleken: Proton Spectroscopic Imaging in Cerebral Ischaemia, Where we Stand and what Can Be Expected.

L. Steiner, C. Lindquist, and M. Steiner: Radiosurgery.

F. Iannotti: Functional Imaging of Blood Brain Barrier Permeability by Single Photon Emission Computerised Tomography and Positron Emission Tomography.

Technical Standards:

B. Jennett and J. Pickard: Economic Aspects of Neurosurgery.

F. Gjerris and S.E. Børgesen: Current Concepts of Measurement of Cerebrospinal Fluid Absorption and Biomechanics of Hydrocephalus.

F. Gentili, M. Schwartz, K. TerBrugge, M.C. Wallace, R. Willinsky, and C. Young: A Multidisciplinary Approach to the Treatment of Brain Vascular Malformations.

Springer-Verlag Wien NewYork

Sachsenplatz 4–6, P.O.Box 89, A-1201 Wien · 175 Fifth Avenue, New York, NY 10010, USA
Heidelberger Platz 3, D-14197 Berlin · 37-3, Hongo 3-chome, Bunkyo-ku, Tokyo 113, Japan

Jeremy C. Ganz

Gamma Knife Surgery
A Guide for Referring Physicians

1993. With 45 figures. XVII, 163 pages.
Soft cover DM 69,–, öS 485,–, US $ 45.00
ISBN 3-211-82476-6

Prices are subject to change without notice.

Radiosurgery has become an established technique, with more than 15000 patients treated world-wide, most of them in the last five years. Yet, there is much uncertainty in the general medical community as to the nature, advantages and limitations of the method. This uncertainty provokes unnecessary debate between colleagues and is a source of avoidable stress to patients.

This book provides an account of the scientific basis of radiosurgery and describes its current applications in respect of the only well established radiosurgical device, the Leksell Gamma Knife. The book assumes the general medical knowledge of a newly qualified medical practitioner.

There are three sections. The first outlines the rationale for radiosurgery and the principles of stereotaxy, radiophysics and radiobiology. The middle section, consisting of a single chapter, describes what a potential patient may expect to experience. In the final section, the current applications are gone through, one by one, indicating what can and what cannot be achieved. The book is intended for neurologists, neurosurgeons, internists, otolaryngologists, oncologists, ophthalmologists, general practitioners, medical students and anyone else who might wish to refer a patient to or advise a patient about Gamma Knife radiosurgery.

 Springer-Verlag Wien New York

Sachsenplatz 4–6, P.O.Box 89, A-1201 Wien · 175 Fifth Avenue, New York, NY 10010, USA
Heidelberger Platz 3, D-14197 Berlin · 37-3, Hongo 3-chome, Bunkyo-ku, Tokyo 113, Japan